Neurodegeneration
Metallostasis and Proteostasis

RSC Drug Discovery Series

Editor-in-Chief
Professor David Thurston, *London School of Pharmacy, UK*

Series Editors:
Dr David Fox, *Pfizer Global Research and Development, Sandwich, UK*
Professor Salvatore Guccione, *University of Catania, Italy*
Professor Ana Martinez, *Instituto de Quimica Medica-CSIC, Spain*
Dr David Rotella, *Northeastern University, USA*

Advisor to the Board:
Professor Robin Ganellin, *University College London, UK*

Titles in the Series:
1: Metabolism, Pharmacokinetics and Toxicity of Functional Groups: Impact of Chemical Building Blocks on ADMET
2: Emerging Drugs and Targets for Alzheimer's Disease; Volume 1: Beta-Amyloid, Tau Protein and Glucose Metabolism
3: Emerging Drugs and Targets for Alzheimer's Disease; Volume 2: Neuronal Plasticity, Neuronal Protection and Other Miscellaneous Strategies
4: Accounts in Drug Discovery: Case Studies in Medicinal Chemistry
5: New Frontiers in Chemical Biology: Enabling Drug Discovery
6: Animal Models for Neurodegenerative Disease
7: Neurodegeneration: Metallostasis and Proteostasis

How to obtain future titles on publication:
A standing order plan is available for this series. A standing order will bring delivery of each new volume immediately on publication.

For further information please contact:
Book Sales Department, Royal Society of Chemistry, Thomas Graham House, Science Park, Milton Road, Cambridge, CB4 0WF, UK
Telephone: +44 (0)1223 420066, Fax: +44 (0)1223 420247
Email: books@rsc.org
Visit our website at http://www.rsc.org/Shop/Books/

Neurodegeneration
Metallostasis and Proteostasis

Edited by

Danilo Milardi
Consiglio Nazionale delle Ricerche, Istituto di Biostrutture e Bioimmagini, Catania, Italy

and

Enrico Rizzarelli
Dipartimento di Scienze Chimiche, Universita degli Studi di Catania, Catania, Italy

RSCPublishing

RSC Drug Discovery Series No. 7

ISBN: 978-1-84973-050-1
ISSN: 2041-3203

A catalogue record for this book is available from the British Library

Published by The Royal Society of Chemistry,
Thomas Graham House, Science Park, Milton Road,
Cambridge CB4 0WF, UK

Registered Charity Number 207890

For further information see our web site at www.rsc.org

Preface

Extensive evidence has been accumulated in recent years indicating that several disorders have the same molecular basis: a change in conformation of a protein. These protein conformational diseases include the neurodegenerative disorders that constitute a worldwide health problem. The conformational modified protein may be implicated in the disease by direct toxic activity, by the lack of the biological function of the normally folded protein, or by improper trafficking: in the cases in which the protein is toxic, it forms fibrillar amyloid aggregates. Most of the members of this group of diseases may be of either inherited or sporadic origin. In familial cases, a mutation induces misfolding of the protein and hence displaces the equilibrium between the normal and the pathological conformational state toward the latter, which accumulates and triggers the disease process. In sporadic cases, the balance is altered by environmental factors (pH, ionic strength, metal ions, free radicals, oxidative stress, *etc.*), the effect of physiological and pathological chaperone proteins, or changes in the protein concentration. The neurodegeneration usually occurs and develops as a result of the imbalance between the production and the clearance of the involved protein. To cope with this alteration of proteostasis that, according to Morimoto and Kelly,[1] refers to control of the concentration, conformation, binding interactions (quaternary structure), and location of individual protein changes, small molecules and/or biological proteostasis regulators have been suggested as agents able to enhance protein homeostasis by binding to and stabilizing specific proteins (pharmacologic chaperones) or by increasing the proteostasis network capacity (proteostasis regulators).[2]

Although transition metals are important for life, there is evidence that they are also involved in neuronal damage in many neurodegenerative disorders. Neurodegenerative diseases associated with the disruption of brain metallostasis include Alzheimer's (AD), Parkinson's (PD) and Huntington's (HD)

RSC Drug Discovery Series No. 7
Neurodegeneration: Metallostasis and Proteostasis
Edited by Danilo Milardi and Enrico Rizzarelli
© Royal Society of Chemistry 2011
Published by the Royal Society of Chemistry, www.rsc.org

diseases as well as amyotrophic lateral sclerosis (ALS). It has been observed that patients with neurodegenerative diseases accumulate metals in their nervous systems,[3,4] suggesting a role of metal ions in those disorders. Metallostasis is frequently altered in neurodegenerative disorders;[5] under certain conditions, redox active metal ions (Cu, Fe) are the most potent pro-oxidants due to their high availability.[6] The excessive production of reactive oxygen species (ROS) involves the oxidation of proteins, DNA and phospholipids leading to structural and functional alterations.[7] Metal-binding proteins and DNA may therefore be vulnerable. Most of the reactions involving Cu and Fe are related to Fenton chemistry, a series of reactions initiated with transition metals and hydrogen peroxide, leading to the formation of highly unstable radicals that affect biological macromolecules.[8] Proteins involved in metal transport and distribution in the nervous system, such as copper transporter protein 1 and ATP7A (copper-transporting P-type ATPase) for Cu,[9,10] transferrin and transferrin receptor (TfR) for Fe, and DMT1 (divalent metal transporter 1) for both Cu and Fe,[11] could be involved in the altered metal homeostasis in the brains of patients with neurodegenerative diseases. Therefore, tight and combined regulation of neuronal protein and metal homeostasis is essential to the integrity of normal brain function, and it has been recently suggested to cope with the cognitive decline in Alzheimer's disease, which involves pathological accumulation of synaptotoxic amyloid-β (Aβ) oligomers and hyperphosphorylated tau.[12] It has been found that ionophore chelating agents greatly elevated cellular levels of copper and zinc but not iron. Ionophore-driven increases in intracellular metal levels resulted in activation of PI3K, leading to downstream phosphorylation of glycogen synthase kinase 3 (GSK3) and potentiation of the mitogen activated protein kinase, JNK. The stimulation of this pathway culminated in up-regulation of cellular metalloprotease activity and degradation of extracellular Aβ peptides.[13] Therapeutic strategies targeting interactions between Aβ, tau and metals to restore proteostasis and metallostasis have been reported. An AD therapeutic strategy has been proposed to target GSK3. Copper-*bis*(thiosemicarbazonate) complexes were employed to increase intracellular copper bioavailability and inhibit GSK3. The lead compound has been reported: i) to significantly inhibit GSK3 in the brains of transgenic AD model mice; ii) to decrease the abundance of Aβ trimers and phosphorylated tau; and iii) to restore performance of AD mice in the Y-maze test to levels expected for cognitively normal animals. Improvement in the Y-maze test correlated directly with decreased Aβ trimer levels. This study demonstrates that increasing intracellular copper bioavailability can restore cognitive function by inhibiting the accumulation of neurotoxic Aβ trimers and phosphorylated tau.[14]

This book is intended to provide a comprehensive review of some of the above-mentioned molecular features related to neurodegenerative disorders. In particular, the first chapter provides a picture of the role of the inorganic elements in different brain areas, thus giving the necessary outline to the problem of the brain activity in terms of chemistry. Chapters 2 and 3 focus on the chemistry and biology of proteostasis by describing the complex network of cellular processes that supervise the maintenance of the proteome. These two chapters are mainly devoted to an overview of those processes that are thought to provide a link between altered proteostasis and neurodegeneration. Specific issues relating to metalloproteins, which are believed to provide links between metallostasis and proteostasis and hence between metallostasis and neurodegeneration, are also addressed. Furthermore, some emerging concepts in the treatment of different severe diseases via the manipulation of proteostasis are reported. Chapter 4 summarizes the role of amyloid peptide channels in the pathophysiology of amyloid diseases by causing depolarization of cell membranes, and redistribution of metal ions across the different compartments. Chapter 5 reviews the most recent observations regarding the effect of an abnormal compartmentalization of metal ions on the efficiency of the protein "quality control" systems (*e.g.* metalloproteases and the ubiquitin proteaseome system) of the cell. Chapter 6 addresses the connections between copper(II) and zinc(II) metallostasis with the dyshomeostais of amyloid-β peptide (Aβ) levels in Alzheimer's disease. A chemical description of these issues is reported, stressing the relationships between speciation of metal complexes with Aβ with the different morphologies and toxicities of metal-loaded Aβ aggregates. In Chapter 7, defects in neurotrophins (NTF) transport or processing, that are normally known to be causally linked to human neurodegenerative disorders, are discussed in terms of the reciprocal interplay between metal ion balance and trophic factors (TF)-dependent signaling. Chapter 8 deals with the delicate equilibrium of metal ions in the central nervous system, essential for the development and maintenance of enzymatic activities, mitochondrial function, and neurotransmission. In particular, the consequences of the disruption of these mechanisms, triggering the cascade of events leading to neurodegeneration, are proposed as the framework of a novel disease-modifying therapeutic strategy for diverse neurodegenerative diseases. Chapter 9 reviews the recent hypotheses on the mechanism of brain iron homeostasis during normal and pathological conditions: how iron is imported from the blood circulation, redistributed through the brain and stored in neurons. In addition, a molecular description of iron accumulation in normal, aged and pathological brain is provided, in order to identify new pharmacological targets related to iron management. Chapter 10 reports the recent findings correlating aluminium to AD, highlighting in particular what remains from the old aluminium hypothesis. In Chapter 11 the role played by an altered metallostasis in the pathology of the debilitating disease amyotrophic lateral sclerosis (ALS) is described. Finally, Chapter 12 elucidates the copper(II) coordination features of the prion protein with emphasis on new concepts relating copper uptake to prion protein function and neuroprotection.

In conclusion, a wide variety of conceptually distinct approaches are presented in an attempt to correlate proteostasis and metallostasis with the molecular features of neurodegenerative disorders whose incidence has dramatically increased. We wish to thank all the contributors for their efforts in summarizing the most relevant results on their topics and we hope that their research can contribute to supporting people involved in finding effective therapies against these diseases.

References

1. W. E. Balch, R. I. Morimoto, A. Dillin and J. W. Kelly, *Science*, 2008, **319**, 916–919.
2. T. Gidalevitz, E. A. Kikis and R. I. Morimoto, *Curr. Op. Struct. Biol.*, 2010, **20**, 23–32.
3. D. Berg and M. B. Youdim, *Top. Magn. Reson. Imag.*, 2006, **17**, 5–17.
4. D. T. Dexter, P. Jenner, A. H. Schapira and C. D. Marsden, *Ann. Neurol.*, 1992, **32**(Suppl.), S94–S100.
5. D. A. Simmons, M. Casale, B. Alcon, N. Pham, N. Narayan and G. Lynch, *Glia*, 2007, **55**, 1074–1084.
6. O. I. Aruoma, H. Kaur and B. Halliwell, *J. R. Soc. Health*, 1991, **111**, 172–177.
7. J. Agar and H. Durham, *Amyotroph. Lateral Scler. Other Motor Neuron Disord.*, 2003, **4**, 232–242.
8. M. B. Youdim, M. Fridkin and H. Zheng, *Mech. Ageing Dev.*, 2005, **126**, 317–326.
9. C. W. Levenson, *Physiol. Behav.*, 2005, **86**, 399–406.
10. E. Madsen and J. D. Gitlin, *Annu. Rev. Neurosci.*, 2007, **30**, 317–337.
11. H. Gunshin, B. Mackenzie, U. V. Berger, Y. Gunshin, M. F. Romero, W. F. Boron, S. Nussberger, J. L. Gollan and M. A. Hediger, *Nature*, 1997, **388**, 482–488.
12. A. I. Bush and R. E. Tanzi, *Neurotherapeutics*, 2008, **5**, 421–432.
13. A. Caragounis, T. Du, G. Filiz, K. M. Laughton, I. Volitakis, R. A. Sharples, R. A. Cherny, C. L. Masters, S. C. Drew, A. F. Hill, Q-X. Li, P. J. Crouch, K. J. Barnham and A. R. White, *Biochem. J.*, 2007, **407**, 435–450.
14. P. J. Crouch, L. W. Hunga, P. A. Adlar, M. Cortes, V. Lal, G. Filiz, K. A. Pereza, M. Nurjono, A. Caragounis, T. Du, K. Laughton, I. Volitakis, A. I. Bush, Q-X Li, C. L. Masters, R. Cappai, R. A. Cherny, P. S. Donnelly, A. R. White and K. J. Barnham, *Proc. Nat. Acad. Sci. USA*, 2009, **106**, 381–386.

Contents

RSC Drug Discovery Series No. 7
Neurodegeneration: Metallostasis and Proteostasis
Edited by Danilo Milardi and Enrico Rizzarelli
© Royal Society of Chemistry 2011
Published by the Royal Society of Chemistry, www.rsc.org

Chapter 7   Zinc, Copper, Neurotrophic Factors and Neurodegeneration   **141**
G. Amadoro and P. Calissano

Chapter 8   Biological Metals: Metallostasis and Alzheimer's Disease   **152**
*A. Rembach, J. A. Duce, L. A. O'Sullivan,
R. E. Tanzi and A. I. Bush*

CHAPTER 1

An Introduction to the Brain and its Biological Inorganic Chemistry

R. J. P. WILLIAMS[a] AND J. J. R. FRAÚSTO DA SILVA[b]

[a] Inorganic Chemistry Laboratory, Oxford University, South Parks Road, Oxford OX1 3QR, UK; [b] Instituto Superior Tecnico, Universidade Tecnica de Lisboa, Lisbon, Portugal

1.1 Introduction

The brain is a complex structure, composed of many zones organised as compartments that are apparently isolated by the manner of folding of the outer structure and by the packing and types of cells in the structures (Figure 1.1 and Table 1.1).[1,2] The functions of the compartments and the chemicals in them, distinguished by staining, apart from their physical characteristics, give ways of delineating them. It is also possible to describe the zones by the size of their differentiated electrical responses stimulated by actual or experimental outside events. A more detailed level of division of the description of the zones is by the cellular and membrane structures and their differences. There are two major classes of cell types in all zones, neurons and glia, and we shall describe them in turn, at first as if they were independent cell types. Each neuron appears to be separated from all others spatially and by several surrounding glial cells except for deliberate connection made between neurons at synaptic junctions. In this general introduction we draw special attention to the part played by metal ions and their enzymes.

RSC Drug Discovery Series No. 7
Neurodegeneration: Metallostasis and Proteostasis
Edited by Danilo Milardi and Enrico Rizzarelli
© Royal Society of Chemistry 2011
Published by the Royal Society of Chemistry, www.rsc.org

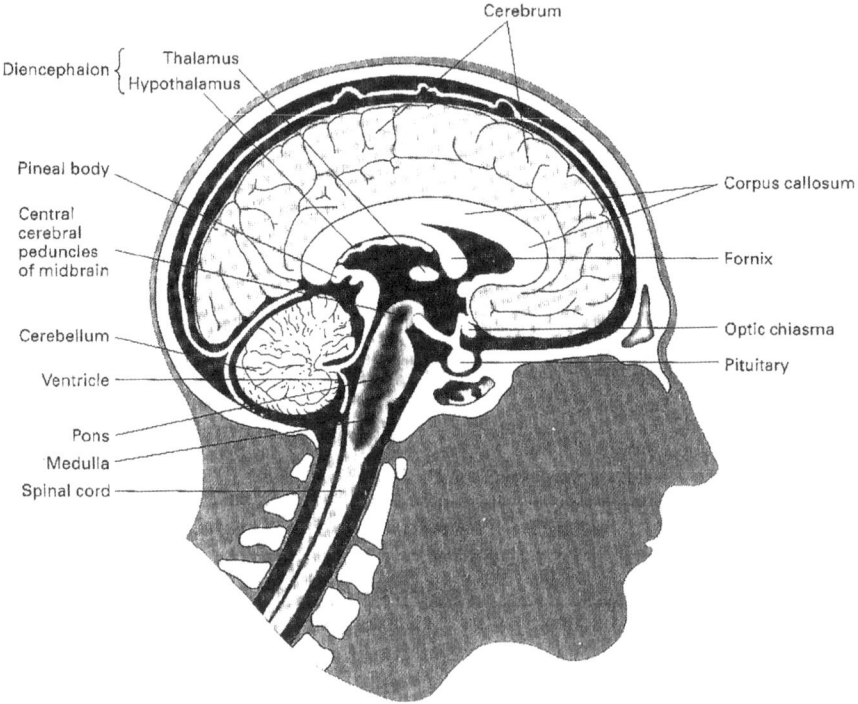

Figure 1.1   A general outline of the zones of the human brain; see Table 1.1 for their situation in the three major parts of the brain.

Table 1.1   Divisions and zones of the brain.

Major Division	Principal Zones	Sub-zone
Forebrain	Cerebral cortex	Olfactory bulb
	Basal ganglia	Caudate nucleus, striatum
	Limbic system	Amygdala, hippocampus
	Thalamus	Connected to eyes
	Hypothalamus	See pituitary
Midbrain	Tectum	Pineal, connected to eyes
		Red nucleus
	Tegmentum	Substantia nigra, raphe nuclei
Hindbrain	Pons	Locus coeruleus, raphe nuclei
	Cerebellum	
	Medulla	

Taken in part from *Physiology of Behaviour*, ed. N. R. Clarkson, 6th edn, 1998, Allyn and Bacon, Boston.

1.2   The Structure of Neurons

The simple description of a neuron is that it is made of a central nuclear body of ill-defined shape but considerable volume, see Figure 1.2(a) and (b), with long very thin tubular extensions called axons. The extensions have termini, which

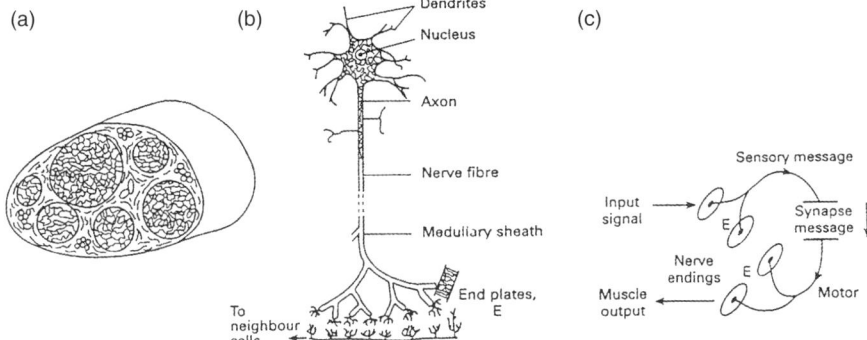

Figure 1.2   An outline of a nerve cell: (a) cross-section, (b) a neuron, (c) sensory and motor connections via synapses.

can act as donors or acceptors of chemicals in the message system of brain states and are seen as somewhat bulbous regions, see Figure 1.2(c). The physical structure of the central region of the neuron (the soma) is that of a eukaryotic cell and has the usual compartments of organelles and vesicles. The axons are structured internally by conventional fibrous proteins, including tubulins capable of allowing transfer of vesicles from the central body to the bulbous termini. The membranes of the axons in the brain cover long stretches between "nodes of Ranvier", Figure 1.3, where there are active channels and pumps for Na and K ions. The protection is provided by myelin proteins produced by oligodendrocytes, a special kind of glial cell. In general it is considered that the axons are just long-range connections between the central region and the bulbous termini. The major components of the liquid in them are generally considered to be of the same ionic content as the cytoplasm, but have few, if any, enzymes and little metabolic activity. However, the nodal membranes have ion gates and ATP-ases as pumps. We return to the chemical composition at termini later since these bulbous zones have a concentration of vesicles of differing chemical contents. The outer membranes here have the usual contents and properties of the eukaryotic cell, being able to exo- and endocytose, and have numerous enzymes on the surfaces able to act in donor or acceptor capacities, especially as channels and pumps. The axons are able to grow independently by cell multiplication or replication. The cells are physically surrounded by extracellular fluid, which in the brain is a special fluid separated from the blood by a blood–brain barrier. Although the whole brain is aerobic and neurons require oxygen they are also supported with some nutrients by glial cells. There is also extensive connective tissue composed of proteins and polysaccharides to maintain structure.

1.3   The Chemical Activity of Neurons

The main chemical activities of the nerve cell are simply divided. The central region is one major supplier of small and large chemicals and energy to its axons and then to the bulbous termini. The axons, at active nodes, seem only to

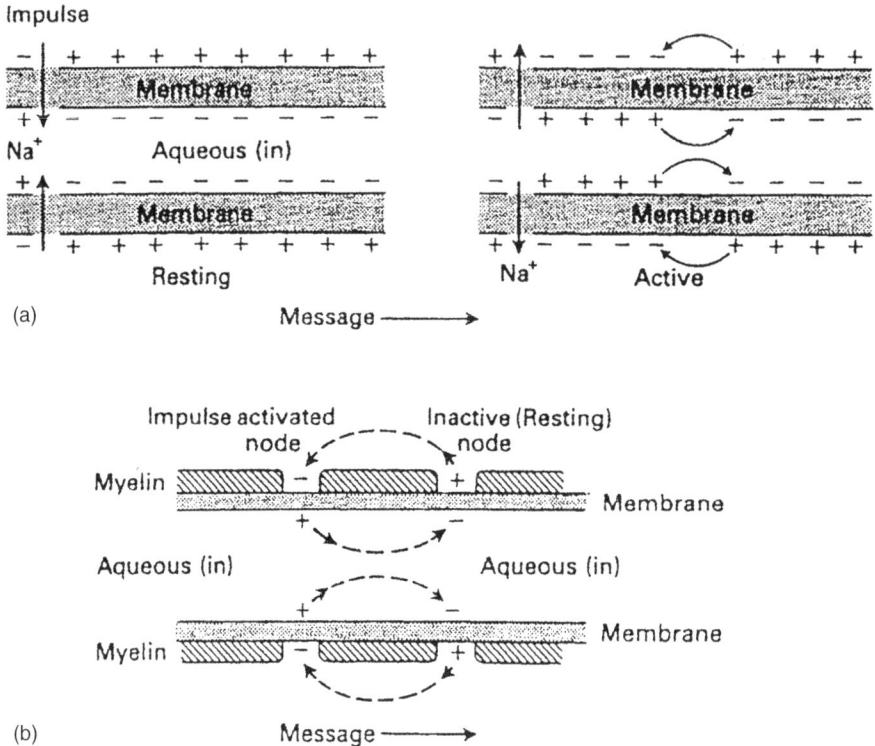

Figure 1.3 The distribution and flow of Na$^+$ ions carrying a message in an axon. The neuron in (a) is not myelinated. Note: K$^+$ flow is in the opposite sense due to a combination of ATP-ases and gates.

require a large amount of energy to maintain their electrical activity, which is their dominant function. The activity is a self-sustaining relay of electrostatic ion flow of such a character that it allows depolarisation and repolarisation due to the flow of Na$^+$/K$^+$ ions from inside to outside, and its reversal, see Figure 1.3. On allowing initial depolarisation through *channels* the wave of depolarisation travels along the axon as an electrical signal, but the axon recovers immediately by pumping the ions back into itself. It is very important, therefore, that both the internal cytoplasmic and the external concentrations of the fluids, see Table 1.2, are very precisely fixed. The maintenance (homeostasis) of Na$^+$ and K$^+$ ions is a critical factor in nerve and brain chemistry. Note that it is standard hospital practice to monitor these levels in humans for any sign of weakness, which could ultimately affect the brain.

The depolarisation wave travels to the termini at the synapse where it activates donor events. The donation is of transmitters, which travel to acceptor centres in the opposing neuron of an adjacent synapse after release from storage vesicles. The chemistry involved thereafter is complex. We shall therefore leave aside the chemical activities in the axons while we describe those

Table 1.2 Elements in rat brain cerebrospinal fluid (CSF) and organs (mEq per kg H_2O).

	Plasma	*CSF*	*Cytoplasm*
Na	148	152	20
K	5.3	3.36	140
Ca	6.14	2.2	10^{-4}
Mg	1.44	1.77	8.0
Cl	106	130	40.5
Glucose	7.2	5.4	
Pyruvate	0.17	0.18	
Lactate	0.7	2.0	
Proteins (mg per 100 ml)	6500	25	10 000

From: R. J. P. Williams and J. J. R. Fraústo da Silva, *Bringing Chemistry to Life*, Oxford University Press, Oxford, 1997, ch. 15.

of the bulbous zones. These terminal zones have outer membranes, which directly or indirectly are stimulated by the depolarisation wave to allow calcium ion entry into the cytoplasm of the bulb. In turn the calcium ions cause a filamentous action to move the vesicles holding transmitters to the cell membrane where they discharge either their small molecules or ions into the extracellular fluid, directed as much as is possible toward a bulbous zone of a receptor cell. This cell then initiates a second Na^+/K^+ wave down its axon, using the ion gradients. The outside concentrations are shown in Table 1.2 while the inside concentrations are: K^+ approx. 100 and Na^+ approx. 5 mEq per kg H_2O. The small molecules and ions, *e.g.* Zn^{2+}, are both called transmitters. The donor bulbous region must now quickly recover its resting chemical content by filling its depleted stores in vesicles; see also Glial Cells. The energy required for the passage of a signal is considerable.

The chemical content of particular interest lies in the packaging of transmitters in the vesicles, which can contain a vast number of different organic and inorganic ions. Different neurons have different vesicle contents. Analysis indicates that these include both positive, *e.g.* adrenaline (epinephrine), and negative, *e.g.* glutamic acid, or even zwitterions, *e.g.* γ-amino butyric acid and small cations such as Zn^{2+}. We have listed some of them in Table 1.3. Note some of the brain amino acids are D- not the conventional L- of the main body. The peptides may not be present in a simple immediately available form, but may be part of larger peptides or proteins, *e.g.* chromogranin A, which are hydrolysed on external release to give statin-like peptides.[3] The differently charged transmitters cannot be stored without molecules carrying the opposite charge. Adrenaline is stored with adenosine triphosphate (ATP), which in this case is free from Mg^{2+}, but there is some Ca^{2+} in the vesicle. The storage of acidic transmitters appears to be with Na^+ (not K^+).

A second activity of the neuron does not concern ions or molecules but a variety of proteins and some enzymes. They include those in the central area, those in the synapse bulbs and those in the axons. The enzymes in the central zone catalyse tubulin synthesis for extension as the axon grows. Axon growth occurs only with repeated electrical activity of the neuron and is therefore associated with

Table 1.3   Messenger compounds between cells in the brain.

Fast Signals

Excitatory (+)/inhibitory (−)
  Glutamate (+)
  Glycine (−)
  GABA (γ-amino butyric acid) (−)
  Acetylcholine

Inorganic signals
  Na^+, K^+, Ca^{2+}, (Zn^{2+})

Intermediate Speed of Signal
  Noradrenaline (norepinephrine)
  Dopamine
  Serotonin

  Nitric oxide
  Carbon monoxide
  Hydrogen sulfide

Slow Signals
  Neuropeptides (large number, > 100)
  Substance P
  Cholecystokinin
  Corticotrophin-releasing factor
  Melatonin

long-term memory. The proteins necessary in the synaptic bulbs for storage of the messenger molecules, transmitters, are transported along the tubulin from the central region to the bulbs. Hence the central region is responsible for transmitter synthesis, but transmitter recovery occurs at the synapses (see Glial Cells). Analysis of a nerve cell is a basic task but in the brain it is essential that, as well as general analysis, we can *image* where any type of neuron including its synapse is located. The study of the locations depends on: (1) direct methods for metal ions contents, including single atom microscopes, assuming that different neurons have different ion contents, (2) use of such tools as direct fluorescence or fluorescence of added dyes which may reflect contents of neurotransmitters directly as they are at high concentration in bulbous regions. Typically the presence of *free* or very weakly bound zinc can be detected by very high resolution electron microscopy or by the use of the dye-stuff dithizone. A very interesting example of a parallel but different direct spectroscopy analysis of vesicles of cells is that of the adrenal gland. The vesicles of the gland release transmitters in much the same fashion as the neurons do. The cells of this gland resemble neurons in that they contain adrenaline, some corticosteroids, adenosine triphosphate and a protein, chromogranin A, which is the source of certain peptide hormonal molecules such as enkephalin. Much of the content of the vesicles of the whole organ can be visualised by nuclear magnetic resonance (NMR),[3] including all four types of messenger molecule. It is possible to use brain slices to perform similar NMR analyses.

1.4 The Enzyme Content of the Neuron

The main enzymes of the neurons are in the three groups of the normal complement of aerobic eukaryotic cells, so they can catalyse the common activities of glycolysis, the Krebs cycle and conventional syntheses. The neurons have mitochondria and the usual vesicles of the endoplasmic reticulum. In addition they must have the ability to synthesise the transmitters. In so far as the transmitters are specialised in different neurons so their enzymes must be present in the specific cell central region. The major separation of transmitters is into those that require specific oxidation, *e.g.* adrenaline and amidated peptides, and those that do not, *e.g.* glutamate. Oxidation enzymes are usually iron-containing in the cytoplasm and copper-containing in vesicles or externally. The analysis locally for copper and iron and their enzymes is therefore very useful. Some of these enzymes should be in higher concentration than in other cells. Zinc enzyme analysis, equally important, is confused by local concentrations of free zinc in vesicles. Enzymes other than those containing metal ions must be recognised by fluorescent products. We return to the distribution in the different neurons when we have described the analysis of the brain's zones.

A peculiarity of nerve cells generally but very importantly in neurons is the synthesis of the myelin sheath. Myelin is 80% lipid and 20% protein, holding the multi-layered membrane rigid. The protein somewhat resembles an outer skin of keratin and its final cross-linked state requires copper oxidases. It is very unusual for an internal cell to have such protection. Note that vitamin B_{12} (cobalamine is the active form of vitamin B_{12}) is necessary to protect the synthesis of myelin.

Neurons in the brain need to grow to make new contacts so as to create long-term memory. Short-term memory may not need such growth. These nerves contain the proteins enabling extension of the axon, the cell. They must therefore have not only tubulins but actomyosin filaments and these proteins must be carried down the axons. The actomyosin proteins are also responsible for the ejection of chemical transmitters from the vesicles as their contractile function, linked to calcium stimulus, moves the vesicles to the outer membrane for discharge. The activity of the actomyosin contraction is mediated by ATP hydrolysis, which is generally of the Mg–ATP complex where Mg acts as a required unit. This activity of a kinase is dependent on calcium entry into the neuron synapse bulk. There are several metal ions of great importance in the brain and nervous system, especially sodium, potassium, magnesium, calcium, iron, copper, zinc, cobalt and manganese. The details of their distribution as free ions or in enzymes, the metallomes, in different brain zones must be a major area for future research.

1.5 Glial Cells

The glial cells (glia)[2] which occur in the central nervous system (CNS) and in the peripheral nervous system (PNS) are much more abundant than the neurons and occupy one half of the volume of the brain. Their number in the human neocortex is about 36–39 000 000 000, and they can reproduce. For many years they were thought to have only the function of supporting the neurons physically (their name derives from the Greek *glia*, glue), but it is now known that besides this function they are essential to maintain and repair the neuronal system, to control the formation of the synapses and to participate in the mechanism of production of energy in the CNS, besides transporting essential ion and organic molecules to the neurons.

In effect there are five different kinds of glial cell: the astrocytes, the oligo-dendrocytes, the Schwann cells, the microglial cells and the satellite cells. The astrocytes (the name is derived from their star shape) are very relevant, linked to the functions of the neurons, and it can be said that they control the formation of the synapses, which they isolate, forming a kind of external barrier. The same happens near the nodes of Ranvier, near where the Na^+ and K^+ channels are concentrated. As mentioned above, these cells also transport ions and other substances required in an intercellular pathway by the neurons and participate in the mechanism of production of energy in the CNS, providing energetic substrates in an activity-dependent manner (pyruvate, lactate, glucose). They also regulate the level of glutamate in the synaptic space, removing K^+ from these spaces, regulate the pH of the cerebrospinal fluid and connect the synaptic activity with the blood flux. Finally they protect against oxidation stress and remove toxic substances (such as ammonia and cell detritus). Note that a Mn^{2+} enzyme (glutamine synthetase) occurs specifically in the astrocytes and converts glutamate and ammonia to gluta-mine. The astrocytes and the neurons communicate via intercellular holes or channels that allow the trafficking of ions and small molecules in the so-called "gap junctions" but they can also act by extracellular trafficking of ions and signalling molecules. In this way they help to produce a kind of network in which information circulates in the relevant areas of the brain as a chain of reactions.

Two other kinds of glial cell are the oligodendrocytes (in the CNS of evolved vertebrates) and the Schwann cells (in the PNS), both of which produce the myelin layers that coat the neuronal axons, isolating them from electronic effects and controlling the concentration of the ionic Na^+/K^+ channels in the nodes of Ranvier. The fourth kind of glial cell is the group of the small microglial cells that have a neuroimmunological function, responding to disease or aggression, phagocytosing cell detritus and providing anti-inflammatory responses; they have, therefore, a function of protection of other cells. The last of the five glial cells, the satellite cells, give physical support to the neurons in the PNS and help in the regulation of the external chemical environment.

All these five kinds of glial cell act as an integrated functional unit, but the production of vasodilators, including nitric oxide, carbon monoxide, ade-nosine, arachidonic acid, *etc.*, derives from the activity of the neurons. The metal ion and metalloenzyme content of the glial cells is not known but their specialist functions will demand a differentiated distribution of ions in them.

1.6 Analysis of Cerebrospinal Fluid and Connective Tissue

The CNS, including the spinal cord, is surrounded by a special fluid separated from the rest of the body by the blood–brain barrier or series of membranes. The fluid differs in content from the normal extracellular fluids in that it contains very few proteins and enzymes. The major mineral content of the cerebrospinal fluid (CSF) is much the same as that outside the brain, although the calcium seems to be somewhat lower, but the absence of proteins and enzymes implies that the extracellular activities are extremely limited. It may be that the demand for growth has to be restricted to zones of intense activity in response to particular external events. It is stimulated then at the cellular level and is not managed as a general widespread development of the whole brain.

The connective tissue appears to be composed of similar proteins and saccharides to this material elsewhere in the body. The synthesis of these often cross-linked polymers requires copper oxides and the breakdown of them to allow growth needs zinc proteases. Connective tissue damage, as elsewhere in the body, can lead to serious disease. It could be that the failing of memory is not only related to difficulties associated with old age; see the following chapters.

1.7 Analysis of the Whole Brain

The analysis of the whole brain is not very meaningful though it is of interest to see how certain elements change during growth (Table 1.4). The table indicates that the growth of cells is very considerable relative to the extracellular fluid. There are indications that this reverses in old age as cells tend to die and are not fully replaced. Nerve–nerve connections increase rapidly immediately after birth. We do not give any detailed data on trace elements since, generally speaking, the whole brain does not differ from several other organs of the body

Table 1.4 The ionic composition during development of the human brain.

	Electrolyte composition (meq kg^{-1})				
			Human brain at given age		
	Prenatal age (Weeks)				
Electrolyte	*13–14*	*20–22*	*Newborn*	*Adult*	*Senescent*
Na$^+$	97.5	91.7	80.9	55.2	Rises
K$^+$	49.6	52.0	58.2	84.6	Falls
Cl$^-$	72.1	72.6	66.1	40.5	Rises
Mg^{2+}		8.4	7.9	11.4	Unknown
Ca^{2+}		4.9	4.8	4.0	Rises
P	57.0	52.2	54.0	109.0	Falls

From: R. J. P. Williams and J. J. R. Fraústo da Silva, *Bringing Chemistry to Life*, Oxford University Press, Oxford, 1997, ch. 15.

in this respect. Details of the free and bound metallomes in the cytoplasm of cells are given in our books and relate to local and general activities. There are however considerable amounts in vesicles and in all cell components, including mitochondria. Special attention has been directed to zinc,[4-6] as it has been correlated with some hormone functions in growth.

1.8 Chemical Analysis of the Brain Regions

The analysis of the brain has to be conducted in defined regions, as it is not a homogeneous organ. The observations so far are still limited despite the progress in imaging, and so we can do no more than give an outline of what is known and what is desirable to know if we are to understand normal and disturbed brains. We shall divide the topic into cell contents and the surrounding cerebrospinal fluids and shall refer mainly to neurons and transmitters, but there are a few general remarks which we can make first with regard to all the different cells, neurons, glia and others. We consider that the *free ions* content of the cytoplasm of the cells, *the free metallome*, is closely similar to that of all aerobic cells (Figure 1.4). The levels of all free ions for K^+, Na^+ and Mg^{2+} are above 10^{-4} M while those for all the other cations and anions, such as Ca^{2+}, are below 10^{-6} M. The free ions of the trace elements are extremely low, sometimes below 10^{-10} M. The levels in the mitochondrial cytoplasm are different, and for example Fe^{2+} is close to 10^{-6} M there. This says nothing about the bound ions, *the bound metallome*, which is low for K^+ and Na^+ but is around 10^{-3} M for Mg^{2+} in all cells. The bound levels of the other trace elements are dependent on the precise zone of the brain under discussion.

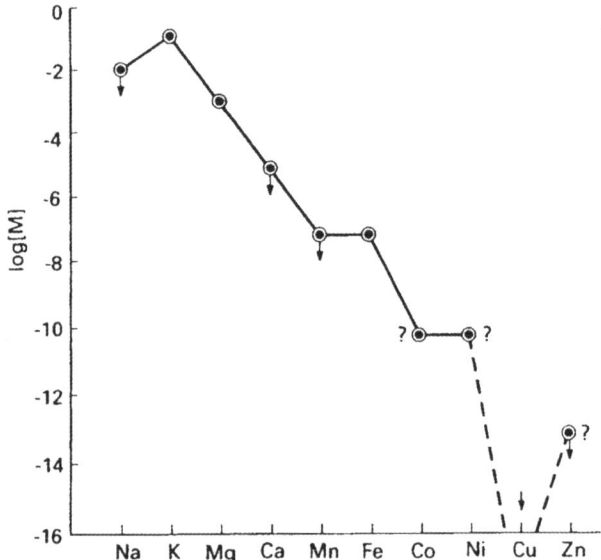

Figure 1.4   The free metal ion content of the cytoplasm of cells, the free metallome.

The important elements selective for particular zones are the content of the neuron vesicles themselves, ions, small or large molecules and the enzymes for producing these small molecules, presumably in the central regions of cells. The free cation contents in these vesicles of special interest are those of zinc and calcium. In the case of zinc the free ion is associated with glutamate transmitters, particularly in mossy fibre vesicles. This association requires that the vesicle content, probably made in the central region of the cell, is pumped into the vesicles up to a total content of close to 10^{-4} M. One pump protein is a known heavy element vesicle membrane ATP-ase. After release of its contents the empty vesicle may return into the synaptic vesicle of the neuron and be refilled. The zinc released is thought to modulate the action of the glutamate in some of its receptors. Apart from the free zinc in vesicles this metal has quite general importance in all cells, particularly in zinc finger transcription factors. Zinc exchange is mediated by special proteins, such as the hydrolases, which are necessary in the hydrolysis of peptides, and carriers and buffers, so that this element controls many functions including growth.

We turn now to the small molecules other than glutamate while noting that those carrying charge need a counter ion. A good example is adrenaline, which requires ATP^{3-} in the ratio of one for three molecules of adrenaline, but the vesicles do not contain high Mg^{2+}, the usual counter ion of ATP^{3-} in enzyme reactions. The interest centres in the synthesis of adrenaline and the uptake mechanisms for both adrenaline and ATP^{3-}. The synthesis is especially intriguing as it involves two steps of oxidation by iron and copper oxidases, the first in the cytoplasm but the second in the vesicle itself. In either case the presence of the two in a specific zone in high concentration should be recognisable by analysis. Table 1.3 gives zones with content of the major hydroxylated transmitters noradrenaline, adrenaline and serotonin, and Tables 1.5 and 1.6 give those with high zinc and copper. The *substantia nigra* and the *locus coeruleus* stand out in both tables, but there are marked differences and also differences of both from the distribution of non-haem iron, Table 1.7. We do not know how this selectivity is achieved but in effect the different neurons of the brain are differentiated cells with different DNA expression.

Table 1.5   Zinc content of brain zones.

Zone	Zinc Content (µg/g dry weight)
Hippocampus	>65
Cerebellum	30–50
Cortex	
Olfactory bulbs	
Thalamus	<25
Hypothalamus	
Medulla	

From: I. E. Dreosti, in *Zinc in Human Biology*, ed. C. F. Mills, Springer-Verlag, New York, 1988, ch. 15. See also Figure 1.1.

Table 1.6   Copper concentration in human brain zones.

Zone	Cu concentration (μg/g wet weight)
Cerebellar cortex	10.4
Hippocampus	6.6
Substantia nigra	18.8
Locus coeruleus	62.0
Others	< 10.0

From: J. R. Prohaska, *Physiol. Rev.*, 1987, **67**, 858–901.

Table 1.7   Non-haem iron content of zones.

Zone	Non-haem Fe mg/100g
Globus pallidus	21[a]
Red nucleus	20[a]
Substantia nigra	18[a]
Putamen	13[a]
Dentate nucleus	10
Caudate nucleus	9
Thalamus	5
Cerebellum	3
Cerebral cortex	1
Medulla	1
Spinal cord	< 1

[a]These zones are centrally placed in the regions of the midbrain and the lower part of the forebrain and above the hindbrain, where there are also Raphe Nuclei, see Figure 1.1 and Table 1.1.
From: M. B. H. Youdin (ed.), *Brain Iron*, Taylor and Francis, London, 1988.

Before we leave the topic of adrenaline vesicles we again refer to these vesicles in the adrenal gland. The study of the whole adrenal gland by nuclear magnetic resonance (NMR) was the first such study of a whole organ. The examination showed that the signals from adrenaline and the ATP, known to be in vesicles, were sharp, indicating that they were freely mobile. Interestingly the study revealed that the vesicles also contained a functional largely unfolded protein, chromogranin A, and certain steroids were present in high amounts in the cytoplasm. Inspection of the unfolded chromogranin and of its hydrolysis revealed that it contained short hormonal peptides. Similar studies of the vesicles of other glands have shown that for example hydroxytryptamine (serotonin) was stored with other pyrophosphates. They require synthesis by iron and copper oxidases. These studies[3] imply that high resolution proton NMR of local brain regions either *in situ* or in slices could reveal much about the vesicles and their content, but while there are also good stains for the zinc content of cells there are only poor methods available as yet for copper imaging.

Another bound metal ion is iron, but it is loosely bound relative to zinc binding so that free Fe^{2+} in the cytoplasm is high, 10^{-6} M, and it speaks directly to enzymes and to transcription factors. Important in the brain are the haem-containing brain-specific neuroglobins and a special cytochrome - P450, the non-haem, non Fe/S, hydroxylases which participate in the early steps of oxidation to give the aromatic transmitters. Just as is the case with other metal ions too much or too little iron is dangerous in the brain. The supply by carrier proteins of iron and other metal ions such as zinc and copper does not occur through the extracellular fluids which have so little protein. The question remains as to whether this is a function of glial cells.

Manganese is a further essential metal ion in glutamine synthetase and in glycosylation of proteins in the endoplasmic reticulum. Excess manganese does affect brain function as has been observed in manganese miners. There is some evidence for the value of molybdenum in sulfite oxidase in some neurons.

1.9 Evolution of the Brain

Great interest centres on how the brain evolved,[6] since it may throw light on how it functions, but we can only make some speculative remarks as so little is certain. We know that the nematode worm and the octopus have very primitive brains (Table 1.8), and probably little more than corresponds to the cerebrum or cerebellum in the earliest brains in invertebrates (Figure 1.5). These zones coordinate movement and short-term memory and are largely associated with what we take to be the earliest transmitters, as shown in Table 1.7. The later developments of the brain are also shown in Figure 1.5 and include the enlargement of the cerebrum and the cerebellum. Quite possibly this corresponds to the introduction of long-term memory capacity and the more complex sets of movements. Most novel development between the brains of the nematode worms and the most advanced animals lies in sense organs and in control of sleep/wake and endocrine systems. They lead to the complicated states of mind

Table 1.8 Possible evolutionary stages of chemical systems in brain.

Animal (Date)	*Innovation*
Nematode neurons (Precambrian)	Na^+/K^+, Ca^{2+}, acetylcholine? Glycine? Butyrate, γ-amino butyric acid (GABA) Recovery by re-entering synapse
Chordate neurons (Early Cambrian)	As above plus first hydroxylations giving Serotonin and dopamine; Iron/pterin Chemistry in cytoplasm; vesicle filled in centre of cell Recovery by amine oxidation (flavo-enzymes)
Jawless fish (Cambrian)	As above; second hydroxylation giving noradrenaline and amidated peptides Copper chemistry in vesicles Recovery additionally by hydrolysis Zinc enzymes
Complete vertebrates (Late Cambrian)	As above plus myelinated neurons ? Use of zinc enzymes in glial cells ? Free zinc
Many animals?	NO/haem chemistry ? in glial cells?

n.b. Note that many transmitters are synthesised in special zones.

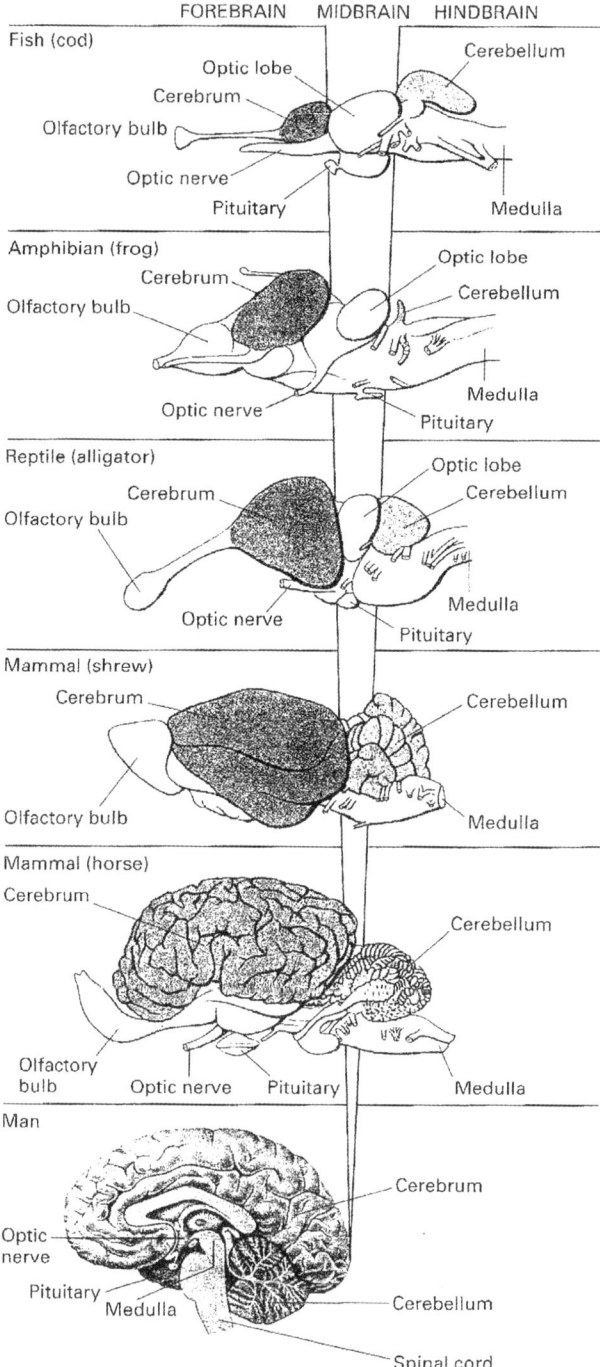

FOREBRAIN MIDBRAIN HINDBRAIN

Fish (cod)
Cerebellum
Optic lobe
Cerebrum
Olfactory bulb
Optic nerve
Pituitary
Medulla

Amphibian (frog)
Optic lobe
Cerebrum
Cerebellum
Olfactory bulb
Medulla
Optic nerve
Pituitary

Reptile (alligator)
Optic lobe
Cerebrum
Cerebellum
Olfactory bulb
Medulla
Optic nerve
Pituitary

Mammal (shrew)
Cerebrum
Cerebellum
Olfactory bulb
Medulla

Mammal (horse)
Cerebrum
Cerebellum
Olfactory bulb
Optic nerve
Pituitary
Medulla

Man
Cerebrum
Optic nerve
Pituitary
Medulla
Cerebellum
Spinal cord

Figure 1.5 A view of the later developments of the brain. A way to follow metal ion involvement during evolution is to analyse these brains, even going back to the octopus and the nematode worm.

and to its disturbances. We may suppose that their evolution was dependent on novel transmitters and novel enzymes for their synthesis. The most obvious likely novelty was through modification of pre-existing amino acids and peptides. They are the hydroxylated amino acids, tyrosine, tryptophan (adrenaline and serotonin), and the amidated peptides prepared by hydroxylation (oxidation) of their carboxyl-terminal glycine. The probability of these transmitters being a late addition is that the enzymes responsible for their production are all copper enzymes. The small molecule transmitter nitric oxide is also prepared by oxidation of another amino acid, arginine, using haem enzymes. It is worth observing that the content of copper is higher in the substantia nigra and locus coeruleus, see Table 1.6. These zones are responsible for motor system connections and awareness (sleep/wake balance) and appear to be a late addition. Metal analysis of primitive brains could be very useful in the understanding of the brain's evolution.

Another metal ion which has a strong locational differentiation in the zones is zinc. It is high in the hippocampus (Table 1.5), which includes the somewhat peculiar neurons, the mossy fibres, which are considered to be functional particularly in spatial resolution and emotional states. We may well think that these are recent developments in the brain. The very presence of considerable copper and zinc in biochemical catalysis of reactions is of relatively recent origin, after 0.75 billion years ago. Thus a consistent story could well be built on the development of brain functions, of novel organic molecules and zinc used as transmitters, and the introduction of copper and zinc in quantity in brain cells. This would suggest that, based on biochemistry, some of the transmitters arose before a billion years ago as the source of the most primitive activity associated with the brain, for example coordinated body movements, while more recently the variety of transmitters has increased.

1.10 Major Functions of the Brain Zones

The general outline of brain functions is known but details are lacking. Table 1.9 lists the possible outstanding functions of brain regions but we must remember that the brain may work as one very large integrated organ with major contributions of particular responses from a local region and with minor contributions from many others. The reason for different transmitters or groups of transmitters in particular zones of neurons appears to be unknown and maybe they are just historical accidents. This possibility is brought out if we look at the evolution of the brain functions from its beginning in nematodes to the sophistication shown in humans (Table 1.10). One approach is to look at brain aberrations from those of a minor kind to those of a major kind including these caused by local injury. Another indication is the effect of drugs that alleviate certain abnormalities in behaviour such as the use of adrenaline (epinephrine) in the treatment of Parkinson's disease. These matters are the subjects of many of the chapters in this volume and we draw attention only to those thought to be related to inorganic elements.

Table 1.9   Some major functions of brain zones.[a]

Amygdala, thalamus (Limbic system)	Emotional and behavioural responses
Hippocampus	Spatial resolution, learning, memory Autobiographical events
Cerebellum	Movement coordination
Tectum (pineal)	Light and auditory responses
Hypothalamus	Controls autonomic nervous and endocrine systems: internal milieu
Red nucleus and substantia nigra	Motor system connections
Locus coeruleus	Vigilance, sleep/wake cycle
Medulla	Control of heart and skeletal muscle

[a]Zones are interconnected but the very fact that they differ in chemical composition, physical appearance and that damage to them causes differential disturbance indicates that they have major functional roles in addition to general cooperative ones.
Taken in part from: J. D. Fix, *High-yield Neuroanatomy*, Lippincott, Williams and Wilkins, Philadelphia, 2nd edn, 2000.

Table 1.10   Possible evolution of the properties of the brain.

Species	*Brain Characteristics*
(1) Nematode ($> 500 \times 10^6$ years)	Simple nerve net from body to head region No senses except touch and general chemotaxis. Ring (of nerve cells) brain.
(2) Amphioxus ($\sim 500 \times 10^6$ years)	As for (1) with an eye spot. Some coordination of sight with the body. No evidence of olfactory or hearing senses. Little or no telencephalox or cerebral cortex. Has forebrain, pineal and hypothalamus-like systems.
(3) Jawless fish (450×10^6 years)	As for (2) with olfactory system and a cerebellum for maintaining posture while viewing. Very small cortex and telencephalox.
(4) Vertebrates with jaws (400×10^6 years)	As for (3) with ever more complex states of awareness. Development of myelin and of mid-brain. Enlarged cortex.
(5) Warm-blooded animals (350×10^6 years)	As for (4) with temperature controls. Enlarged cortex and states resembling emotions and consciousness.

Taken in part from: J. M. Allman, *Evolving Brains*, Scientific American Library, New York, 2000.

1.11   The Chemistry of Brain Damage

Many chapters of this book are devoted to injury to the brain and drugs used to combat brain damage. We can only draw attention to the frequent concentration on the oxidised transmitters and related drugs and on the problems caused by the presence of excesses of metal ions such as Mn, Cu, Fe and Zn. We are still at some distance from understanding the bioinorganic chemistry of the brain, although excessive or aberrant oxidation is thought to be a major problem. Of major interest are Parkinson's disease, related in some way to adrenaline (epinephrine), and Alzheimer's disease, associated with the formation of β-amyloid protein plaques, which can bind both copper and zinc.[4] Is the problem of the plaque not just its presence but its ability to collect metal ions and become an active enzyme?

1.12 Conclusions

This introduction to the role of metal ions in the brain has required a presentation of many general features of this complex organ. Against this complexity it is clear that much detail is lacking so that any definite conclusions as to the exact roles of the metal ions are not yet available. We have therefore attempted no more than an approach showing the problems. Subsequent chapters describe the present state of our knowledge of various normal and abnormal conditions of the brain in the context of this general outline. It is shown that, as elsewhere in the body of organisms, several metal ions have major roles both as free ions and in combination with proteins.

References

Note. Much of the material of this chapter is taken from previous articles and books. We draw attention especially to references 1 and 2. Other major references are included in the Tables where they are most relevant.

There have been several publications recently on brain chemistry related to this chapter. In particular the article by M. J. Pushie, I. J. Pickering, G. R. Morton, S. Tsutsui, F. R. Jink and G. N. George, Prion protein expression levels atters regional copper, iron and zinc content in the mouse brain, *Metallomics*, 2010, **3**, 206–214.

1. R. J. P. Williams, *J. Inorg. Chim. Acta*, 2003, **356**, 27–40.
2. J. J. R. Fraústo da Silva and J. Armando L. da Silva, *The Inorganic Chemistry of the Brain* (in Portuguese), Gradiva, Lisbon, 2008.
3. A. Daniels, R. J. P. Williams and P. E. Wright, *Nature*, 1976, **261**, 321–323.
4. A. Kretzel and W. Maret, *J. Amer. Chem. Soc.*, 2007, **129**, 10911–10921.
5. R. J. P. Williams, *Endeavour*, 1984, **8**, 65–70.
6. J. J. R. Fraústo da Silva and R. J. P. Williams, *The Biological Chemistry of the Elements*, Oxford University Press, Oxford, 2001.

CHAPTER 2

Chemistry and Biology of Proteostasis

M. J. SAARANEN AND L. W. RUDDOCK*

Department of Biochemistry, University of Oulu, Oulu, Finland

2.1 Introduction

Proteostasis is, at its most simple, defined as the regulation of the level of individual proteins that make up the proteome. However, it is often extended beyond the regulated control of the concentration of individual proteins to include modulation of functionality and/or localization of these proteins.

The formation of the composite word proteostasis from the phrase protein-homeostasis sometimes leads to misconceptions as stasis leads to the concept that the levels are fixed or stationary *i.e.* they are invariant. This could not be further from the truth. The levels of different proteins that make up the proteome vary widely, while the level of an individual protein will vary between different organisms, between different tissues within a single organism, between different cells within a single tissue and with different developmental or cell-cycle states or environmental conditions experienced by an individual cell. Cells are constantly adapting to their individual needs and the primary method for doing this is by alterations in their proteome. Hence individual protein levels are not static, they are constantly changing and adapting to the needs of the individual cell. Due to this the proteome of a single cell at any point is probably unique, with similar cells in similar environments showing a clustering of the proteome state without being identical.

RSC Drug Discovery Series No. 7
Neurodegeneration: Metallostasis and Proteostasis
Edited by Danilo Milardi and Enrico Rizzarelli
© Royal Society of Chemistry 2011
Published by the Royal Society of Chemistry, www.rsc.org

At the most simplistic level the level of any individual protein can be considered to arise due to the relative rate of synthesis versus the rate of degradation. However, each of these is a composite of multiple inter-dependent cellular processes (Figure 2.1). Indeed there are no cellular processes which do not, directly or indirectly, modulate proteostasis. Hence when examining the effects of modulators of proteostasis it can be difficult to dissect out all of the processes which contribute to the net change in the system.

Considering proteostasis also leads to issues relating to what should be considered as a single protein. There are an estimated 20 000–25 000 genes in the human genome,[1] but the estimates of the number of individual human proteins have reached as high as nearly 2 million.[2] This difference arises primarily due to differences in mRNA splicing and protein post-translational modification, but other factors such as alternative initiation or protein splicing also contribute. While mRNA splicing and post-translational modifications undoubtedly result in the generation of distinct protein species with different biophysical properties, the question arises whether this defines the whole proteome. Two further issues arise (Figure 2.2).

Firstly, proteins are not conformationally static entities; rather the native state of a protein is best described as a probabilistic ensemble of inter-converting structures. While these structures do not chemically differ from one another, with the possible exception of the protonation states of individual side chains or the N- or C-terminus, they may vary considerably with respect to biophysical properties and especially with respect to biological activity. With the exception of gross changes in structure, for example those experienced by prion proteins,[3,4] these different structural states are often not considered. Furthermore, again with the exception of gross changes in structure, they are exceedingly difficult to either study or even define under physiological conditions and yet such conformational exchange or flexibility is essential for enzyme function.[5,6] The population of different conformational states for a given protein

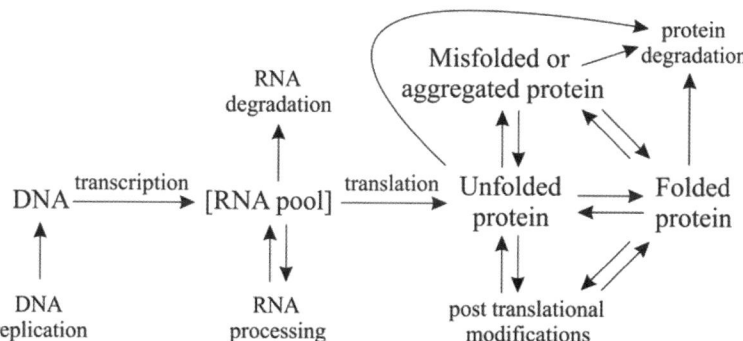

Figure 2.1   Overview of proteostasis pathways. The amount of an individual folded protein in the cell depends on multiple pathways. All of these pathways interconnect with each other and with all other cellular processes *e.g.* metabolism, signaling, *etc.* All of the pathways shown can be modulated by metallostasis.

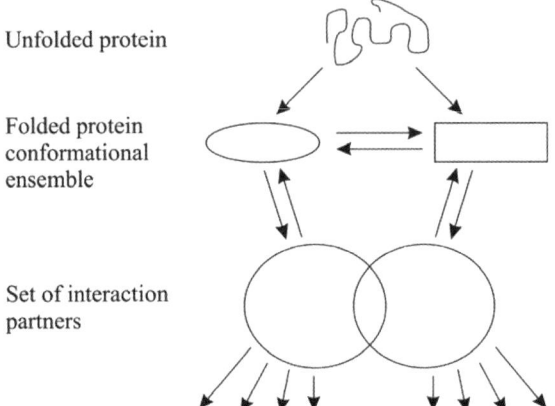

Unfolded protein

Folded protein
conformational
ensemble

Set of interaction
partners

Each complex has different biological properties

Figure 2.2   Conformational and complex formation, inter-conversion and modulation. Folded proteins are not a single structural entity, rather they are in
dynamic equilibrium between different states (for simplicity only two are
shown), which will have different biological properties. In addition, biological function is modulated by the formation of transient interaction
complexes between different proteins, some of which are mutually
exclusive. Interaction with other proteins (or other small molecules) will
modulate the equilibria between the different conformational states, and
different conformational states may have a different set of interaction
partners. Both conformational exchange and the formation of transient
interaction complexes are highly dynamic allowing rapid changes in
response to the needs of the cell without the absolute requirement for
changes in the cellular concentration of the protein. Both can also be
highly regulated, for example by post-translational modifications.

and the rates of inter-conversion between them are extremely sensitive to changes
in environment and hence functional changes in the activities of a proteome
may result from just about any change to the cell or its environment, and this
may not be observed in quantitative changes in protein concentrations.

The second issue that arises with respect to defining the entire proteome is
that proteins, especially under physiologically relevant protein concentrations
of up to 300 mg/ml of protein in a typical cell, are constantly forming interaction complexes. These complexes, also known as the interactome,[7] in turn
either regulate or define protein function. These complexes may often be very
transient and one protein may have multiple possible interaction partners at
least some of which will be mutually exclusive. The interactome is extremely
difficult to characterize or define, especially under physiological conditions, but
conceptually it is easy to hypothesize how functional changes in the activities of
a proteome related to one protein may result without quantitative changes in
the concentration of that protein (see Figure 2.2).

While the complications of protein flexibility and proteome ensemble
(and their inter-dependence) are often not fully considered due to the difficulties
of quantification of either under physiological conditions, even examining

proteostasis at the less complex level of protein concentrations is not simple. Many methodologies exist for the quantification of individual proteins in a complex mix, for example western blotting, but these often lack sensitivity and/or cannot distinguish between different post-translational modifications. These methodologies may also result in a lack of appreciation of the complexity of the *in vivo* situation where usually there are multiple changes in the proteome induced by any change in the state of the cell such as changes in metallostasis. On a global level proteomic approaches such as two-dimensional gels and/or mass spectrometry are often used. However, there are over one million human proteins and these global techniques maximally look at a small percentage that are the most abundant,[8,9] unless sub-fractionation is used. Since global quantification of the human proteome has not been achieved and would be prohibitively costly, such pre-fractionation is often based on pre-conceived ideas regarding which system(s) or types of protein may be altered under varying physiological conditions, something which introduces an intrinsic bias to the data set obtained. In addition, questions can be raised about the effects on quantification of individual proteins of such sub-fractionation and/or on quantification of the different sub-populations of an individual protein, for example post-translational modifications. Hence, powerful though these techniques are, they should be viewed as being semi-quantitative. Finally, while micro-array analysis can be used to quantify global mRNA levels it does not give a quantitative picture of the proteome, even when validated by quantitative reverse transcriptase (RT)-PCR, since mRNA levels often do not correlate well with protein levels and they give no information on which post-transcriptional modification variant(s) are present.

Many factors affect proteostasis. Due to space limitations each area will be discussed only briefly with our main focus being on those that are widely acknowledged to be significant with respect to a link between proteostasis and neurodegeneration. In addition to generic mechanisms that modulate proteostasis, we will also consider later in this chapter specific issues relating to metalloproteins which provide further strong links between metallostasis and proteostasis. First though the generic mechanisms for regulation of proteostasis are described (see also Figure 2.1).

2.2 Cellular Processes affecting Proteostasis

2.2.1 DNA Replication

The fidelity of DNA replication is sometimes overlooked when considering proteostasis as it is not a mechanism by which an individual organism regulates the levels of individual proteins within their proteome. However, proteins are translated from mRNA which is transcribed from DNA and so errors in DNA replication may directly result in mutations in genes which either alter the sequence of the encoded protein or change its expression level due to alterations in the rate of transcription or translation or alterations in mRNA processing or stability.

Replication of DNA is a multi-step process,[10] which has numerous minor variations. Initiation of DNA replication requires the formation of a pre-replicative complex, marking the origin of replication and the establishment of replication forks. These are associated with the replisome, a multi-component protein complex that includes factors for unwinding the coiled DNA and elongation factors, such as DNA polymerases, that carry out replication. Replication proceeds in a semi-discontinuous manner with nucleotides being added only to the $3'$ end of the synthesized strand. Under normal physiological conditions the error rates of DNA replication are normally very low in comparison with other related processes,[11] with editing and repair mechanisms giving rise to error rates of 1 in 10^8–10^{10}. While this error rate is very small, it is essential for evolution to proceed and has implications for proteostasis.

Given the fidelity of replication it is now believed that an individual will have 150–200 changes in nuclear encoded DNA compared with the DNA inherited from their parents.[12] This is a global average since errors in replication occur with each cell division and hence different cells within the same person may have differences in their DNA. Furthermore, significant differences at the DNA level arise in the human population as it is estimated that between 1 and 4% of people in Eurasia show genetic descent from Neanderthals.[13] In extreme cases errors in DNA replication may directly lead to a disease state, including cancer, but more often the effects of these mutations are more subtle; they may result in a small change to a single protein or system, but the cell adapts to this by other, often subtle, changes in the proteome.

Many of the enzymes involved in DNA and RNA metabolism, including DNA polymerases involved in DNA replication, are magnesium dependent and hence levels of this metal can modulate proteostasis (see Section 2.3.2). Additionally, metals such as nickel and chromium result in alterations in the fidelity of replication,[14,15] with concomitant changes in proteostasis. Other enzymes which feed into the process also contain structural or catalytic metal ions, for example ribonucleotide reductase, which is required to generate deoxyribonucleotides, has a di-iron tyrosyl radical cofactor essential for catalysis,[16] providing further links between proteostasis and metallostasis.

In addition to nuclear DNA, eukaryotic organisms have two other distinct repositories of DNA, in mitochondria and for plants and phototropic protists in chloroplasts. Given its localization this DNA requires alternative routes for DNA replication. Just like the replication of nuclear DNA, the replication of mitochondrial DNA is not a mechanism by which an individual organism regulates the levels of individual proteins, however mitochondrial DNA replication has a number of unique features which may result in unique proteostasis modulation. Mitochondrial DNA is replicated by a nuclear encoded DNA polymerase, the *POLG* and *POLG2* gene products in humans.[17] These proteins are translated in the cytoplasm and then transported into mitochondria presumably by the TOM (transport outer membrane) and TIM (transport inner membrane) pathways.[18] Due to a combination of differences in the available repair mechanisms and damage due to reactive oxygen species produced during aerobic respiration the mutation rate of mitochondrial DNA replication is ten fold greater than that of nuclear encoded DNA.[19] This implies a greater chance in variation of the mitochondrial proteome between individuals and between cells within an individual. One further consideration with respect to proteostasis for an individual is that mitochondrial DNA is maternally inherited, in contrast to nuclear DNA which is inherited from both parents.

2.2.2 Transcription

Transcription is the process by which the genetic code is copied from the storage state (DNA) to the cellular functional state (RNA). There are significant differences in the mechanisms of transcription in prokaryotes and eukaryotes, especially in the initiation and termination steps, and in prokaryotes transcription is linked to translation. The primary regulation of proteostasis by biological systems is via modulation of transcription, the levels of individual cellular proteins being adjusted by the up- or down-regulation of the synthesis of the corresponding mRNAs. Transcription is controlled by complex pathways that depend upon the presence of elements in the promoter region of each gene. In eukaryotes the core promoter sequence is found within 100 base pairs upstream of the transcription initiation site, though other promoter or enhancer elements are typically found with five kilobases of this site and may lie upstream or downstream. These elements are bound by transcription factors which, in complex with other proteins, may either activate or repress transcription. One defining feature of transcription factors is that they must be able to bind to specific sequences of DNA; they do this by having one or more DNA-binding domains.

Transcription is linked to metallostasis at multiple levels. For example, in eukaryotes several RNA polymerases (pol) share the task of DNA transcription. Pol I is responsible for more than half of all nuclear transcription, including the synthesis of 28 S, 18 S and 5.8 S ribosomal RNA, while pol II synthesizes mRNA and many non-coding RNAs and pol III produces 5 S rRNA, tRNA and other short non-coding RNAs.[20] All of these RNA polymerases are divalent metal ion dependent enzymes and therefore their activity is linked to metallostasis. Another direct example is that one class of DNA-binding domain used by transcription factors is the so called zinc-finger domain, which actually consists of several subclasses of zinc–cysteine motifs.[21] A search of the human genome revealed the presence of circa 900 C2H2 zinc finger transcriptional activators out of a total of 2000 hypothetical transcription activators.[22] This reveals both the complexity of transcriptional regulation and the intimacy of the link between metallostasis and transcription and hence proteostasis. The converse is also true; metallostasis is regulated primarily by transcriptional regulation due to the presence of metal-responsive transcription factors.[23]

While transcription is often considered to be a nuclear process in eukaryotes, transcription must also occur in other cellular compartments in which DNA is stored, *i.e.* in mitochondria and chloroplasts. Human mitochondrial DNA encodes 37 genes of which 24 encode RNA molecules (tRNA and rRNA) involved in the process of translation. Other components of the transcription apparatus, for example mitochondrial DNA directed RNA polymerase,[24] are nuclear encoded and must be imported into mitochondria by the same mechanisms as those proteins involved in DNA replication (see Section 2.2.1).

2.2.3   mRNA Processing

While many of the processes involved in transcription are similar in eukaryotes and prokaryotes one notable difference is that eukaryotic RNA polymerase associates with mRNA processing enzymes during transcription. This is the start of hugely differing pathways for mRNA processing, with eukaryotic mRNAs undergoing a much wider variety of mRNA processing, including mRNA splicing which allows one eukaryotic gene to encode multiple mature mRNAs and hence multiple different proteins. In eukaryotes the first processing event that happens to the nascent mRNA transcript is methylation of the 5′ guanosine to generate a 7-methylguanosine cap required for ribosome binding. Three enzymatic activities are needed, a triphosphatase, a guanyl transferase, and a methyl transferase. The first two may be present either in one enzyme as in mammals or in two separate enzymes as in yeast, while all eukaryotes have a separate methyl transferase.[25] The next step in mRNA processing is splicing, where introns, the gene-interrupting non-coding sequences, are removed by a macromolecular machine called the spliceosome. The spliceosome consists of five small nuclear ribonucleoprotein particles and a large number of protein splicing factors.[25,26] The transcripts generated are then cleaved from the 3′ end through the recognition of *cis*-acting elements by an endonuclease complex and then have a polyadenyl tail added. While the cleavage and polyadenylation reactions *per se* are simple, the machinery required to do it is complex. Polyadenylation protects mRNA from degradation by exonucleases, and facilitates mRNA release from the transcription site and its ultimate export through the nuclear pore complex to cytoplasmic translation.[25,27] Modulation of any of the processes involved in mRNA processing will have very significant effects on proteostasis.

A large number of the proteins involved in mRNA processing are metalloproteins. For example, the spliceosome employs at least one divalent metal ion in catalysis in a conserved region of U6 small nuclear RNA.[28] Similarly, the cleavage and polyadenylation specificity factor CPSF-73 has a metallo-β-lactamase domain that binds divalent cations such as zinc and iron, CPSF-30 has five zinc-finger domains and poly(A) polymerase contains an N-terminal domain that coordinates the two metal ions (magnesium or manganese) that are required for catalysis.[29]

RNA processing not only occurs for nuclear encoded RNA molecules, but also for mitochondrial and chloroplast encoded RNAs. The 37 mitochondrial encoded genes are transcribed from just three promoters, resulting in the formation of polycistronic transcripts which are subsequently processed into mRNA, tRNA and rRNA molecules.[24,30] Similar to modulation of mRNA processing of nuclear encoded genes, an alteration in mRNA processing of mitochondrial encoded genes will have significant effects on proteostasis and such alterations can arise from changes in metallostasis.

2.2.4 mRNA Stability

The lifetime of an mRNA will directly influence the number of protein molecules which can be made from it. mRNA stability depends on a number of factors, for example in eukaryotes it is promoted by 5' end capping and 3' end processing *e.g.* the binding of poly(A) binding protein (PABP) to polyadenylated mRNA inhibits degradation in mammalian cells.[29] While the lifetime of the mRNA influences the degree of translation possible, translation in turn influences mRNA stability. mRNAs that are being translated are held in a 'closed loop' conformation, where the 5' cap is kept in close proximity to the polyadenylated tail and thus protected from exonucleolytic degradation. Decapping, de-adenylation and/or endonucleolytic cleavage of mRNA renders it more susceptible to degradation by exonuclease complexes, called exosomes, or discrete cellular foci called cytoplasmic processing bodies (P-bodies).[31,32] The half-lives of mRNAs vary very significantly from the long and constant half-lives of those for housekeeping genes to a group of mRNAs with a very short half-life and fast induction and repression rates after transcriptional activation or arrest. The half-life of a particular mRNA depends on sequences in the transcript itself, generally located in the 3' untranslated region (3' UTR) of the mRNA (*e.g.* AUrich element, ARE) and on RNA-binding proteins that bind to these sequences. mRNA degradation can be also triggered by the binding of specific micro-RNAs to the 3' UTR of target mRNAs.[32] As for mRNA synthesis and processing, many of the enzymes connected with mRNA stability are metalloproteins and hence can be deleteriously affected by changes in metallostasis.

Metal ions are also used in biological systems to modulate mRNA stability, especially as feedback systems for metallostasis. For example, the binding of iron by iron-responsive proteins results in their interaction with iron-responsive elements in the 3' untranslated region of the transferrin receptor mRNA which stabilizes the mRNA, resulting in increased levels of the protein.[33]

2.2.5 Translation

Translation is the conversion of the genetic code stored in mRNA to the amino acid code used in proteins. It comprises three distinct phases, initiation, elongation and termination, with charging of the tRNA molecules with the correct amino acids as an essential feed-reaction. The exact mechanisms used vary between species, with sharp differences between translation in prokaryotes and eukaryotes.

The fidelity of translation is cited as being 1 in 1000 to 1 in 10 000 and as such translation has a higher error rate than either DNA replication or transcription.[11] These errors are often stated to arise either from misacylation of the tRNA *i.e.* loading in the incorrect amino acid on a tRNA, or from the incorrect use of a tRNA during transcription, but other errors during translation such as frameshifting or incorrect termination can also occur.[34] The aminoacyl-tRNA synthetases use a combination of methods to ensure increased fidelity of aminoacylation,[35] including having a second active site which acts as a proof-reading mechanism, similar to that used by many DNA polymerases. These result in an error rate of around 1 in 10^4–10^5 for acylation. The fidelity of extension of the amino acid chain depends on a number of factors, including kinetic partitioning.[34] The fidelity of both steps in translation is dependent on metallostasis, for example many of the aminoacyl-tRNA synthetases require a metal ion cofactor for catalysis.

The rate of translation is dependent on many factors. Since partially translated proteins cannot fold to their biologically active native state and are more prone to aggregation and/or degradation events the rate of translation affects proteostasis. The k_{cat} for the ribosome is often cited[36] as being around 50 s^{-1} but has recently been shown to be faster[37] and to be independent of the aminoacyl-tRNA used.[38] However even after initiation, translation is slower than this and can be limited by the availability of amino acid-charged tRNAs which in turn is linked to the availability of each mature tRNA, the aminoacyl-tRNA synthetases and amino acid concentrations. Hence many processes modulate the rate of translation and metallostasis feeds into translation on many levels.

In addition to the non-physiological effects of metal ions on translation, metal ions are used by biological systems to regulate translation of some genes. For example, iron binding by iron-responsive protein 1 results in it binding to iron response elements in the 5′ untranslated region of the ferritin mRNA, inhibiting translation of the ferritin protein.[33] As for the use of iron response elements in modulating mRNA stability such systems are used in feedback mechanisms in metallostasis.

Usually the initiation codon for translation is AUG (encodes methionine). However, there are examples of other codons *e.g.* CUG (encodes leucine) or GUG (encodes valine) also being used.[39,40] It is currently unclear how such alternative initiation occurs or what the potential regulatory mechanisms are, and hence it is not known how this aspect of proteostasis is or can be modulated.

While translation is often referred to as occurring in the cytoplasm, this is only true for prokaryotic proteins or for nuclear encoded eukaryotic proteins. Translation also occurs in mitochondria and chloroplasts with both compartments having distinct translational apparatus and regulatory mechanisms, with 22 tRNA molecules and two ribosomal RNA subunits being encoded by mitochondrial DNA.[41] The differences extend to codon usage with mitochondrial codons showing species-specific differences compared with nuclear encoded proteins. In vertebrate mitochondria the codon AUA encodes methionine instead of isoleucine, UGA encodes tryptophan instead of being a stop codon, and AGA and AGG encode stop codons instead of arginine.[41]

As for transcription, translation in these compartments requires mitochondrial import and hence translation may be modified directly or indirectly, either of which will modulate proteostasis.

2.2.6 Protein Folding

The rate of translation is around five amino acids per second in mammalian cells and hence a large protein such as apolipoprotein B may take more than 10 minutes to be synthesized,[42] and this imposes a limitation on the time it takes for a protein to adopt its native conformation. Since unfolded or partially folded intermediates on the folding pathway have a higher propensity to have exposed hydrophobic side chains than folded proteins, they are much more prone to the formation of aggregate, which results in partitioning of proteins to an amorphous inactive state. To avoid this, two broad mechanisms are adopted. Firstly, many proteins are formed from distinct modules, or domains, which are able to fold autonomously from the rest of the protein *i.e.* the N-terminal region of the protein may be able to fold without having to wait until the C-terminal region is synthesized. Secondly, distinct cellular factors have evolved to aid protein folding. These fall into two general categories: i) protein folding catalysts such as peptidyl-prolyl *cis–trans* isomerases or catalysts of disulfide bond formation and thiol-disulfide isomerization which catalyze steps on the productive folding pathway, and ii) molecular chaperones, such as the small heat shock proteins or Hsp60, Hsp70, Hsp90, Hsp100 families which either inhibit off-pathway reactions or return off-pathway folding states to the productive pathway. Protein folding is complex, may often be protein specific and the results may be contradictory and/or dependent on the methods used to examine folding. However, there is growing convergence between *in vitro* and *in vivo* concepts.[43]

As would be expected from the central importance of protein folding in proteostasis, protein folding is highly regulated with complex feedback mechanisms, the most well characterized of which are the heat shock response[44–46] and the unfolded protein response (UPR).[47–51] These regulatory pathways often operate on multiple levels. For example the UPR, which is linked to protein folding in the endoplasmic reticulum (ER), is known to have at least three parallel pathways in humans, acting via the transmembrane proteins IRE1, PERK and ATF6, and activation of these pathways results in increased transcription of protein folding catalysts and molecular chaperones, decreased rates of general protein translation, enlargement of the ER, increased rates of ER-associated protein degradation, alterations in metabolism and if all else fails the UPR triggers cellular apoptosis.

There is a large number of disease states associated with defects in protein folding. These include loss of function diseases such as Fabry, Gaucher and cystic fibrosis and gain of toxic function diseases such as Parkinson's, Huntington's and Creutzfeldt–Jakob.[11] Due to the combination of numerous disease states and cellular mechanisms that can be manipulated, protein folding is a major target for manipulating proteostasis.[52–54] Such manipulation includes the counter-intuitive approach of using enzyme inhibitors to recover functional levels of the enzyme being inhibited, for example in the treatment of lysosomal storage diseases,[55] an approach based on so called pharmacological chaperones or small molecules *e.g.* ligands or inhibitors, which bind to and stabilize the native state and/or folding intermediates.

It should be remembered that not all proteins attain a stable tertiary structure. Instead some have limited regular secondary structure and significant conformational flexibility. While these have been called natively unfolded proteins this name is potentially misleading and they should be more appropriately known as intrinsically disordered proteins (IDPs).[56–58] While IDPs may be relatively rare, many proteins may have IDP regions, with an estimated 28% of mammalian proteins containing extensive IDP regions.[59]

Metal ions are known to significantly affect protein folding in many systems; for examples see Chapters 6, 11 and 12 of this book. The interaction between folding proteins and metals is similar to that between metals and folded proteins except that the polypeptide chain has not yet reached a native conformation and hence residues which are normally buried in the core of the protein are also accessible in the non-natively folded state. Metal ions may also interact with molecular chaperones and protein folding catalysts. For example, the ER-resident molecular chaperone uses adenosine triphosphate (ATP), and ATP binding is enhanced by calcium or magnesium,[60] while the activity of the ER-resident protein folding catalyst PDI is inhibited by zinc.[61]

Much of the focus of the effects of metal ions on protein folding has been on their influence on oxidative protein folding. Oxidative protein folding occurs in the endoplasmic reticulum and inter-membrane space of the mitochondria of humans as well as in the periplasm of bacteria and the cytoplasm of some archaebacteria. It is differentiated from other protein folding by the formation of disulfide bonds between the thiol side chains of cysteine residues, a process which increases the oxidation state of the sulfur atoms involved from -2 to -1, and hence the name. Native disulfide bond formation is a relatively slow and complex process, being comprised of least two distinct steps, disulfide bond formation (or oxidation) and disulfide bond rearrangement (or isomerization) to the native state.[62–64] Metal ions have a significant effect on oxidative protein folding due to the significant interaction of the thiol side chains of cysteines with metal ions, an interaction which generally inhibits catalyzed oxidative protein folding. While around 2.5% of amino acids found in human proteins are cysteines,[65] relatively few reactive thiol groups can be found on the surface of natively folded proteins and hence unfolded proteins and/or protein folding intermediates represent a significant proportion of the pool of metal binding protein thiols.

2.2.7 Post-Translational Modifications

As described in the introduction to this chapter, 20 000–25 000 human genes are thought to give rise to around two million different proteins. While splicing, alternative initiation of translation, *etc.* contribute to this, the largest single category of processes which contribute to this difference in the number of genes *vs.* proteins is post-translation modification or PTM. PTMs are covalent modifications of the amino forming the primary structure of the protein that occur after translation. The name itself is potentially misleading as many PTMs can occur co-translationally, for example disulfide bond formation or N-glycosylation.[66,67] However, in these cases the amino acids being modified have been translated even if the whole of the polypeptide chain has not yet been completed.

More than 300 different types of PTM are known,[2] the most common of which include phosphorylation, disulfide bond formation, N-glycosylation, proteolytic processing, hydroxylation, ubiquitinylation, acylation, alkylation, amidation, nitrosylation and sulfation. Not all PTMs are controlled by cellular mechanisms. For example, oxidation of the side chains of amino acids such as methionine or glycation of amino acid side chains occurs spontaneously inside (and outside) the cell. For any given protein the PTMs may be homogeneous, for example a folded protein will usually always have the same cysteine residues linked together to form disulfide bonds. However, PTMs for a given protein are often heterogeneous, for example N-glycans may be variably added to consensus N-glycosylation sites in proteins in the ER and then these may be further heterologously modified in the ER and Golgi to give tens or even hundreds of different protein species from the same gene product.[68,69] PTMs may be synergistic or they may be mutually antagonistic, for example phosphorylation at one site in a protein may either increase or inhibit the likelihood of phosphorylation at another site. This complex interplay between PTMs is combined with the large number of proteins involved in the process; for example there are over 800 human known and putative kinases and phosphatases[70,71] each with unique specificities and mechanisms of regulation. The ensemble of different possible PTMs that a protein has may be unique for a given protein and for many will be dynamic, showing temporal-dependent changes with varying cellular environment. Hence proteostasis depends on PTMs and PTMs depend on proteostasis.

While PTMs may result in only very small changes in mass or pI of the protein, they can result in very large changes in function – a good example of which is phosphorylation, a reversible PTM widely used in cellular signaling cascades which can in some cases switch activity between inactive and active states, or in other cases can cause more subtle changes in function. PTMs can also change the stability of a protein, its interactome, its folding and its sub-cellular localization.

PTMs regulate most cellular processes, for example phosphorylation, methylation, ubiquitinylation and acetylation of histones regulate access to DNA and hence can modulate DNA replication and transcription and, along with DNA methylation, play a role in epigenetics[72] which in turn modulates proteostasis.

PTMs are sensitive to changes in metallostasis. For example, kinases use magnesium in their catalytic cycle, while many phosphatases are also metalloproteins, for example with zinc and/or magnesium as a cofactor, and in addition many proteins involved in phosphorylation are activated by calcium.[73,74] Similarly, prolyl hydroxylases, both those involved in protein biosynthesis *e.g.* collagen hydroxylation and those involved in the hypoxia response, use iron as a cofactor.[75,76] A third example would be methylation of both DNA and proteins which is linked to metallostasis at a variety of levels. For example, S-adenosylmethionine required for methylation is synthesized and regenerated by methionine synthase, an enzyme which has a cobalt-containing cofactor.[77] Numerous other examples could be given, but since PTMs and their effects can be so protein specific and since PTM pathways are so numerous and complex, it is sufficient to say that any cellular pathway may be modulated by changes in proteostasis that could be linked to metal-dependent changes in PTMs.

2.2.8   Protein Transport

The correct sub-cellular localization of a protein is essential for its function. There are several mechanisms for ensuring correct localization. These include cleavable signal sequences,[78] for example targeting to the ER, mitochondria or chloroplasts, and other targeting sequences within the protein, for example the C-terminal KDEL motif required for ER-retention or C-terminal peroxisomal signaling motifs.[79,80] Other mechanisms are based on PTMs, for example membrane association due to prenylation, palmitoylation, *etc.*[81,82] or transport from the Golgi to lysozomes based on the presence of mannose 6-phosphate.[83] The interactome of a protein may also play a role in sub-cellular localization, for example the prolyl-4-hydroxylase α-subunit is retained in the ER by physical association with the KDEL-containing protein PDI.[84] All of these mechanisms may be modulated by changes in proteostasis as outlined in the other sections, for example the insulin-degrading enzyme (IDE), a zinc-metallopeptidase, undergoes alternative initiation of translation which results in translation of a protein either with a mitochondrial targeting signal or without one, which results in its cytoplasmic location.[85]

2.2.9   Protein Degradation

Protein turnover occurs in most cells, exceptions being red blood cells which have no mechanisms for making new proteins, and compartments such as the lens of the eye whose proteins do not change during the course of the life of the organism.[86] Protein turnover requires degradation of proteins such that the amino acids can be released and re-used for new protein synthesis. The rate of degradation of different proteins varies widely from minutes to several days[87] with generic rules, such as the N-end rule in prokaryotes, giving indications of the likely half-life of any individual protein. Many of the best known proteases, such as trypsin, chymotrypsin and pepsin, are secreted from the cell, but others are cell associated, including lysosomal resident proteases and the proteasome, both involved in protein turnover, and a variety of proteases involved in protein maturation and/or activation. Complex mechanisms have evolved connected with individual cellular processes to ensure correct degradation of proteins. For example, ER-associated degradation (ERAD) is linked to the selective proteolytic removal of proteins that misfold in the ER.[88,89] Similarly, mechanisms exist to link degradation with transcription[90] to interlink different aspects of proteostasis.

Proteases are split into different categories dependent on their catalytic mechanisms. Zinc proteases, which include the matrix metalloproteases (MMPs) involved in tissue remodeling, as their name suggests contain a catalytic zinc atom.[91,92] There is a strong link between MMP activity and neurodegeneration.[93] Other proteases respond to changes in metallostasis, for example calpains are activated by calcium and this may be linked to neurodegeneration.[94] The proteasome, a large complex found in eukaryotes and archaebacteria but not in eubacteria, is central to regulated intracellular protein turnover[95] and is also sensitive to metallostasis. This sensitivity arises by multiple pathways including direct inhibition of the proteasome by metal containing compounds,[96,97] and also because one of the pathways that targets proteins for degradation is based on ubiquitinylation and some ubiquitin ligases contain RING finger domains which have structural zinc ions.[98]

2.2.10 Biophysical Conditions

Protein structure and function are often sensitive to even minor changes in biophysical conditions, for example pH, temperature, ionic strength. While the cell has responses to these changes, for example the heat shock response, even minor variations in these conditions are likely to result in changes in proteostasis. In addition to changes in physical conditions, the cell also often uses feedback inhibition mechanisms to regulate protein function and hence changes in the concentration of any metabolite in the cell or organism will usually result in the alteration of metabolic processes which link to that metabolite, with potential concomitant changes in proteostasis. With the exception of very extreme conditions which will inactivate the majority of proteins, the effects of all of these changes in environment on individual proteins in the proteome will be very dependent on the protein.

2.2.11 Flora and Fauna

It must be remembered that we do not live in isolation, human somatic and germ cells are outnumbered by a factor of ten by microorganisms living on and in humans.[99] The mutalistic, commensal or parasitic interactions between the human host and these organisms will result in changes in proteostasis in both. Furthermore there may be genetic transfer of material between organisms which further modulates interactions *e.g.* the transfer of genes encoding carbohydrate metabolizing enzymes from marine bacteria to gut microbiota of

Japanese individuals is connected with eating sushi.[100] Recent studies suggest that the population of bacteria growing on our skin may be more unique to the individual than fingerprints.[101] Hence each of us will have unique changes in proteostasis resulting from these unique colonizations, changes that are constantly modulating as our associated flora and fauna alter. With the exception of associations which result in such a significant change that we register it as a disease state, these changes are rarely considered to the extent which they possibly should be.

2.3 Metalloproteins

While biochemistry is essentially the application of organic chemistry by living systems, and hence the most abundant elements in all living organisms are carbon, hydrogen, nitrogen and oxygen, there are also a large number of metals that are found playing a functional role in all organisms. These include the highly abundant elements calcium, magnesium, potassium and sodium as well as the essential, but less abundant, metals cobalt, copper, iron, manganese and zinc. In addition to these some organisms have a specific requirement for trace amounts of arsenic, chromium, gallium, molybdenum, nickel, tungsten and vanadium.

While some metal ions are required for functions such as signaling within a cell or across a membrane, the major function of metals within biological systems, especially for transition metals, is as structural or functional elements of proteins. In particular metal ions provide alternative reaction mechanisms to those available through the use of amino acid side chains alone: nearly half of all classes of enzymes for which a structure has been solved have a functional or structural metal ion.[102]

Due to their requirement for metal ions, proteostasis of metalloproteins has a direct link to metallostasis. Any decrease in the concentrations of metals, either globally within the organism or locally within the cell, for example due to changes in diet or malfunction or dysregulation of carrier or transporter molecules, may result in suboptimal production of the metal ion-containing cofactor or loading of the metal ion into the protein. For some proteins this will result in the formation of a stably folded, but inactive, protein while for others the result will be severe misfolding resulting in enhanced rates of degradation. Since feedback systems for regulating proteostasis are linked mainly to the activity of proteins either of these situations will result in increased protein production, but the former will lead to increased levels of protein (mainly inactive), while the latter will not.

As well as the direct effect on the proteostasis of metalloproteins by changes in metallostasis, decreases in the intracellular concentrations of metals will lead to direct and indirect effects on global cellular proteostasis. This may be a direct effect, for example the loss of activity of a metalloprotein may result in global changes in transcription, translation or metabolism, *etc.* For example arginine-tRNA ligase is a magnesium dependent enzyme, with concomitant global changes in proteostasis. The effect may also be more indirect. Since the cell has limited capacity to synthesize proteins any requirement to substantially increase the level of transcription, translation or protein folding for one protein or sub-class of proteins will result in a net decrease in the capacity of the system for the production of other proteins, with a potential concomitant decrease in their levels. Similarly, increases in the amounts of metals in the system will result in significant changes in proteostasis. It is especially worth noting that elevated levels of one metal often result in biological deficiency in another due to inhibition of uptake or recycling systems and that elevated levels of one metal can result in the inappropriate loading of that metal ion into the active site or metal cofactor of an enzyme, usually resulting in substantially altered biophysical properties. In addition, the non-physiological interaction of proteins with metals, especially during protein folding, can have a large effect on proteostasis. Hence, metallostasis is dynamic, with the levels of individual metals influencing the levels of other metals, proteostasis is dynamic, with the levels of individual proteins influencing the levels of other proteins, and the interplay between metallostasis and proteostasis is dynamic, with variations in metal concentrations impacting on individual and global protein levels and *vice versa*. This complex interplay makes the elucidation of specific effects *in vivo* very challenging, but this is essential for the development of novel proteostasis regulators.

Given the large number of metalloproteins it is worth *briefly* reviewing the role(s) of individual metals in proteins so that potential global changes in cellular proteostasis can be borne in mind before considering the specific links between metallostasis, proteostasis and neurodegeneration that are discussed in Chapters 4 through 13. So-called second and third tier elements, those found in all organisms include nine metals with biological importance.

2.3.1 Calcium

Calcium is the most abundant metal in the human body. It is used in multiple secondary messenger systems, as a major component of bones, in muscle contraction and in other physiological functions such as blood clotting. Calcium is bound by many proteins with low affinity, with many ER-located proteins having high-capacity, low-affinity calcium binding as a secondary function, to allow the ER to act as a calcium store.[103] In addition, some secreted proteins such as Factor X undergo a gamma-carboxylation post-translational modification to allow higher affinity calcium binding.[104] Calcium is also used as a cofactor by a number of enzymes including phospholipases, mannosidases, transketolase and peroxidases.[105–108] Calcium deficiencies are mainly initially manifested in symptoms related to its role in muscle contraction, including paresthesia, spasms and cardiac arrhythmia,[109] and changes to proteostasis may be linked to any of its physiological functions. Calcium homeostasis shows age dependent changes with implications for function in peripheral neurons.[110] Hypercalcemia is a relatively rare condition in isolation and is usually linked to other disease states.

2.3.2   Magnesium

Magnesium is the most widely occurring catalytic metal ion in enzymes with known structures, with nearly 250 different enzyme classifications utilizing it.[111] Whereas many metalloenzymes have a tightly bound metal ion, or metal-containing cofactor, magnesium is often found bound to substrates, especially phosphate groups, and so only transiently associates with the enzyme. Due to the association between magnesium and phosphate nearly all enzymes which synthesize or utilize ATP or which use other nucleotides, including the synthesis of DNA and RNA, are magnesium dependent. In addition to the activation of P–O bonds through polarization, magnesium is also used by many enzymes to polarize C–O bonds, for example isocitrate dehydrogenase, involved in the Krebs cycle. Given the critical role of magnesium-dependent enzymes in DNA replication, transcription, translation and metabolism it is unsurprising that changes in magnesium levels will have very significant biological effects. In addition to its role in these processes, magnesium regulates many other systems including sodium–potassium exchange, calcium release from the sarcoplasmic reticulum and neurological effects such as blocking NMDA glutamate receptors and release of acetylcholine.[112] Given these combined effects it is not surprising that magnesium deficiency and hypomagnesemia exhibit many clinical features, many of which are neurological.[113] Excess magnesium is effectively excreted by the kidney[114] and hence hypermagnesemia is rare.

2.3.3   Manganese

Manganese is an unusual metal in metalloproteins. It is an essential element and is an important catalytic element in a large number of enzymes. However, in many of these manganese is apparently interchangeable with either magnesium or zinc.[111] Manganese is an essential component of human Mn–superoxide dismutase, pyruvate decarboxylase, glutamine synthase and arginase and has other essential roles in biological systems such as the plant photosystem II and in reverse transcriptases of many retroviruses. As manganese is readily available from a wide variety of foods manganese deficiency has not been reported in the normal population, but enforced deficiency or other disturbances in manganese metabolism result in a range of symptoms including skeletal abnormalities and alterations in high density lipoprotein homeostasis.[115] Manganese toxicity is also rare and is often known as manganese-induced Parkinsonism or manganism, characterized by neurological deterioration.[115] The mechanisms by which this occurs are unknown, but they may be linked to alterations in homeostasis of other metals, especially iron; for example manganese appears to inhibit mitochondrial aconitase, an iron–sulfur cluster protein involved in the tricarboxylic acid (TCA) cycle.[116]

2.3.4   Potassium and Sodium

The biological roles of these two very abundant alkali metals are mainly linked to their selective transport across biological membranes rather than a direct functional or structural role in metalloproteins. These roles are often coupled to the transport of another species, for example Na^+/H^+, K^+/H^+, Na^+/K^+, Na^+/Ca^{2+}, Na^+/bicarbonate transporters or Na^+/glucose. The biological role of such transporters is either linked to the co-transported molecule *e.g.* modulating pH gradients across membranes, while the Na^+/K^+ ATPase is involved in signal transduction and osmoregulation. Generally Na^+ is the major cation in blood plasma, while K^+ is the major intracellular cation. Modulations in the level of sodium in the blood are usually manifested in symptoms connected with its role in osmoregulation. Hypernatremia[117] results from elevated sodium levels in the blood, but since excess sodium is effectively excreted by the kidney this condition usually results from decreased levels of water *i.e.* dehydration, with symptoms associated with changes in osmoregulation. Hyponatremia,[117] or decreased level of sodium, is commonly associated with acute diarrhea or vomiting. Again the symptoms are linked to changes in osmoregulation, with a shift of movement of water into cells. In contrast, modulations in the level of potassium are seen mainly in modulating its role in signal transduction, with hypokalemia[118] resulting in numerous symptoms arising from impairment of muscle function. Any changes in either signal transduction or osmoregulation will have knock-on effects that alter proteostasis as will changes associated with Na^+-dependent transport of organic molecules such as glucose.

2.3.5 Iron

Iron is the major redox active metal used in biological systems, with around 120 iron dependent enzyme classifications for which a structure is known, representing 18% of all catalytic metals found in enzymes.[111] Iron is found in a wide range of states including being an integral component of heme prosthetic groups as well as iron–sulfur clusters.

Heme is formed from an iron atom at the center of a porphyrin ring. It is not a single chemical moiety but rather the name for a cluster of closely related, but chemically and biologically distinct, species.[119] Heme is best known for being the oxygen carrier in proteins such as hemoglobin and forming redox active sites in cytochromes, but it is also found in other proteins such as the myeloperoxidase used by neutrophils to generate cytotoxic species as part of their antimicrobial function,[120] sulfite oxidase, catalase and prostaglandin–endoperoxide synthase.

The second widespread use of iron in redox centers is as iron–sulfur clusters. Again these are not a single chemical species, but rather the name for a cluster of related chemical species with 2Fe–2S and 4Fe–4S being common, but other states are also used in biological systems. Iron–sulfur containing proteins include ferredoxins, found in a range of processes including acting as redox carriers in photsynthesis, aconitase, a component of the TCA cycle, 3-isopropylmalate dehydratase, involved in leucine biosynthesis, glutamate synthase, succinate dehydrogenase, amidophosphoribosyltransferase involved in purine nucleotide synthesis, xanthine dehydrogenase, NADH dehydrogenase, nitrogenase and some hydrogenases (others contain iron, but no iron–sulfur cluster). In eukaryotes iron–sulfur clusters are synthesized in mitochondria and, where needed, these are exported out. The biogenesis of iron–sulfur clusters is complex and highly regulated and disruption of the process causes a range of disease states including the neurodegenerative disorder Friedreich's ataxia.[121]

A wide range of non-heme, non-iron–sulfur cluster iron-containing enzymes exist which link into a correspondingly wide range of biological processes. These enzymes include ribonucleotide reductase and prolyl-hydroxylases.

Given the importance of iron it is not a surprise that iron homeostasis is tightly regulated. Iron deficiency,[122,123] or hypoferremia, results in a range of symptoms; the most obvious one is iron deficiency anemia. Iron homeostasis is also tightly linked to other homeostatic mechanisms, one of which is unsurprisingly oxygen homeostasis.[76,124]

More details of the links between iron and neurodegeneration can be found in this book, in particular in Chapters 4, 9 and 11.

2.3.6 Zinc

Zinc is the second most abundant metal ion found in enzymes after magnesium, with around 130 enzyme classifications known with solved structures that contain a zinc ion.[111] Zinc is found in the active site of many proteases, such as MMPs and carboxypeptidase A, as well as wide range of other enzymes such as carbonic anhydrase, required for CO_2 transport in mammals and anaerobic respiration in plants, alcohol dehydrogenase, cytidine deaminase, involved in the formation of uridine and deoxyuridine, some alkaline phosphatases and some enzymes involved in glutathione metabolism such as lactoylglutathione lyase and hydroxyacylglutathione hydrolase.

In addition to its catalytic function zinc plays a structural role in many proteins, the most well known of which are zinc-finger containing proteins including many transcription factors. The term zinc-finger covers a class of related chemical structures of all which use the side chain of cysteine and histidine residues to bind a zinc atom. They consist of several groups including Cys_2His_2, gag-knuckle, treble-clef, Zn_2/Cys_6 and zinc ribbon groupings.[125] Many zinc-finger proteins are involved in binding to DNA or RNA, such as transcription factors including ligand activated factors such as nuclear hormone receptors. Zinc-finger structures may also act as biological redox switches.[126]

As well as a functional role in proteins zinc can also be involved in cell signaling in the brain, immune system, salivary glands, pancreas and prostate, and zinc signaling is linked to oncogenesis.[127]

Zinc deficiency is a very widespread syndrome associated with many disease states, including growth retardation and infection susceptibility.[128,129] Similarly excess zinc causes a range of symptoms, including inhibition of copper (and iron) uptake which leads to copper deficiency.[130]

More details of the links between zinc and neurodegeneration can be found in this book, in particular in Chapters 4, 7, 8 and 12.

2.3.7 Copper

Copper is an essential trace metal with a significant function in redox chemistry. It is found in a number of redox enzymes including cytochrome c oxidase in the respiratory electron transport chain of mitochondria,[131] plastocyanin which acts as an electron shuffle in chloroplasts between components of plant photosystems,[132] azurins involved in electron transport in some bacteria,[133] eukaryotic superoxide dismutase which has an associated copper chaperone,[134] ascorbate oxidases involved in vitamin C metabolism in plants,[135] laccases from fungi and plants,[136] lysyl oxidase required for elastin and collagen maturation,[137] nitrite reductases and arsenate reductase, among others. In some animals hemocyanin, a copper-containing protein,[138] is used in place of iron–phorphyrin complexes in hemoglobin for oxygen transport. Copper is also intimately linked to iron homeostasis via its role in the function of hephaestin and ceruloplasmin,[139,140] with ceruloplasmin being implicated in a variety of neurodegenerative diseases.[141]

Due to its key role in many core redox processes it is unsurprising that copper deficiency has a number of severe phenotypes associated with it.[142,143] Similarly an excess of copper is highly toxic due to its redox activity and copper has been linked to a number of neurological conditions. More details of the links between copper and neurodegeneration can be found in this book, in particular in Chapters 7, 8, 12 and 13.

2.3.8 Cobalt

The major biological role of cobalt is usually listed as being linked to vitamin B12-dependent enzymes. Vitamin B12, or cobalamin, is a water soluble coenzyme that cannot be synthesized by humans. It contains a substituted corrin macrocycle. Cobalamin is used as a cofactor by a variety of enzymes including many involved in the transfer of methyl groups between molecules. Since it cannot be synthesized by humans cobalamin deficiency should be treated as being separate from cobalt deficiency. Two human enzymes, methylmalonyl coenzyme A mutase and 5-methyltetrahydrofolate-homocysteinemethyl transferase, use cobalamin as an essential cofactor. The latter enzyme is not only involved in methionine synthesis but also in the regeneration of folate required for the production of thymine and hence DNA synthesis. Hence cobalamin deficiency impacts proteostasis on multiple levels. Unsurprisingly the clinical symptoms of cobalamin deficiency are highly polymorphic.[144,145] While the essential role of cobalt is sometimes listed as being only linked to vitamin B12-dependent enzymes, cobalt is also found in at least eight other classes of enzymes,[146] including methionine aminopeptidase, found in a wide range of species including humans.[147] This enzyme removes N-terminal methionine residues from newly translated proteins and is required for cell cycle progression[148] and hence cobalt deficiency can have a substantial effect on proteostasis. Excess cobalt can also inhibit aminopeptidase activity by binding in the active site.[147] Excess cobalt also has similar effects to other carcinogenic metals such as arsenic, lead, nickel and vanadium, inhibiting DNA repair systems, modulating cellular redox regulation and deregulating cellular proliferation pathways,[149,150] all of which have major effects on proteostasis.

2.3.9 Fourth Tier Elements

Fourth tier elements are those which are not essential for all organisms, but for which some organisms have a specific requirement for trace amounts. These include a number of metal ions of which we will discuss only a few briefly due to their connection with proteostasis in humans.

2.3.9.1 Molybdenum

Molybdenum is a critical component of nitrogenase, a key enzyme in nitrogen fixation in some bacteria.[151] However, in most enzymes that utilize molybdenum it is found in a complex with one or two pterin molecules.[152] Three human enzymes are known that contain a molybdopterin cofactor, xanthine dehydrogenase, aldehyde oxidase and sulfite oxidase. Molybdenum deficiency in humans results in a reduction in the activity of these enzymes which in turn results in high levels of sulfite and urate in the blood along with neurological damage,[153] while high levels disrupt copper uptake resulting in copper deficiency,[154] and hence both significantly affect proteostasis.

2.3.9.2 Nickel

There are currently eight known classes of enzyme that contain nickel, namely urease, NiFe-hydrogenases, methyl coenzyme M reductase, acireductone dioxygenase, Ni-superoxide dismutases, carbon monoxide dehydrogenase, acetyl CoA-synthase and glyoxylase I,[155] which are found in microbes and plants. While there are no known biological functions for nickel in vertebrates, there have been a variety of reports that nickel deficiency causes a variety of physiological abnormalities in rats.[156] In addition, there is a direct link between nickel-containing enzymes and human proteostasis, due to the presence of such enzymes in bacteria that form part of our normal (or abnormal) microflora. For example, there may be a correlation between nickel levels and growth of *Helicobacter pylori*, which causes ulcers and is implicated in cancer development, as it contains a nickel-dependent urease which is required for colonization.[157] Nickel toxicity and carcinogenicity are complex multi-factorial events which include DNA damage, inhibition of DNA repair, induction of oxidative stress and induction of a pseudo-hypoxic state,[156] all of which trigger multiple changes in proteostasis.

2.3.9.3 Cadmium

To date only a single enzyme has been reported that contains cadmium, a carbonic anhydrase from the marine diatom *Thalassioira weissflogii*.[158]

Cadmium is however a potent modulator of proteostasis in humans. It is able to replace zinc in many biological systems,[159] in particular proteins that use cysteine side chains to bind the zinc atom, such as zinc-finger protein structures. It can also replace magnesium, calcium and iron in other systems[159] and can have very significant effects on protein folding by chelation of thiol groups, for example catalyzed oxidative folding is completely inhibited by 200 µM cadmium.[160] Cadmium is also reported to have hormone-like effects in a range of biological systems.[161] Ingestion of cadmium leads to Itai-itai disease, respiratory tract and kidney problems such as glucosuria, hypophosphatemia and proteinuria, as well as a weakening of bones due to osteomalacia and osteoporosis.[162,163] However, the major route for cadmium exposure is through inhalation and particularly from tobacco smoking, with smokers having up to 5 times higher blood cadmium concentrations than non-smokers.[164] Some more details of the modulation of amyloid peptides can be found in Chapter 4.

2.3.9.4 Vanadium

The role of vanadium in biological systems is controversial. While a few vanadium containing enzymes, such as some nitrogenases, have been reported[165] and the presence of vanadium in the diet of rats has been reported to enhance growth by 40%,[166] it has yet to be established as an essential element for humans. However, it is known to have a wide range of biological effects and to affect a wide range of enzymatic systems including phosphatases, ATP-ases, peroxidases, ribonucleases, protein kinases and oxidoreductases,[167] and hence it will trigger multiple independent changes in proteostasis.

In addition to those metals which play a productive role in biological systems there are many that do not but which can have severe effects on proteostasis. The most abundant of these is aluminium, which plays a role in a number of neurodegenerative diseases (see Chapter 10).

2.4 Conclusions

In conclusion, proteostasis depends on a complex network of inter-connected pathways. It is defined in response to the current state of the cell and in turn defines or modulates all cellular processes. As we have tried to demonstrate, all of the main pathways involved in proteostasis can be linked to metallostasis, either directly through metalloproteins or indirectly. Since metallostasis depends on proteostasis, they should be seen as being intimately entwined, with any variation in one resulting in expected and sometimes unexpected variations in the other. It is a fine balancing act and so it should not be surprising that any perturbations may have very significant impact on the organism, including neurodegeneration which forms the basis for the rest of this book.

References

1. International Human Genome Sequencing Consortium, *Nature*, 2004, **431**, 931.
2. O. N. Jensen, *Curr. Opin. Chem. Biol.*, 2004, **8**, 33.
3. K. M. Pan, M. Baldwin, J. Nguyen, M. Gasset, A. Serban, D. Groth, I. Mehlhorn, Z. Huang, R. J. Fletterick, F. E. Cohen and S. B. Prusiner, *Proc. Natl. Acad. Sci. USA*, 1993, **90**, 10962.
4. S. Colacino, G. Tiana, R. A. Broglia and G. Colombo, *Proteins*, 2006, **62**, 698.
5. E. L. Kovrigin and J. P. Loria, *Biochemistry*, 2006, **45**, 2636.
6. E. D. Watt, H. Shimada, E. L. Kovrigin and J. P. Loria, *Proc. Natl. Acad. Sci. USA*, 2007, **104**, 11981.
7. D. Plewczyński and K. Ginalski, *Cell. Mol. Biol. Lett.*, 2009, **14**, 1.
8. P. G. Righetti, A. Castagna, P. Antonioli and E. Boschetti, *Electrophoresis*, 2005, **26**, 297.
9. B. Wittmann-Liebold, H. R. Graack and T. Pohl, *Proteomics*, 2006, **6**, 4688.
10. D. Branzei and M. Foiani, *Nature*, 2010, **11**, 208.
11. D. N. Hebert and M. Molinari, *Physiol. Rev.*, 2007, **87**, 1377.
12. Y. Xue, Q. Wang, Q. Long, B. L. Ng, H. Swerdlow, J. Burton, C. Skuce, R. Taylor, Z. Abdellah, Y. Zhao, Asan, D. G. MacArthur, M. A. Quail, N. P. Carter, H. Yang and C. Tyler-Smith, *Curr. Biol.*, 2009, **19**, 1453.
13. R. E. Green, J. Krause, A. W. Briggs, T. Maricic, U. Stenzel, M. Kircher, N. Patterson, H. Li, W. Zhai, M. H. Fritz, N. F. Hansen, E. Y. Durand, A. S. Malaspinas, J. D. Jensen, T. Marques-Bonet, C. Alkan, K. Prüfer, M. Meyer, H. A. Burbano, J. M. Good, R. Schultz, A. Aximu-Petri, A. Butthof, B. Höber, B. Höffner, M. Siegemund, A. Weihmann, C. Nusbaum, E. S. Lander, C. Russ, N. Novod, J. Affourtit, M. Egholm, C. Verna, P. Rudan, D. Brajkovic, Z. Kucan, I. Gusic, V. B. Doronichev, L. V. Golovanova, C. Lalueza-Fox, M. de la Rasilla, J. Fortea, A. Rosas, R. W. Schmitz, P. L. Johnson, E. E. Eichler, D. Falush, E. Birney, J. C. Mullikin, M. Slatkin, R. Nielsen, J. Kelso, M. Lachmann, D. Reich and S. Pääbo, *Science*, 2010, **328**, 710.
14. J. Singh and E. T. Snow, *Biochemistry*, 1998, **37**, 9371.
15. W. Hu, Z. Feng and M.-S. Tong, *Carcinogenesis*, 2005, **25**, 455.
16. J. Stubbe, *Curr. Opin. Chem. Biol.*, 2003, **7**, 183.
17. W. C. Copeland, *Annu. Rev. Med.*, 2008, **59**, 131.
18. W. Neupert and J. M. Herrmann, *Annu. Rev. Biochem.*, 2007, **76**, 723.
19. W. M. Brown, M. George and A. C. Wilson, *Proc. Natl. Acad. Sci. USA*, 1979, **76**, 1967.
20. R. J. White and A. D. Sharrocks, *Trends Genet.*, 2010, **26**, 214.
21. R. O. Emerson and J. H. Thomas, *PLoS Genetics*, 2009, **5**, e1000325.
22. R. Tupler, G. Perini and M. R. Green, *Nature*, 2001, **409**, 832.
23. J. C. Rutherford and A. J. Bird, *Euk. Cell*, 2004, **3**, 1.

24. M. Falkenberg, N. G. Larsson and C. M. Gustafsson, *Annu. Rev. Biochem.*, 2007, **76**, 679.
25. M. J. Moore and J. N. Proudfoot, *Cell*, 2009, **136**, 688.
26. J. R. Sanford and J. F. Caceres, *J. Cell Sci.*, 2004, **117**, 6261.
27. S. Millevoi and S. Vagner, *Nucleic Acids Res.*, 2010, **38**, 2757.
28. E. J. Sontheimer, *Nat. Struct. Biol.*, 2001, **8**, 11.
29. C. R. Mandel, Y. Bai and L. Tong, *Cell. Mol. Life Sci.*, 2008, **65**, 1099.
30. P. Fernández-Silva, J. A. Enriquez and J. Montoya, *Exp. Physiol.*, 2003, **88**, 41.
31. S. F. Newbury, *Biochem. Soc. Trans.*, 2006, **34**, 30.
32. F. Bolognani and N. I. Perrone-Bizzozero, *J. Neurosci. Res.*, 2008, **86**, 481.
33. K. J. Waldron, J. C. Rutherford, D. Ford and N. J. Robinson, *Nature*, 2009, **460**, 823.
34. L. Cochella and R. Green, *Curr. Biol.*, 2005, **15**, R540.
35. J. Ling, N. Reynolds and M. Ibba, *Annu. Rev. Microbiol.*, 2009, **63**, 61.
36. M. Lovmar and M. Ehrenberg, *Biochimie*, 2006, **88**, 951.
37. M. Johansson, E. Bouakaz, M. Lovmar and M. Ehrenberg, *Mol. Cell*, 2008, **30**, 589.
38. S. Ledoux and O. C. Uhlenbeck, *Mol. Cell*, 2008, **31**, 113.
39. C. Touriol, S. Bornes, S. Bonnal, S. Audigier, H. Prats, A. C. Prats and S. Vagner, *Biol. Cell*, 2003, **95**, 169.
40. J. L. Wegrzyn, T. M. Drudge, F. Valafar and V. Hook, *BMC Bioinformatics*, 2008, **9**, 232.
41. J. W. Taanman, *Biochim. Biophys. Acta*, 1999, **1410**, 103.
42. S. O. Olofsson, K. Boström, P. Carlsson, J. Borén, M. Wettesten, G. Bjursell, O. Wiklund and G. Bondjers, *Am. Heart J.*, 1987, **113**, 446.
43. F. U. Hartl and M. Hayer-Hartl, *Nat. Struct. Mol. Biol.*, 2009, **16**, 574.
44. S. D. Westerheide and R. I. Morimoto, *J. Biol. Chem.*, 2005, **280**, 33097.
45. M. V. Powers and P. Workman, *FEBS Lett.*, 2007, **581**, 3758.
46. A. Shamovsky and E. Nudler, *Cell. Mol. Life Sci.*, 2008, **65**, 855.
47. M. Schröder and R. J. Kaufman, *Annu. Rev. Biochem.*, 2005, **74**, 739.
48. S. J. Marciniak and D. Ron, *Physiol. Rev.*, 2006, **86**, 1133.
49. J. D. Malhotra and R. J. Kaufman, *Sem. Cell Dev. Biol.*, 2007, **18**, 716.
50. M. Schröder, *Cell. Mol. Life Sci.*, 2008, **65**, 862.
51. K. Zhang and R. J. Kaufman, *Nature*, 2008, **454**, 455.
52. W. E. Balch, R. I. Morimoto, A. Dillin and J. W. Kelly, *Science*, 2008, **319**, 916.
53. F. Hatahet and L. W. Ruddock, *Curr. Pharm. Des.*, 2009, **15**, 2488.
54. E. T. Powers, R. I. Morimoto, A. Dillin, J. W. Kelly and W. E. Balch, *Annu. Rev. Biochem.*, 2009, **78**, 959.
55. J.-Q. Fan, *Biol. Chem.*, 2008, **389**, 1.
56. H. J. Dyson and P. E. Wright, *Nat. Rev. Mol. Cell Biol.*, 2005, **6**, 197.
57. K. Dunker, I. Silman, V. N. Uversky and J. L. Sussman, *Curr. Opin. Struc. Biol.*, 2008, **18**, 756.

58. P. E. Wright and H. J. Dyson, *Curr. Opin. Struc. Biol.*, 2009, **19**, 31.

59. C. J. Oldfield, Y. Cheng, M. S. Cortese, C. J. Brown, V. N. Uversky and A. K. Dunker, *Biochemistry*, 2005, **44**, 1989.

60. H. K. Lamb, C. Mee, W. Xu, L. Liu, S. Blond, A. Cooper, I. G. Charles and A. R. Hawkins, *J. Biol. Chem.*, 2006, **281**, 8796.

61. A. Solovyov and H. F. Gilbert, *Protein Sci.*, 2004, **13**, 1902.

62. F. Hatahet and L. W. Ruddock, *Antioxid. Redox Signal.*, 2009, **11**, 2807.

63. J. M. Herrmann, F. Kauff and H. E. Neuhaus, *Biochim. Biophys. Acta*, 2009, **1793**, 71.

64. J. Reimer, N. Bulleid and J. M. Herrmann, *Science*, 2009, **324**, 1284.

65. R. E. Hansen, D. Roth and J. R. Winther, *Proc. Natl. Acad. Sci.*, 2009, **106**, 422.

66. M. Molinari and A. Helenius, *Nature*, 1999, **402**, 90.

67. L. W. Ruddock and M. Molinari, *J. Cell Sci.*, 2006, **119**, 4373.

68. A. Dell and H. R. Morris, *Science*, 2001, **291**, 2351.

69. B. Tissot, S. J. North, A. Ceroni, P.-C. Pang, M. Panico, F. Rosati, A. Capone, S. M. Haslam, A. Dell and H. R. Morris, *FEBS Letts.*, 2009, **583**, 1728.

70. G. Manning, D. B. Whyte, R. Martinez, T. Hunter and S. Sudarsanam, *Science*, 2002, **298**, 1912.

71. J. P. MacKeigan, L. O. Murphy and J. Blenis, *Nat. Cell Biol.*, 2005, **7**, 591.

72. U. Khan and S. Krishnamurthy, *Front. Biosci.*, 2005, **10**, 866.

73. D. Barford, A. K. Das and M.-P. Egloff, *Annu. Rev. Biophys. Biomol. Struct.*, 1998, **27**, 133.

74. J. A. Adams, *Chem. Rev.*, 2001, **101**, 2271.

75. J. Myllyharju and K. I. Kivirikko, *EMBO J.*, 1997, **16**, 1173.

76. D. R. Mole, *Antioxid. Redox Signal.*, 2010, **12**, 445.

77. R. V. Banerjee and R. G. Matthews, *FASEB J.*, 1990, **4**, 1450.

78. R. S. Hegde and H. D. Bernstein, *Trends Biochem. Sci.*, 2006, **31**, 563.

79. H. R. Pelham, *Trends Biochem. Sci.*, 1990, **15**, 483.

80. C. Brocard and A. Hartig, *Biochim. Biophys. Acta*, 2006, **1763**, 1565.

81. J. T. Dunphy and M. E. Linder, *Biochim. Biophys. Acta*, 1998, **1436**, 245.

82. S. M. Miggin, O. A. Lawler and B. T. Kinsella, *J. Biol. Chem.*, 2003, **278**, 6947.

83. K. Kollmann, S. Pohl, K. Marschner, M. Encarnac, I. Sakwa, S. Tiede, B. J. Poorthuis, T. Lubke, S. Müller-Loennies, S. Storch and T. Braulke, *Eur. J. Cell Biol.*, 2010, **89**, 117.

84. R. Noiva, *Semin Cell Dev. Biol.*, 1999, **10**, 481.

85. M. A. Leissring, W. Farris, X. Wu, D. C. Christodoulou, M. C. Haigis, L. Guarente and D. J. Selkoe, *Biochem. J.*, 2004, **383**, 439.

86. U. P. Andley, *Prog. Retin. Eye Res.*, 2006, **26**, 78.

87. Y. Ohsumi, *IUBMB Life*, 2006, **58**, 363.

88. J. Hoseki, R. Ushioda and K. Nagata, *J. Biochem.*, 2010, **147**, 19.

89. B. Meusser, C. Hirsch, E. Jarosch and T. Sommer, *Nat. Cell Biol.*, 2005, **7**, 766.

90. M. Muratani and W. R. Tansey, *Nat. Rev. Mol. Cell Biol.*, 2003, **4**, 192.

91. N. M. Hooper, *FEBS Lett.*, 1994, **354**, 1.

92. D. F. Seals and S. A. Courtneidge, *Genes Dev.*, 2003, **17**, 7.

93. G. A. Rosenberg, *Lancet Neurol.*, 2009, **8**, 205.

94. P. S. Vosler, C. S. Brennan and J. Chen, *Mol. Neurobiol.*, 2008, **38**, 78.

95. M. H. Glickman and A. Ciechanover, *Physiol. Rev.*, 2002, **82**, 373.

96. K. G. Daniel, D. Chern, B. Yan and Q. P. Dou, *Front. Biosci.*, 2007, **12**, 135.

97. L. Li, H. Yang, D. Chen, C. Cui and Q. P. Dou, *Toxicol. Appl. Pharmacol.*, 2008, **229**, 206.

98. C. T. Chasapis and G. A. Spyroulias, *Curr. Pharm. Des.*, 2009, **15**, 3716.

99. P. J. Turnbaugh, R. E. Ley, M. Hamady, C. M. Fraser-Liggett, R. Knight and J. I. Gordon, *Nature*, 2007, **449**, 804.

100. J.-H. Hehemann, G. Correc, T. Barbeyron, W. Helbert, M. Czjzek and G. Michel, *Nature*, 2010, **464**, 908.

101. N. Fierera, C. L. Lauberb, N. Zhoub, D. McDonaldc, E. K. Costelloc and R. Knight, *Proc. Natl. Acad. Sci. USA*, 2010, **107**, 6477.

102. C. Andreini, I. Bertini, G. Cavallaro, G. L. Holliday and J. M. Thornton, *J. Biol. Inorg. Chem.*, 2008, **13**, 1205.

103. A. Gorläch, P. Klappa and T. Kietzmann, *Antioxid. Redox Signal.*, 2006, **8**, 1391.

104. J. Oldenburg, M. Marinova, C. Müller-Reible and M. Watzka, *Vitam. Horm.*, 2008, **78**, 35.

105. T. Pawelczyk and A. Matecki, *Eur. J. Biochem.*, 1998, **257**, 169.

106. S. Kouichirou, H. Hayasawaa and B. Lönnerdalb, *Biochem. Biophys. Res. Commun.*, 2001, **281**, 1024.

107. A. Sevostyanova, V. A. Yurshev, O. N. Solovjeva, S. V. Zabrodskaya and G. A. Kochetov, *Proteins*, 2008, **71**, 541.

108. Y. Zhu, M. D. Suits, A. J. Thompson, S. Chavan, Z. Dinev, C. Dumon, N. Smith, K. W. Moremen, Y. Xiang, A. Siriwardena, S. J. Williams, H. J. Gilbert and G. J. Davies, *Nat. Chem. Biol.*, 2010, **6**, 125.

109. M. S. Cooper and N. J. Gittoes, *BMJ*, 2008, **336**, 1298.

110. J. N. Buchholz, E. J. Behringer, W. J. Pottorf, W. J. Pearce and C. K. Vanterpool, *Aging Cell*, 2007, **6**, 285.

111. C. Andreini, I. Bertini, G. Cavallaro, G. L. Holliday and J. M. Thornton, *J. Biol. Inorg. Chem.*, 2008, **13**, 1205.

112. H.-T. Liu, M. W. Hollmann, W.-H. Liu, C. W. Hoenemann and M. E. Durieux, *Anesth. Analg.*, 2001, **92**, 1173.

113. R. Swaminathan, *Clin. Biochem. Rev.*, 2003, **24**, 47.

114. J. M. Topf and P. T. Murray, *Rev. Endocr. Metab. Disord.*, 2003, **4**, 195.

115. J. W. Finley and C. D. Davis, *BioFactors*, 1999, **10**, 15.

116. W. Zheng, S. Ren and J. H. Graziano, *Brain Res.*, 1998, **799**, 334.

117. S. M. Bagshaw, D. R. Townsend and R. C. McDermid, *Can. J. Anesth.*, 2009, **56**, 151.

118. M. Greenlee, C. S. Wingo, A. A. McDonough, J.-H. Youn and B. C. Kone, *Ann. Int. Med.*, 2009, **150**, 619.

119. S. Tsiftsoglou, A. I. Tsamadou and L. C. Papadopoulou, *Pharmacol. Ther.*, 2006, **111**, 327.

120. M. B. Hampton, A. J. Kettle and C. C. Winterbourn, *Blood*, 1998, **92**, 3007.

121. R. Lill and U. Mühlenhoff, *Annu. Rev. Biochem.*, 2008, **77**, 669.

122. J. Umbreit, *Am. J. Hemat.*, 2005, **78**, 225.

123. M. B. Zimmermann and R. F. Hurrell, *Lancet*, 2007, **370**, 511.

124. A. Salahudeen and R. K. Bruick, *Ann. N.Y. Acad. Sci.*, 2009, **1177**, 30.

125. S. S. Krishna, I. Majumdar and N. V. Grishin, *Nucleic Acids Res.*, 2003, **31**, 532.

126. K. D. Kröncke and L. O. Klotz, *Antioxid. Redox Signal.*, 2009, **11**, 1015.

127. M. Maurakami and T. Hirano, *Cancer Sci.*, 2008, **99**, 1515.

128. K. M. Hambidge and N. F. Krebs, *J. Nutr.*, 2007, **137**, 1101.

129. M. J. Turek and N. Fazel, *Curr. Opin. Gastroenterol.*, 2009, **25**, 136.

130. G. J. Fosmire, *Am. J. Clin. Nutr.*, 1990, **51**, 225.

131. T. Tsukihara, H. Aoyama, E. Yamashita, T. Tomizaki, H. Yamaguchi, K. Shinzawa-Itoh, R. Nakashima, R. Yaono and S. Yoshikawa, *Science*, 1995, **25**, 1069.

132. N. Melkozernov, J. Barber and R. E. Blankenship, *Biochemistry*, 2006, **45**, 331.

133. O. Farver and I. Pecht, *Proc. Natl. Acad. Sci. USA*, 1989, **86**, 6968.

134. J. M. Leitch, P. J. Yick and V. C. Culotta, *J. Biol. Chem.*, 2009, **284**, 24679.

135. N. Smirnoff, *Curr. Opin. Plant Biol.*, 2000, **3**, 229.

136. P. Baldrian, *FEMS Microbiol. Rev.*, 2006, **30**, 215.

137. H. A. Luvero and H. M. Kagan, *Cell Mol. Life Sci.*, 2006, **63**, 2304.

138. K. E. Van Holde and K. I. Miller, *Adv. Protein Chem.*, 1995, **47**, 1.

139. C. D. Vulpe, Y. M. Juo, T. L. Murphy, L. Cowley, C. Askwith, N. Libina, J. Gitschier and G. J. Anderson, *Nat. Genet.*, 1999, **21**, 195.

140. N. E. Hellman and J. D. Gitlin, *Annu. Rev. Nutr.*, 2002, **22**, 439.

141. S. J. Texel, X. Xu and Z. L. Harris, *Biochem. Soc. Trans.*, 2008, **36**, 1277.

142. J. Bertinato and M. R. L'Abbé, *J. Nutr. Biochem.*, 2004, **15**, 316.

143. E. Madsen and J. D. Gitlin, *Curr. Opin. Gastroenterol.*, 2007, **23**, 187.

144. R. Carmel, *Annu. Rev. Med.*, 2000, **51**, 357.

145. N. Dali-Youcef and E. Andrès, *Q. J. Med.*, 2009, **102**, 17.

146. M. Kobayashi and S. Shimizu, *Eur. J. Biochem.*, 1999, **261**, 1.

147. A. Addlagatta, X. Hu, J. O. Liu and B. W. Matthews, *Biochemistry*, 2005, **44**, 14741.

148. X. Hu, A. Addlagatta, J. Lu, B. W. Matthews and J. O. Liu, *Proc. Natl. Acad. Sci. USA*, 2006, **103**, 18148.

149. A. Hartwig and T. Schwerdtle, *Toxicol. Lett.*, 2002, **127**, 47.

150. D. Beyersmann and A. Hartwig, *Arch. Toxicol.*, 2008, **82**, 493.

151. L. C. Seefeldt, B. M. Hoffman and D. R. Dean, *Annu. Rev. Biochem.*, 2009, **78**, 701.

152. G. Schwarz, R. R. Mendel and M. W. Ribbe, *Nature*, 2009, **460**, 839.

153. G. Schwarz, *Cell Mol. Life Sci.*, 2005, **62**, 2792.

154. A. Vyskocil and C. Viau, *J. Appl. Toxicol.*, 1999, **19**, 185.

155. S. W. Ragsdale, *J. Biol. Chem.*, 2009, **284**, 18571.

156. E. Denkhaus and K. Salnikow, *Crit. Rev. Oncol. Hematol.*, 2002, **42**, 35.

157. R. J. Maier, S. L. Benoit and S. Seshadri, *Biometals*, 2007, **20**, 655.

158. T. W. Lane, M. A. Saito, G. N. George, I. J. Pickering, R. C. Prince and F. M. M. Morel, *Nature*, 2005, **435**, 42.

159. A. Martelli, E. Rousselet, C. Dycke, A. Bouron and J. M. Moulis, *Biochimie*, 2006, **88**, 1807.

160. R. Noiva, R. B. Freedman and W. J. Lennarz, *J. Biol. Chem.*, 1993, **268**, 19210.

161. C. Byrne, S. D. Divekar, G. B. Storchan, D. A. Parodi and M. B. Martin, *Toxicol. Appl. Pharmacol.*, 2009, **238**, 266.

162. M. H. Bhattacharyya, *Toxicol. Appl. Pharmacol.*, 2009, **238**, 258.

163. G. F. Nordberg, *Toxicol. Appl. Pharmacol.*, 2009, **238**, 192.

164. L. Järup and A. Akesson, *Toxicol. Appl. Pharmacol.*, 2009, **238**, 201.

165. D. Rehder, *J. Inorg. Biochem.*, 2000, **80**, 133.

166. K. Schwarz and D. B. Milne, *Science*, 1971, **174**, 426.

167. A. Bishayee, A. Waghray, M. A. Patel and M. Chatterjee, *Cancer Lett.*, 2010, **294**, 1.

CHAPTER 3

Proteostasis as a Signaling Device and Therapeutic Target

S. CENCI, M.D. AND R. SITIA, M.D.

Division of Genetics and Cell Biology, San Raffaele Scientific Institute and Università Vita-Salute San Raffaele, Via Olgettina 58, 20132 Milano, Italy

3.1 Introduction

Proteins are the main components of the cell's hardware. They come in different shapes and sizes, hence their name is derived from *Proteus*, the Greek mythic character who constantly changed his looks. Proteins have also extremely diverse half-lives, their lifespan ranging from minutes to years. Aaron Ciechanover often tells that if a good physicist were to describe one human being, (s)he should list the precise arrangement of all of his/her atoms. Ten years after, however, only a fraction of the atoms are still there, although the identity of the individual remains the same. So, proteins come and go all the time.

Protein homeostasis, or *proteostasis*, refers to the capacity of eukaryotic cells to control the concentration, conformation and localization of proteins, as well as their interactions, thereby shaping and maintaining a healthy and functional proteome. Even staying away from the Ciechanover identity paradox, our cells continuously readapt their shapes and functions, according to the physiological needs of the moment. Often they differentiate or get infected with pathogens: both events entail an even more drastic reshaping of the proteome. These changes challenge the fidelity of the proteome and hence represent great risks, as excessive production, insufficient degradation, or unbalanced subunit bio-synthesis can cause the dispatch of erroneous biological messages or the

RSC Drug Discovery Series No. 7
Neurodegeneration: Metallostasis and Proteostasis
Edited by Danilo Milardi and Enrico Rizzarelli
© Royal Society of Chemistry 2011
Published by the Royal Society of Chemistry, www.rsc.org

accumulation of toxic species. Over recent years, the concept of proteotoxicity, following the older paradigms of glycotoxicity and lipotoxicity, became extremely popular, as many degenerative diseases are caused by the accumulation of aberrant proteins that create damage.

A plastic capacity, proteostasis is therefore crucial for eukaryotic cells to survive and adapt to changing intrinsic and environmental conditions. Proteostasis ensures a number of vital functions, from the cellular to the organismal level, including differentiation, metabolic control, and resistance to stress, crucial to maintain the health of the proteome and to protect the organism against aging and diseases.

3.2 The Proteostasis Integrated Networks

In a cell at steady state, proteostasis implies that every protein that is synthesized is either released in the environment or degraded to make room for a new one. Proteostasis is maintained through intertwined networks (Figure 3.1) that ultimately ensure the integrated control of protein synthesis, folding, trafficking, and degradation.[1-4] If proteostasis is insufficient, protein accumulation, condensation and/or aggregation ensue, damaging vital cell functions, and ultimately resulting in cell death (Figure 3.2).

The first level of control operates at the level of protein synthesis (Figure 3.1). Besides the abundance and stability of the corresponding mRNA, multiple

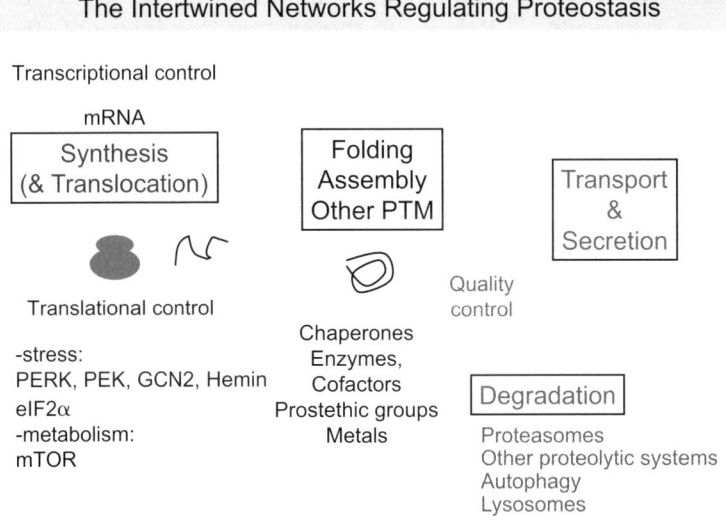

Figure 3.1 The intertwined networks regulating proteostasis. Proteostasis can be regulated at four main levels: 1. Synthesis (and in the case of secretory and mitochondrial proteins, translocation), 2. Folding, assembly and other posttranslational mechanisms, 3. Intracellular transport and sequestration, and 4. Degradation.

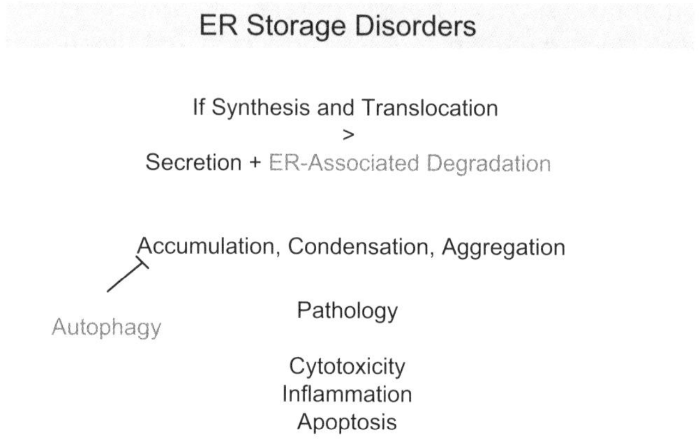

ER Storage Disorders

If Synthesis and Translocation
>
Secretion + ER-Associated Degradation

Accumulation, Condensation, Aggregation

Autophagy   Pathology

Cytotoxicity
Inflammation
Apoptosis

Figure 3.2   ER storage disorders. When the synthesis of a secretory protein that co-
translationally translocates into the ER exceeds its exit (that is, the sum of
secretion and degradation), accumulation (and often condensation and/or
aggregation) inevitably ensues. Dilated cisternae of the early secretory
compartments accumulate in cells, implying the existence of sorting
mechanisms to sort aggregation-prone molecules. Whilst undoubtedly
pathogenic for organisms (though the exact mechanisms of toxicity remain
largely unknown), these aberrant cisternae are often well tolerated by cell
lines, possibly because cell division generates space for them (see also
Figure 3.4).

mechanisms control translation initiation, for which the reader is referred to
comprehensive reviews.[5-10] Relative to this chapter, conditions of stress activate kinases (such as Perk, Pek or hemin) that phosphorylate eIF2α (eukaryotic
Initiation Factor 2α), hence inhibiting translation.

For proteins destined for the secretory pathway that enter co-translationally
into the endoplasmic reticulum (ER), translocation appears also to be regulated. In conditions of ER stress, 'difficult to fold' proteins such as the prion
protein P_1P,[11,12] or regulatory chaperones, such as calreticulin,[13] are denied
entry into the ER and shunted to proteasomal degradation, in a process defined
as pre-emptive ER quality control.[14-16]

Once a protein is made, a major proteostatic control is exerted at the level of
its folding and assembly. Whatever the cell compartment the protein attempts
to achieve its native state in, a cohort of chaperones and enzymes come into
play, helping its structural maturation and performing a stringent quality
control. Proteins that fail to reach the native state within due time are arrested
and sent to degradation.

Chaperones and folding enzymes that interact with nascent polypeptides to
minimize aggregation and ensure protein quality control are only part of the
proteostasis network. Indeed, a large variety of enzymes indirectly influence the
physical chemistry of protein folding by altering noncovalent (*e.g.* hydrophobic) forces. Metabolic pathways also produce small-molecule ligands and/
or metals that can bind to and stabilize the folded state of specific proteins,

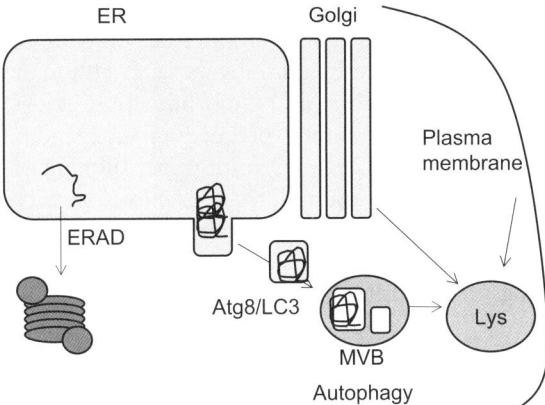

Figure 3.3 Degradation of secretory proteins. Secretory proteins can be eliminated in several ways. ER-associated degradation entails the recognition, partial unfolding and retrotranslocation of the substrate for disposal by cytosolic proteins. When aggregates form that are too stable or large to be retro-translocated, autophagy (or ERphagy) contributes to maintain proteostasis. At present, it is not known how autophagic vacuoles recognize the ER subregions that contain protein aggregates. Some proteins can be shunted to lysosomes from later compartments of the secretory pathway, including the plasma membrane.

enhancing folding by shifting folding equilibria.[17] Moreover, protein folding depends on the distribution, concentration, and subcellular localization of chaperones and folding assistants.[18]

Cells use stress sensors and inducible pathways to respond to a loss of proteostatic control, including the heat shock response (HSR) that regulates cytoplasmic proteostasis,[19,20] and the unfolded protein response (UPR) that maintains protein homeostasis in the secretory pathway.[2,21,22]

As making and breaking form a perfect equilibrium, degradative machineries are essential components of the proteostatic network (Figure 3.3). A dedicated paragraph in this Chapter (Section 3.5) provides an overview of the main proteocatabolic routes, the ubiquitin proteasome system (UPS) and autophagy, their functions and interplay.

3.3 Proteostasis and Aging

Among the numerous definitions of aging proposed by gerontologists, a comprehensive one defines senescence as an *age-dependent* series of *cumulative*, *progressive*, *intrinsic* and *deleterious* changes occurring after reproductive maturity that culminate in death.[23] This definition contains a number of hallmarks of aging, *in primis* its detrimental effects on tissue and organ function, which accumulate gradually, causing increased fragility of the individual. Another key feature is that aging is *universal* among all members of the same species. At the same time, it is also hallmarked by a high degree of intrinsic *heterogeneity*: a common experience is that different individuals age differently, and senescence increases inter-individual diversity.[24] Although aging enhances susceptibility to extrinsic factors and environmental stress, senescence is dictated by an intrinsic pace, thereby proceeding independent of modifiable agents.[25] The environment is extremely important, though. In environmental conditions that compromise reproduction – *e.g.* caloric restriction – organisms can reallocate energy to maximize somatic maintenance and longevity, waiting for reproduction-permissive conditions. Hence, *plasticity* is another key feature of aging, with longevity resulting from a plastic balance of reproduction and somatic maintenance.[26,27] Besides its teleological implications, this concept raises the idea that, in principle, aging can be manipulated. Along this line, caloric restriction was the first and most consistent measure capable of prolonging lifespan in vertebrate laboratory animals.[28–32]

Aging also occurs at the cellular and molecular levels, and proteins are no exception. Indeed, the intracellular environment is a tough place for a protein to maintain its native structure and function, because generation of reactive oxygen species, the presence of reactive sugars, proteases, and detergents, and the continuous emergence of partially folded nascent polypeptide chains acting as aggregation seeds constantly expose proteins to damage, a process referred to as protein aging.[33] This is more evident in senescent cells and tissues, which tend to accumulate oxidized and damaged proteins, with increased propensity to aggregate, thereby challenging the maintenance of proteostasis.[4,34–37] Collectively, this evidence led to the hypothesis that proteostasis plays a role in the molecular mechanisms of aging. In more recent studies, an age-associated decline in proteostatic control has been proposed based on correlative evidence in different models, offering a potential explanation for why aging predisposes to many diseases. A reduced capacity to adapt to stress and maintain homeostasis, referred to as *homeostenosis*, is a common feature of aging across species. By analogy, a progressive loss of the proteostatic capacity could be referred to as *proteostenosis*.

Recent elegant studies exploiting metastable reporters of proteostasis revealed a generalized collapse of the proteostatic capacity preceding aging and tissue damage in *Caenorhabditis elegans*. Such loss of proteostasis was causal to the aging process, because restoring the folding machinery by inducing cytosolic and ER chaperones rescued tissue degeneration.[38] Should mammals also experience such *proteopause* early in senescence, agents manipulating proteostasis would hold great promise against aging.

The potential importance of proteostasis in longevity is also supported by emerging evidence suggesting that signaling pathways involved in metabolic control, resistance to stress, or genomic stability, recently shown to control lifespan, also strongly influence proteostasis. For example, a decrease in insulin growth factor signaling has a profound beneficial effect on protein aggregation and toxicity in models of Alzheimer's disease in *C. elegans*[39] and in mice.[40,41] The insulin/insulin growth factor (IGF)-1 signaling (IIS) pathway is a prominent regulator of lifespan and youthfulness in worms, flies, and mammals, and reduced IIS has been shown consistently to extend lifespan in different organisms.[42–45] Studies in worms revealed a tight link between IIS-dependent longevity and the central HSR transcriptional activator heat shock factor-1 (HSF-1), with increased HSF-1 extending lifespan through DAF-16, a transcription factor crucial to IIS.[46] The observation that the DAF-16 and HSF-1 transcriptomes include numerous chaperones[46,47] suggests that the integrity of protein folding could play a key role in lifespan determination and amelioration of aggregation-associated proteotoxicity.

In other words, the proteostasis network is likely to be tightly interconnected with the major biological pathways demonstrated to play a role in longevity, further strengthening the rationale to improve proteostasis to maintain youthfulness.[4,48–50]

3.4 Stemness and Proteostasis

Parents are known to (hyper)protect their offspring. It came however as a great surprise, even in Israel and Italy, that unicellular organisms also show motherly behavior (Figure 3.4). In the budding yeast *Saccharomyces cerevisiae* cell division is asymmetrical, and this asymmetry is believed to be linked to the aging process.[51] Some genetic material has long been known to be unequally distributed: for example, the daughter cell does not inherit extrachromosomal ribosomal DNA circles (ERCs), which have been suggested to participate in replicative senescence, and accumulate in mother cells during growth.[52,53] Since

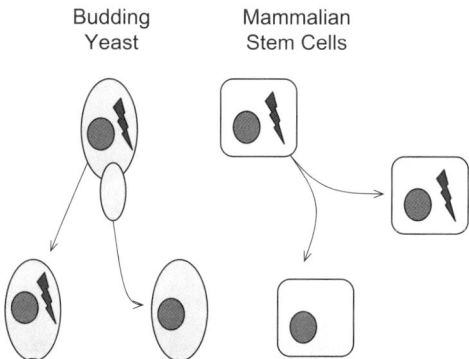

Figure 3.4 Asymmetric inheritance of proteotoxic species. In many unicellular organisms, including *E. coli* and *S. cerevisiae*, carbonylated or aggregated proteins can be retained by the mother cell to yield a daughter cell free of proteotoxic species. This ancient motherly mechanism seems to be active in metazoan stem cells. Therefore, cell division becomes a way to dilute toxic species.

damaged components, *e.g.* oxidized proteins, accumulate in yeast mother cells, an asymmetric segregation of damaged molecules, particularly proteins, has been proposed to be crucial to determine age asymmetry.[54,55] A seminal paper demonstrating that damaged proteins are asymmetrically segregated during mitosis was produced by Thomas Nystrom and coworkers, who showed that carbonylated proteins, a product of irreversible oxidative damage, accumulate as a function of replicative age, being retained by mother cells during cytokinesis. This active process requires Sir2, a lifespan determinant, raising the possibility that irreversibly damaged proteins may act as timers of age.[56] In subsequent studies, the same scientists were able to film the active process of retention of protein aggregates, spontaneously arisen with aging, or generated by stress (heat shock), identifying the responsible mother-directed transport mediated by actin filaments.[57,58] These studies exploited the capacity of the chaperone HSP104 to hallmark protein aggregates. This precious disaggregating chaperone functions with Hsp70 and Hsp40 in solubilizing intracellular aggregates of heat-damaged proteins.[59,60] Present in fungi, plants, and bacteria, HSP104 has great therapeutic potential, but whether mammalian cells possess any such endogenous activity remains unclear.[61–64]

More recently, an asymmetric partition of protein aggregates has been reported also in unstressed bacteria (*Escherichia coli*), otherwise known to divide symmetrically. During mitosis, protein aggregates segregate with the old cell pole, and their accumulation throughout rounds of cell division is associated with decreased replicative potential, a hallmark of senescence.[65] The data suggest that this bacterium uses polarity to ensure that deleterious material be differentially inherited by aging cells,[66] and further links protein aggregates with senescence.[67]

Recent work has elaborated mathematical models of symmetric versus asymmetric segregation of damaged cellular components.[68–70] These *in silico* studies predict that a small asymmetry may be sufficient to protect daughter cells, in the presence of low damage rates. Hence, the researchers hypothesize that asymmetric division may offer a growth advantage for the progeny, in the absence of 100% efficient defense mechanisms.[68 70] These predictions are in agreement with initial experimental evidence demonstrating a proliferative advantage in cells that inherited the aggregate-cleared new pole.[71]

In more complex organisms, asymmetric division is a crucial feature of stem cells (Figure 3.4). It is tempting to speculate that cell polarization may have primarily evolved in unicellular organisms to restrict senescence to one daughter cell during division by enabling the differential segregation of damaged material. Metazoans may have co-opted asymmetric division to permit the differential inheritance of centrosomes, RNAs, proteins, and membranes, essential for cell fate diversification, lineage commitment and morphogenesis.[67]

Whatever the teleology, proteotoxicity-clearing mechanisms hold great therapeutic promise, although the presence of asymmetric division of protein aggregates in higher eukaryotes awaits solid demonstration. In its support, a seminal study has shown that aggregated proteins are asymmetrically distributed in mitosis, leaving one daughter cell free of accumulated protein damage. This evidence was generated in two distinct cell models, human epithelial crypts of the small intestine of patients with a protein-folding disease, and neural precursors from *Drosophila melanogaster*.[72] Should this evidence be confirmed, mitotic divisions previously thought to be symmetrical may turn out to be asymmetrical, at least with respect to protein aggregates.[73] Cell division could thus serve an important clearing function, whose loss in post-mitotic terminally differentiated cells such as neurons, along with their long lifespan, may explain their propensity to proteotoxicity and conformational diseases.

This new frontier raises a multitude of exciting questions, with profound biological and therapeutic implications. An anthology is presented as follows.

In principle, the asymmetric partition of protein aggregates requires a strict spatial quality control. The storage of damaged or misfolded proteins into aggresomes, large cytosolic aggregates, by the microtubule-organizing center, or Russell bodies, subregions of the early secretory pathway,[74] may serve this purpose. Understanding the signals driving cellular decisions on their fate, *i.e.* to repair, degrade, or asymmetrically segregate and deliver to the older daughter cell, is of great importance.

Could Russell bodies also be asymmetrically segregated in dividing cells? Are certain ER cisternae somehow labeled as 'old' and kept away from the new pole as the cell prepares to divide? Feroz Papa and coworkers recently provided evidence that *S. cerevisiae* experiencing ER stress generated daughter cells that were essentially free of proteostatic problems.[75]

The transport mechanism responsible for the asymmetric partitioning of damaged proteins that do not show an aggregate pattern, like in the case of carbonylated proteins, is more challenging. Are they transported one by one? How are they recognized? This implies huge energetic costs for the cell.

Stem cells are the paradigmatic example of asymmetric division in complex multicellular organisms. How is this asymmetry lost in the non-stem progeny? Is asymmetric division somehow extrinsically enabled by niche-shaping factors? In the case of regenerative stimuli, stem cells proliferate symmetrically to maintain their compartment. Does such homeostatic proliferation result in decreased regenerative potential of the stem pool?

Does asymmetric segregation of protein aggregates also occur in cancer stem cells? If this were the case, biomarkers could be devised to identify such compartments, with valuable therapeutic implications.

3.5 Degradation Machineries: Proteasomes and Autophagy

Protein degradation is an essential step to maintain proteostasis, as demonstrated by a wealth of observations that genetic or pharmacologic blockade of proteocatabolic pathways disrupts protein homeostasis. Protein degradation is a highly selective and regulated process, and depends on two main pathways: the UPS and the autophagy-lysosomal system.

3.5.1 Proteasomes

In eukaryotic cells, most proteins destined for degradation are first labeled with ubiquitin, and then digested to small peptides by the 26S proteasome, a large (2.4 MDa) multi-meric ($\geq$ 50 subunits), multi-catalytic, multi-functional complex that consumes energy to operate. It consists of a core 20S particle, in which proteins are digested to short peptides, and one or two 19S regulatory particles, responsible for substrate recognition and transport into the core particle.[76] The 20S proteasome is composed of four eptameric-stacked rings, forming a barrel surrounding a central cavity. The two inner β rings form a central proteolytic chamber isolated from the surrounding cytosol. Of the six β peptidases, two cleave after hydrophobic residues, two after acidic residues and two after basic residues, enabling proteasomes to cut most peptide bonds. The α rings form a gated channel, whose access is regulated by the associated ATPases in the 19S particle.[77,78] Inaccessible to globular proteins, this architecture requires ATP-dependent mechanisms to recognize, unfold, linearize, and inject substrates into the 20S particle. The base of the 19S consists of six ATPases that unfold proteins, translocate them and trigger gate opening into the 20S.[78]

The essential features of the cellular degradative machinery emerged early in evolution, mainly to serve the fundamental homeostatic function as a quality-control system capable of rapidly eliminating misfolded or damaged proteins whose accumulation would hamper cell function and viability.[33,79,80] The complex structure of the proteasome reflects the necessity to select targets with absolute specificity, avoiding degradation of essential cellular constituents. Ubiquitin conjugation is a feature of eukaryotes, further ensuring that only unwanted proteins are selectively degraded. Adenosine triphosphate (ATP) is initially essential to activate ubiquitin, which is then transferred to one of the cell's ubiquitin-carrier proteins (*circa* 40, named E2). The exquisite selectivity of this pathway resides in the ubiquitin ligases (E3), which are specific for different protein substrates.[81] Mammalian cells contain hundreds of different ubiquitin ligases, which, together with a specific E2, catalyze the formation of the ubiquitin chain on a limited number of protein substrates, triggering their rapid degradation by the 26S proteasome.[33,76]

The last ~15 years of research have unveiled many additional roles of the ubiquitin–proteasome pathway, from fundamental homeostatic cell functions (regulation of cell division, signal transduction, gene expression) to intercellular integrative functions in metazoa, with complexity correlating with the evolutionary scale (lineage differentiation, tissue homeostasis, self recognition, immune surveillance).[81]

3.5.2 The Mystery of Rapidly Degraded Proteins

Upon ribosomal release, polypeptides face a kinetic competition between successful folding and rapid hydrolysis of unfolded species. Recent experiments showed that a sizeable fraction of newly synthesized proteins undergo degradation within minutes of their synthesis in eukaryotic cells. In aneuploid cells, such as cancers and the common lines utilized in laboratories, rapidly degraded proteins (RDP) may rise up to two-digit percentages.[82,83] Initially attributed to errors in translation (hence the acronym DRiPs, for defective ribosomal products), most short-lived species may result from unsuccessful folding or multimer assembly. Rapid degradation of newly made proteins also reflects the metabolic state of the cell, representing an adaptive measure to maintain the pool of free amino acids during starvation.[84] Whatever their origin, RDP provide a rich source of immunogenic peptides to be presented on the surface in association with major histocompatibility complex class I molecules. The notion that the peptidome presented on the cell's surface provides a real-time representation of the cell's *transductome* has an intriguing teleological explanation. RDP may provide our immune system with an efficient antiviral strategy, enabling our cells to rapidly display viral peptides when viruses hijack the cell's proteosynthetic apparatus, before viruses can assemble and propagate.

The exploitation of protein degradation to provide real-time information on what protein species are synthesized so as to inform lymphocytes roaming the body about cell identity and warn them against intracellular parasites is an admirable invention of evolution. An ecological paradigm, the cellular society utilizes the requirement for single cells to dispose of the waste and maintain proteostasis in order to develop superior functions, such as immune recognition and memory.

A neuropsychological parallel could also be drawn. In psychoanalysis, sensorial β elements are either transformed in abstract thinking – α elements – or eliminated, as they would turn toxic to our mind, representing an unthinkable state of mind – the 'nameless dread'.[85] The proteasome serves a similar function: it copes with toxic proteins, transforming them into signals that our immune system is able to decode, generating knowledge and memory. Just like our mind, our cells require a digestive–transforming capacity – the α function – which is carried out by the proteasome.

3.5.3 Regulating Proteasome Biogenesis

Like many crucial homeostatic mechanisms, the cellular proteasome pool is not fixed, but rather adapts to heterogeneous conditions that demand increased levels of proteasome activity, generally referred to as *proteasome stress*. This can be mimicked *in vitro* by applying pharmacological or genetic proteasomal inhibition, which activates the coordinated synthesis of all proteasomal subunits in the proper stoichiometry,[86,87] thereby defining a *proteasome stress response*.[88] Pathophysiological causes of proteasome stress are conditions that weaken protein folding or cause protein damage, including oxidative stress or intense synthesis of proteins – especially secretory ones, generally rich in disulfide bonds.[89] Proteasomes can also be overburdened by genetic and environmental factors (*i.e.* mutations or lack of cofactors), increasing the frequency of aberrant proteins. Such situations are encountered in many conformational diseases, a broad category encompassing neurodegeneration, diabetes, genetic disorders, and others.[90] Deciphering the adaptive strategy regulating *de novo* proteasome biogenesis is of great importance to understand proteostasis. An elegant mechanism has been discovered in yeast. In *Saccharomyces cerevisiae* the transcription factor Rpn4 controls all the genes encoding proteasome subunits by activating a common *cis*-acting element, named PACE for proteasome-associated control element.[91] Feed-back regulation is guaranteed by Rpn4 being itself a short-lived proteasome substrate.[92,93] Less clear are the mechanisms that regulate proteasome biogenesis in metazoans, apart from the well described effects of interferon-γ on the three subunits specifically found as part of the immunoproteasomes of professional antigen-presenting cells.[94] Although all mammalian 26S subunit RNAs increase in a concerted fashion upon proteasomal blockade, neither an ortholog of Rpn4 nor a universal conserved DNA regulatory element has been identified so far in the promoter of mammalian proteasome genes. Intriguingly, the 5′ untranslated regions (UTR) are conserved in mammals and required for mRNA induction of proteasome genes in *Drosophila melanogaster*, although no obvious consensus sequence emerges from comparing different regulated genes.[87] Consistent with the notion that a weak folding capacity may activate proteasome biogenesis, HSF1, the transcriptional inducer of the HSR, has been shown to indirectly induce the coordinated expression of many proteasome subunits in yeast.[95] Oxidative stress may increase proteasome levels through the canonical antioxidant Nrf2–Keap1 pathway, as antioxidant response elements (ARE) present in the regulatory sequences of many proteasome subunits seem to guide proteasome synthesis in response to indirect antioxidants.[96] However, in mouse embryonic fibroblasts, the related transcription factor Nrf1 – but not Nrf2 – has recently been found to be necessary for the concerted transcriptional induction of proteasome subunits after proteasome inhibition, a mechanism dependent on ARE *cis* elements.[97]

3.5.4   Autophagy

Besides the Ub–proteasome system, autophagy is the second main proteolytic strategy in eukaryotic cells, with a clear role against proteotoxicity.[98] Autophagy is a highly conserved mechanism consisting in the formation of double-membrane autophagic vacuoles, also known as autophagosomes, which transport cytoplasmic cargo to the lysosome for enzymatic digestion.[99–101] Identified through genetic screens in yeast, the autophagy genes (*Atg*) are largely conserved in structure and function in animals, including worms, flies, and mammals.[102] The most ancient function of autophagy is the degradation of cytoplasmic components to generate oxidizable substrates when nutrients are scarce. Autophagy is regulated by the class I and class III phosphatidylinositol 3-kinase (PI3K) signaling pathways. While class III PI3K stimulates autophagic vacuole formation, class I PI3K, crucial in the control of protein synthesis, cell growth, and apoptosis, acts as a negative regulator of autophagy. Being activated by the insulin receptor, class I PI3K provides a link to nutrient availability, inhibiting autophagy through Akt/protein kinase B (PKB). The tumor suppressor phosphatase and tensin homolog (PTEN) instead prevents the accumulation of phosphates on lipids and the activation of Akt/PKB, thereby stimulating autophagy. The tuberous sclerosis complex 1 (TSC1) and TSC2 proteins are positive regulators of autophagy repressed by Akt. TSC1 and TSC2 inhibit the small G protein Rheb, which regulates the mammalian target of rapamycin (mTor), a central signal integrator that blocks autophagy.[101,103]

Atg genes are required for autophagic vacuole formation. Among them are Atg6/Beclin1, and the Atg12 and Atg8 ubiquitin-like conjugation pathways. Atg4 is a cysteine protease that cleaves Atg8. Atg7 is an E1-like protein, and Atg10 and Atg3 encode E2-like proteins. Atg5, Atg12 and Atg16 are physically associated with the isolation membrane, whereas Atg8/LC3, the only Atg factor consumed during the process, is directly conjugated to phosphatidylethanolamine and inserted in the isolation membrane.[104,105] Autophagy has been proposed both as a death-mediating strategy and a protective strategy, depending on the cellular context.[98]

3.5.5 Integrating Protein Degradation

Proteasomes and autophagy have long been viewed as independent pathways with no cross talk or point of intersection, but this view has been strongly challenged by recent reports. As a result, these two degradation systems are now known to cooperate in complementarity, possibly compensating for each other's insufficiency. First, the discovery of a role for autophagy in survival of post-partum starvation in mice[106] changed the previous view of autophagy as an extreme measure in pathological conditions, *e.g.* to face proteotoxicity by polyglutamine diseases, or in UPS insufficiency. The following reports that conditional ablation of autophagy in the central nervous system causes spontaneous neurodegeneration at multiple cerebral sites with ubiquitin positive pathology demonstrated that autophagy cooperates with proteasomes to maintain cerebral homeostasis.[107,108] Moreover, autophagy is activated in conditions of impaired proteasomal degradation, as a compensatory response to accumulation of protein aggregates.[109,110] Molecular sensors, transducers and executors are necessary to activate one system upon insufficiency of the other, defining a biological black box that only recently began to be opened, unveiling proteins capable of recognizing ubiquitinated proteins (*e.g.* p62/sequestosome-1), cytoskeletal modifiers (*e.g.* HDAC6) and molecular motors involved in storing malfolded proteins into a few large aggresomes, so as to enable the autophagic machinery to engulf them.[111] As a result, autophagy can compensate for insufficient proteasomes via HDAC6[110] and protect from the toxicity induced by proteasome inhibitors.[112]

Recently, an autophagic clearance of polymeric aggregates has been proposed to contribute to proteostasis in the secretory pathway (ERAD II) in a model of osteogenesis imperfecta, a conformational monogenic disease caused by mutations of collagen type I, the main bone matrix component secreted by osteoblasts. *In vitro* studies on different collagen mutants suggest that polymeric aggregates are disposed of by a lysosomal pathway that utilizes the autophagic machinery, centered on Atg8/LC3, while monomeric aggregates are degraded by the UPS upon retro-translocation to the cytosol – the canonical ER-associated degradation (ERAD) pathway.[113] Since the latter strategy requires unfolding of the doomed protein, a mechanism devoted to store large multimeric ER aggregates and dispatch them to the lysosome is an attractive strategy for the cell's economy. Such a mechanism further expands the knowledge of autophagy and its cross talk with the proteasome, raising a number of exciting questions: What molecular ruler sorts ER aggregates? How are ER aggregates compartmentalized? How does the autophagic machinery in the cytosol recognize aggregate-storing ER cisternae? How are these cisternae discontinued from the ER and engulfed by the phagophore?

These and many questions on the molecular circuit bi-directionally linking the UPS and autophagy remain to be addressed. Learning how the two main proteocatabolic pathways complement each other to maintain cellular homeostasis may offer strategies to manipulate them against proteotoxicity. A major limitation is that most cellular models generally exploited utilize the genetic or pharmacological blockade of the proteasome and autophagy. More natural models are thus needed that recapitulate physiological or pathological conditions in which one system is insufficient or overwhelmed by excessive functional demands, thereby activating the complementary proteocatabolic route. The development of such cellular models holds great promise towards the identification of specific molecular targets against conformational diseases.

3.6   Learning from Development and Disease

Studies on normal and neoplastic plasma cells recently contributed promising cellular models of naturally challenged proteostasis showing restricted protein homeostatic capacity, or proteostenosis: another healthy lesson against our inclination to set borders between disciplines.

Plasma cells are terminally differentiated B lymphocytes specialized in antibody (Ab) production, a key arm of adaptive immunity. Their differentiation entails a spectacular genetic and functional metamorphosis from small resting long-lived cells into huge protein factories, specialized in intense Ab synthesis at due time.[21,114–116] The vast majority of plasma cells are short-lived to limit the Ab response, but some of them return to the bone marrow where they enjoy a long survival, producing low titers of protective antibodies even for a lifetime.[117,118] The mechanisms leading to programmed death of plasma cells are still not completely understood.[115,119] Specific molecular timers or counters are likely to set apoptotic programs after a terminally differentiated plasma cell has produced enough Ab. Recently, we proposed that the balance between protein synthesis and degradation contributes to sensitize plasma cells to spontaneous and pharmacologic apoptosis, ultimately triggering spontaneous cell death.[120,121] This model, referred to as the proteasome *load vs. capacity* concept, stemmed from the observation that while protein synthesis and hence the demand for proteasomal degradation increase geometrically, proteasome abundance and activity paradoxically decrease in activated B cells. The resulting overload of the UPS causes accumulation of ubiquitinated proteins and consumption of the pool of free ubiquitin, resulting in stabilization of a number of proteins, including pro-apoptotic factors, conferring exquisite apoptotic sensitivity to proteasome inhibitors, powerful anti-cancer agents.[120,121] Also the fraction of rapidly degraded proteins increases dramatically in B lymphoma cells following mitogen stimulation.[122] These RDP could further burden a reduced pool of proteasomes, likely contributing to proteasomal overload and impairment.[120,121] The situation resembles that of an inefficient bureau at rush hour: despite an increasing affluence of customers (protein synthesis), many desks unexpectedly close. Respectful customers (long-lived proteins) face another problem, as many unkind clients (RDP) skip the line and greatly prolong the average serving time: a long queue inevitably forms.

The decrease of proteasome capacity following B cell activation is a challenging paradox. First, it contradicts the established capability of eukaryotic cells to adapt to proteasome stress by increasing *de novo* proteasome biogenesis.[88] Second, it fails to match the increasing metabolic demands of the Ab factory.[116] This apparent paradox of proteasome insufficiency in plasma cells may acquire sense in view of the short lifespan of most Ab secretors. The natural load *vs.* capacity imbalance may serve as an inbuilt intracellular counter of the work accomplished, capable of transducing intrinsic signals and launching death when a functional threshold is met, offering a sensible strategy to end the humoral response.[89] In other words, by engineering a bottleneck in an otherwise efficient quality control strategy devoted to Ig folding and assembly, plasma cells may have evolved a mechanism to count the work accomplished, and predispose themselves to die.

Formal evidence that proteasomal overload sensitizes to apoptosis came from studies on human multiple myelomas (MM), the malignant transformation of plasma cells, which provided another model of proteostenosis. This tumor displays exquisite sensitivity to proteasome inhibitors, but the molecular bases are currently under investigation.[123,124] Moreover, a substantial proportion of patients fail to respond, the reasons for the diverse susceptibility remaining largely unexplained. Attesting to a key role for protein synthesis in sensitizing MM cells to death, manipulation of Ig synthesis in MM lines resulted in altered sensitivity to bortezomib, a first-in-class proteasome inhibitor in clinical use against MM.[125] The differential sensitivity of MM lines to proteasome inhibitors was found to be determined by the load *vs.* capacity ratio, with the most proteostenotic tumors revealing the highest vulnerability.[122]

An important implication of these studies is that proteotoxic stress and apoptotic sensitivity to proteasome inhibitors could represent general features of Ig-secreting plasma cells, making Ab responses amenable to treatment with proteasome inhibitors so as to lower Ab titers.[121,126] Proteotoxicity thus proves useful, at least in the case of proteostenotic plasma cells.

Together, normal and neoplastic plasma cells offer valuable models in which the proteasome workload exceeds the cell's degradative capacity.

Most interestingly for cancer biology, the stressful condition unveiled in MM by basal accumulation of endogenous ubiquitin conjugates is still compatible with apparently normal cell functions, suggesting that selected substrates are degraded normally. However, extreme proteasome stress is a situation of precarious equilibrium, that can be easily pushed towards ill-omened consequences. Accordingly, the sensitivity of MM cells to proteasome inhibitors can be exacerbated simply by increasing the rates of RDP generation with chemicals that cause severe protein misfolding and ultimately enhance proteasomal load, or can *vice versa* be mitigated by up-regulating the proteasomal capacity with repeated administration of very low doses of reversible proteasome inhibitors resulting in *de novo* proteasome biogenesis.[122] This observation opens the possibility to exploit the striking pro-apoptotic synergism observed *in vitro* by combined administration of proteasome inhibitors and ER stressors for therapeutic purposes.

This model nicely fits with the general, novel idea that cytotoxic stress can be exploited against cancer.[127–131] Due to deregulated growth, cancer cells generally experience more cytotoxic stress than normal counterparts (*e.g.* hypoxia, nutrient deprivation, pH changes, oxidative stress). As a result, adaptive responses to stress are often activated to higher levels, providing therapeutic specificity. Most stress responses entail apoptotic pathways that can be activated if stress duration or intensity increase, turning these responses maladaptive. In cancer, these responses are generally intact, with lower apoptotic thresholds, while physiological apoptotic responses to genotoxic stress are often disabled. Thus, strategies exploiting cytotoxic stress hold therapeutic promise against cancer.

3.7   Targeting Proteostasis

Manipulating proteostasis is emerging as a new powerful therapeutic opportunity against many apparently heterogeneous conditions, ranging from infectious diseases, to cancer, to conformational disorders.

A high proteostatic capacity may be critical for infections and cancer. Viruses are obligate intracellular parasites that monopolize the cellular proteosynthetic apparatus and require high protein folding and trafficking capacity for replication and assembly. Proteostasis downregulators that selectively target the folding of viral proteins may provide a general antiviral strategy to combat pathogen propagation and resistance.[132] Recently, negative regulators of proteostasis, including inhibitors of HSP90 and of the proteasome, proved effective against not only highly proliferating tumors, but also plasma cell dyscrasias, indicating efficient proteostasis as a critical requirement in cancer biology.[124,133–135] More recently, inhibitors of autophagic aggresome degradation unveiled an additional anti-proteostatic strategy to fight cancer.[136]

On the other side, deficiencies in proteostasis are causal to many inherited and age related diseases, including genetic, metabolic, neurodegenerative, and cardiovascular disorders that strongly impact on morbidity and mortality. Currently available approaches are generally disease-specific and include protein replacement, pharmacologic chaperoning, and kinetic stabilizers. More general therapeutic strategies may aim at restoring the proteostatic capacity by small molecules or biologicals (small interfering RNAs, cDNAs, proteins) capable of modulating the proteostasis network, including protein synthesis, folding, trafficking, and degradation.[4] Positive proteostasis regulators could in principle tackle multiple diseases involving protein misfolding or aggregation, ranging from rare enzymatic defects, *e.g.* Gaucher's and Fabry's, to frequent neurodegenerative diseases such as Alzheimer's, Huntington's, and Parkinson's. Potential new compounds are expected to increase the cellular endogenous folding capacity by empowering the HSR, and by manipulating ER homeostasis, including calcium levels, and the UPR.[4,137,138]

Encouraging experimental results have been achieved with cellular models of loss of function diseases such as cystic fibrosis and lysosomal storage disorders.[137,139,140] At the organismal level, RNAi-based proteostasis regulators targeting the insulin pathway ameliorated gain of toxic function neurodegeneration in nematode models of Alzheimer's and Huntington's diseases.[47,141] Interestingly, small-molecule positive proteostasis regulators may include natural compounds such as celastrol, a HSF-1 activator known to up-regulate the cytosolic chaperoning capacity.[142,143]

Like in all novel biomedical fields, druggability is a key issue, and side effects pose serious concerns. In the case of positive proteostasis regulators, improved adaptation to cytotoxic stress, including proteotoxicity, may boost viral infections or cancer. Encouraging are observations that non-steroidal anti-inflammatory drugs, medicines of established safety, do increase ER and cytosolic chaperones.[144–146] On the other hand, negative proteostasis regulators may accelerate aging and age-related neurodegenerative diseases. This risk is made acceptable by the fact that the short life expectancy typical of the diseases for which these drugs are currently utilized, as in the case of MM, greatly reduces the lifetime risk for developing dementia and other diseases of the elderly. If up to the late 1990s it may have seemed unfeasible to draw therapeutic benefits from targeting protein degradation – a process essential for cell function and viability – proteasome inhibitors taught us that drugging the undruggable is not an act of *hubris*.

3.8 Conclusions

Recent advances in the general understanding of cell physiology and adaptive responses to stress are unveiling complex networks of signaling pathways that influence virtually all cell functions through the integrated control of protein homeostasis. Proteostasis now appears to be involved in the aging process and in a wide variety of inherited and acquired fatal and disabling diseases. This offers a novel post-genomic opportunity to design therapeutic strategies against diseases. Proteostasis regulators are being actively screened and tested for their properties of manipulating the load or the catabolic pathways that defend our cells against proteostenosis and proteotoxicity, with the potential to treat or even cure some of the most challenging diseases of our era. As time does not stop, we all have great hopes in their activities against aging and degeneration.

Abbreviations

Ab   Antibody
ARE   Antioxidant response element
Atg   Autophagy gene
DRiP   Defective ribosomal product
ER   Endoplasmic reticulum
ERAD   ER associated degradation
ERCs   Extrachromosomal ribosomal DNA circles
HSF-1   Heat shock factor-1
HSR   Heat shock response
Ig   Immunoglobulin
IGF-1   Insulin growth factor-1
IIS   Insulin/IGF-1 signaling
mTor   mammalian target of rapamycin
MM   Multiple myeloma
PTEN   Phosphatase and tensin homologue
PI3K   phosphatidylinositol 3-kinase
PKB   Protein kinase B
RDP   Rapidly degraded polypeptides
TSC   Tuberous sclerosis complex
UPS   Ubiquitin proteasome system
UPR   Unfolded protein response
UTR   Untranslated region

Acknowledgements

We thank Pietro Calissano, Francesca Fontana and Niccolò Pengo for helpful suggestions and discussions, Ana Fella and Raffaella Brambati for skilful and patient secretarial assistance and AIRC, Cariplo, MIUR, Telethon for generously supporting our work.

References

1. Y.-C. Tang, H.-C. Chang, M. Hayer-Hartl and F. U. Hartl, *Cell*, 2007, **128**, 412.
2. D. Ron and P. Walter, *Nat. Rev. Mol. Cell Biol.*, 2007, **8**, 519.
3. C. Queitsch, T. A. Sangster and S. Lindquist, *Nature*, 2002, **417**, 618.
4. W. E. Balch, R. I. Morimoto, A. Dillin and J. W. Kelly, *Science*, 2008, **319**, 916.

5. R. J. Jackson and M. Wickens, *Curr. Opin. Genet. Dev.*, 1997, **7**, 233.
6. T. V. Pestova, V. G. Kolupaeva, I. B. Lomakin, E. V. Pilipenko, I. N. Shatsky, V. I. Agol and C. U. Hellen, *Proc. Natl. Acad. Sci. USA*, 2001, **98**, 7029.
7. A. Miluzio, A. Beugnet, V. Volta and S. Biffo, *EMBO Rep.*, 2009, **10**, 459.
8. T. Nakamoto, *Gene*, 2009, **432**, 1.
9. G. Hernandez, M. Altmann and P. Lasko, *Trends Biochem. Sci.*, 2010, **35**, 63.
10. R. J. Jackson, C. U. T. Hellen and T. V. Pestova, *Nat. Rev. Mol. Cell Biol.*, 2010, **11**, 113.
11. N. S. Rane, J. L. Yonkovich and R. S. Hegde, *EMBO J.*, 2004, **23**, 4550.
12. A. Orsi, L. Fioriti, R. Chiesa and R. Sitia, *J. Biol. Chem.*, 2006, **281**, 30431.
13. K. L. Shaffer, A. Sharma, E. L. Snapp and R. S. Hegde, *Dev. Cell*, 2005, **9**, 545.
14. S.-W. Kang, N. S. Rane, S. J. Kim, J. L. Garrison, J. Taunton and R. S. Hegde, *Cell*, 2006, **127**, 999.
15. D. T. Rutkowski, S.-W. Kang, A. G. Goodman, J. L. Garrison, J. Taunton, M. G. Katze, R. J. Kaufman and R. S. Hegde, *Mol. Biol. Cell*, 2007, **18**, 3681.
16. R. S. Hegde and S.-W. Kang, *J. Cell Biol.*, 2008, **182**, 225.
17. P. Hammarstrom, R. L. Wiseman, E. T. Powers and J. W. Kelly, *Science*, 2003, **299**, 713.
18. R. L. Wiseman, E. T. Powers, J. N. Buxbaum, J. W. Kelly and W. E. Balch, *Cell*, 2007, **131**, 809.
19. R. I. Morimoto, *Genes Dev.*, 1998, **12**, 3788.
20. E. A. A. Nollen and R. I. Morimoto, *J. Cell Sci.*, 2002, **115**, 2809.
21. R. Sitia and I. Braakman, *Nature*, 2003, **426**, 891.
22. M. Schroder and R. J. Kaufman, *Annu. Rev. Biochem.*, 2005, **74**, 739.
23. R. Arking, *The Biology of Aging: Observations and Principles*, Oxford University Press, New York, 3rd edn, 2005.
24. C. E. Finch and T. B. L. Kirkwood, *Chance, Development, and Aging*, Oxford University Press, New York, 9th edn, 2000.
25. B. L. Strehler, *Exp. Gerontol.*, 1986, **21**, 283.
26. R. Arking, S. Buck, V. N. Novoseltev, D.-S. Hwangbo and M. Lane, *Ageing Res. Rev.*, 2002, **1**, 209.
27. R. Arking, J. Novoseltseva, D.-S. Hwangbo, V. Novoseltsev and M. Lane, *J. Gerontol. A Biol. Sci. Med. Sci.*, 2002, **57**, B390.
28. B. P. Yu, E. J. Masoro and C. A. McMahan, *J. Gerontol.*, 1985, **40**, 657.
29. E. C. Hadley, C. Dutta, J. Finkelstein, T. B. Harris, M. A. Lane, G. S. Roth, S. S. Sherman and P. E. Starke-Reed, *J. Gerontol. A Biol. Sci. Med. Sci.*, 2001, **56**, Spec No 1, 5.
30. R. Weindruch, K. P. Keenan, J. M. Carney, G. Fernandes, R. J. Feuers, R. A. Floyd, J. B. Halter, J. J. Ramsey, A. Richardson, G. S. Roth and S. R. Spindler, *J. Gerontol. A Biol. Sci. Med. Sci.*, 2001, **56**, Spec No 1, 20.
31. J. E. Morley, *J. Gerontol. A Biol. Sci. Med. Sci.*, 2002, **57**, M2.
32. J. Koubova and L. Guarente, *Genes Dev.*, 2003, **17**, 313.

33. A. L. Goldberg, *Nature*, 2003, **426**, 895.
34. Q. Zhang, E. T. Powers, J. Nieva, M. E. Huff, M. A. Dendle, J. Bieschke, C. G. Glabe, A. Eschenmoser, P. Wentworth, R. A. Lerner and J. W. Kelly, *Proc. Natl. Acad. Sci. USA*, 2004, **101**, 4752.
35. A. C. Massey, R. Kiffin and A. M. Cuervo, *Cell Cycle*, 2006, **5**, 1292.
36. R. Kiffin, S. Kaushik, M. Zeng, U. Bandyopadhyay, C. Zhang, A. C. Massey, M. Martinez-Vicente and A. M. Cuervo, *J. Cell Sci.*, 2007, **120**, 782.
37. R. R. Erickson, L. M. Dunning and J. L. Holtzman, *J. Gerontol. A Biol. Sci. Med. Sci.*, 2006, **61**, 435.
38. A. Ben-Zvi, E. A. Miller and R. I. Morimoto, *Proc. Natl. Acad. Sci. USA*, 2009, **106**, 14914.
39. E. Cohen, J. Bieschke, R. M. Perciavalle, J. W. Kelly and A. Dillin, *Science*, 2006, **313**, 1604.
40. E. Cohen, J. F. Paulsson, P. Blinder, T. Burstyn-Cohen, D. Du, G. Estepa, A. Adame, H. M. Pham, M. Holzenberger, J. W. Kelly, E. Masliah and A. Dillin, *Cell*, 2009, **139**, 1157.
41. E. Cohen, D. Du, D. Joyce, E. A. Kapernick, Y. Volovik, J. W. Kelly and A. Dillin, *Aging Cell*, 2010, **9**, 126.
42. C. Kenyon, J. Chang, E. Gensch, A. Rudner and R. Tabtiang, *Nature*, 1993, **366**, 461.
43. M. Holzenberger, J. Dupont, B. Ducos, P. Leneuve, A. Geloen, P. C. Even, P. Cervera and Y. Le Bouc, *Nature*, 2003, **421**, 182.
44. C. Kenyon, *Cell*, 2005, **120**, 449.
45. L. Kappeler, C. De Magalhaes Filho, J. Dupont, P. Leneuve, P. Cervera, L. Perin, C. Loudes, A. Blaise, R. Klein, J. Epelbaum, Y. Le Bouc and M. Holzenberger, *PLoS Biol.*, 2008, **6**, e254.
46. A.-L. Hsu, C. T. Murphy and C. Kenyon, *Science*, 2003, **300**, 1142.
47. J. F. Morley and R. I. Morimoto, *Mol. Biol. Cell*, 2004, **15**, 657.
48. R. I. Morimoto, *Genes Dev.*, 2008, **22**, 1427.
49. R. I. Morimoto and A. M. Cuervo, *J. Gerontol. A Biol. Sci. Med. Sci.*, 2009, **64**, 167.
50. L. Partridge, *Philos. Trans. R. Soc. Lond. B Biol. Sci.*, 2010, **365**, 147.
51. M. Kaeberlein, *Nature*, 2010, **464**, 513.
52. L. H. Hartwell and M. W. Unger, *J. Cell Biol.*, 1977, **75**, 422.
53. D. A. Sinclair and L. Guarente, *Cell*, 1997, **91**, 1033.
54. P. Laun, A. Pichova, F. Madeo, J. Fuchs, A. Ellinger, S. Kohlwein, I. Dawes, K. U. Frohlich and M. Breitenbach, *Mol. Microbiol.*, 2001, **39**, 1166.
55. C.-Y. Lai, E. Jaruga, C. Borghouts and S. M. Jazwinski, *Genetics*, 2002, **162**, 73.
56. H. Aguilaniu, L. Gustafsson, M. Rigoulet and T. Nystrom, *Science*, 2003, **299**, 1751.
57. N. Erjavec, L. Larsson, J. Grantham and T. Nystrom, *Genes Dev.*, 2007, **21**, 2410.
58. N. Erjavec and T. Nystrom, *Proc. Natl. Acad. Sci. USA*, 2007, **104**, 10877.
59. J. R. Glover and S. Lindquist, *Cell*, 1998, **94**, 73.

60. S. Diamant, A. P. Ben-Zvi, B. Bukau and P. Goloubinoff, *J. Biol. Chem.*, 2000, **275**, 21107.
61. J. Shorter and S. Lindquist, *EMBO J.*, 2008, **27**, 2712.
62. C. Lo Bianco, J. Shorter, E. Regulier, H. Lashuel, T. Iwatsubo, S. Lindquist and P. Aebischer, *J. Clin. Invest.*, 2008, **118**, 3087.
63. P. Wendler, J. Shorter, D. Snead, C. Plisson, D. K. Clare, S. Lindquist and H. R. Saibil, *Mol. Cell*, 2009, **34**, 81.
64. S. M. Doyle and S. Wickner, *Trends Biochem. Sci.*, 2009, **34**, 40.
65. A. B. Lindner, R. Madden, A. Demarez, E. J. Stewart and F. Taddei, *Proc. Natl. Acad. Sci. USA*, 2008, **105**, 3076.
66. T. Nystrom, *PLoS Genet.*, 2007, **3**, e224.
67. I. G. Macara and S. Mili, *Cell*, 2008, **135**, 801.
68. M. Watve, S. Parab, P. Jogdand and S. Keni, *Proc. Natl. Acad. Sci. USA*, 2006, **103**, 14831.
69. S. N. Evans and D. Steinsaltz, *Theor. Popul. Biol.*, 2007, **71**, 473.
70. N. Erjavec, M. Cvijovic, E. Klipp and T. Nystrom, *Proc. Natl. Acad. Sci. USA*, 2008, **105**, 18764.
71. J. Winkler, A. Seybert, L. Konig, S. Pruggnaller, U. Haselmann, V. Sourjik, M. Weiss, A. S. Frangakis, A. Mogk and B. Bukau, *EMBO J.*, 2010, **29**, 910.
72. M. A. Rujano, F. Bosveld, F. A. Salomons, F. Dijk, M. A. W. H. van Waarde, J. J. L. van der Want, R. A. I. de Vos, E. R. Brunt, O. C. M. Sibon and H. H. Kampinga, *PLoS Biol.*, 2006, **4**, e417.
73. L. C. Fuentealba, E. Eivers, D. Geissert, V. Taelman and E. M. De Robertis, *Proc. Natl. Acad. Sci. USA*, 2008, **105**, 7732.
74. R. R. Kopito and R. Sitia, *EMBO Rep.*, 2000, **1**, 225.
75. P. I. Merksamer, A. Trusina and F. R. Papa, *Cell*, 2008, **135**, 933.
76. D. Voges, P. Zwickl and W. Baumeister, *Annu. Rev. Biochem.*, 1999, **68**, 1015.
77. M. Groll, M. Bajorek, A. Kohler, L. Moroder, D. M. Rubin, R. Huber, M. H. Glickman and D. Finley, *Nat. Struct. Biol.*, 2000, **7**, 1062.
78. N. Benaroudj, P. Zwickl, E. Seemuller, W. Baumeister and A. L. Goldberg, *Mol. Cell*, 2003, **11**, 69.
79. A. L. Goldberg and J. F. Dice, *Annu. Rev. Biochem.*, 1974, **43**, 835.
80. M. Y. Sherman and A. L. Goldberg, *Neuron*, 2001, **29**, 15.
81. M. H. Glickman and A. Ciechanover, *Physiol. Rev.*, 2002, **82**, 373.
82. U. Schubert, L. C. Anton, J. Gibbs, C. C. Norbury, J. W. Yewdell and J. R. Bennink, *Nature*, 2000, **404**, 770.
83. J. W. Yewdell, *Trends Cell Biol.*, 2001, **11**, 294.
84. R. M. Vabulas and F. U. Hartl, *Science*, 2005, **310**, 1960.
85. W. R. Bion, *Learning from Experience*, William Heinemann, London, 1962.
86. S. Meiners, D. Heyken, A. Weller, A. Ludwig, K. Stangl, P. M. Kloetzel and E. Kruger, *J. Biol. Chem.*, 2003, **278**, 21517.
87. J. Lundgren, P. Masson, Z. Mirzaei and P. Young, *Mol. Cell Biol.*, 2005, **25**, 4662.

88. J. Hanna and D. Finley, *FEBS Lett.*, 2007, **581**, 2854.
89. S. Cenci and R. Sitia, *FEBS Lett.*, 2007, **581**, 3652.
90. R. J. Kaufman, *J. Clin. Invest.*, 2002, **110**, 1389.
91. G. Mannhaupt, R. Schnall, V. Karpov, I. Vetter and H. Feldmann, *FEBS Lett.*, 1999, **450**, 27.
92. Y. Xie and A. Varshavsky, *Proc. Natl. Acad. Sci. USA*, 2001, **98**, 3056.
93. J. A. Fleming, E. S. Lightcap, S. Sadis, V. Thoroddsen, C. E. Bulawa and R. K. Blackman, *Proc. Natl. Acad. Sci. USA*, 2002, **99**, 1461.
94. A. Macagno, M. Gilliet, F. Sallusto, A. Lanzavecchia, F. O. Nestle and M. Groettrup, *Eur. J. Immunol.*, 1999, **29**, 4037.
95. J. S. Hahn, D. W. Neef and D. J. Thiele, *Mol. Microbiol.*, 2006, **60**, 240.
96. M. K. Kwak, N. Wakabayashi, J. L. Greenlaw, M. Yamamoto and T. W. Kensler, *Mol. Cell Biol.*, 2003, **23**, 8786.
97. S. K. Radhakrishnan, C. S. Lee, P. Young, A. Beskow, J. Y. Chan and R. J. Deshaies, *Mol. Cell*, 2010, **38**, 17.
98. B. Levine and G. Kroemer, *Cell*, 2008, **132**, 27.
99. D. J. Klionsky and S. D. Emr, *Science*, 2000, **290**, 1717.
100. J. J. Lum, R. J. DeBerardinis and C. B. Thompson, *Nat. Rev. Mol. Cell Biol.*, 2005, **6**, 439.
101. E. H. Baehrecke, *Nat. Rev. Mol. Cell Biol.*, 2005, **6**, 505.
102. B. Levine and D. J. Klionsky, *Dev. Cell*, 2004, **6**, 463.
103. D. J. Klionsky, *J. Cell Sci.*, 2005, **118**, 7.
104. Y. Ohsumi, *Nat. Rev. Mol. Cell Biol.*, 2001, **2**, 211.
105. D. J. Klionsky, *Nature*, 2004, **431**, 31.
106. A. Kuma, M. Hatano, M. Matsui, A. Yamamoto, H. Nakaya, T. Yoshimori, Y. Ohsumi, T. Tokuhisa and N. Mizushima, *Nature*, 2004, **432**, 1032.
107. M. Komatsu, S. Waguri, T. Chiba, S. Murata, J.-i. Iwata, I. Tanida, T. Ueno, M. Koike, Y. Uchiyama, E. Kominami and K. Tanaka, *Nature*, 2006, **441**, 880.
108. T. Hara, K. Nakamura, M. Matsui, A. Yamamoto, Y. Nakahara, R. Suzuki-Migishima, M. Yokoyama, K. Mishima, I. Saito, H. Okano and N. Mizushima, *Nature*, 2006, **441**, 885.
109. A. Iwata, B. E. Riley, J. A. Johnston and R. R. Kopito, *J. Biol. Chem.*, 2005, **280**, 40282.
110. U. B. Pandey, Z. Nie, Y. Batlevi, B. A. McCray, G. P. Ritson, N. B. Nedelsky, S. L. Schwartz, N. A. DiProspero, M. A. Knight, O. Schuldiner, R. Padmanabhan, M. Hild, D. L. Berry, D. Garza, C. C. Hubbert, T.-P. Yao, E. H. Baehrecke and J. P. Taylor, *Nature*, 2007, **447**, 859.
111. U. B. Pandey, Y. Batlevi, E. H. Baehrecke and J. P. Taylor, *Autophagy*, 2007, **3**, 643.
112. D. C. Rubinsztein, *Neuron*, 2007, **54**, 854.
113. Y. Ishida and K. Nagata, *Autophagy*, 2009, **5**, 1217.
114. J. W. Brewer and L. M. Hendershot, *Nat. Immunol.*, 2005, **6**, 23.
115. M. Shapiro-Shelef and K. Calame, *Nat. Rev. Immunol.*, 2005, **5**, 230.

116. E. van Anken, E. P. Romijn, C. Maggioni, A. Mezghrani, R. Sitia, I. Braakman and A. J. Heck, *Immunity*, 2003, **18**, 243.
117. R. A. Manz, A. E. Hauser, F. Hiepe and A. Radbruch, *Annu. Rev. Immunol.*, 2005, **23**, 367.
118. A. Radbruch, G. Muehlinghaus, E. O. Luger, A. Inamine, K. G. C. Smith, T. Dorner and F. Hiepe, *Nat. Rev. Immunol.*, 2006, **6**, 741.
119. S. Masciarelli, A. M. Fra, N. Pengo, M. Bertolotti, S. Cenci, C. Fagioli, D. Ron, L. M. Hendershot and R. Sitia, *Mol. Immunol.*, 2010, **47**, 1356.
120. S. Cenci, A. Mezghrani, P. Cascio, G. Bianchi, F. Cerruti, A. Fra, H. Lelouard, S. Masciarelli, L. Mattioli, L. Oliva, A. Orsi, E. Pasqualetto, P. Pierre, E. Ruffato, L. Tagliavacca and R. Sitia, *EMBO J.*, 2006, **25**, 1104.
121. P. Cascio, L. Oliva, F. Cerruti, E. Mariani, E. Pasqualetto, S. Cenci and R. Sitia, *Eur. J. Immunol.*, 2008, **38**, 658.
122. G. Bianchi, L. Oliva, P. Cascio, N. Pengo, F. Fontana, F. Cerruti, A. Orsi, E. Pasqualetto, A. Mezghrani, V. Calbi, G. Palladini, N. Giuliani, K. C. Anderson, R. Sitia and S. Cenci, *Blood*, 2009, **113**, 3040.
123. A. L. Goldberg and K. Rock, *Nat. Med.*, 2002, **8**, 338.
124. S. V. Rajkumar, P. G. Richardson, T. Hideshima and K. C. Anderson, *J. Clin. Oncol.*, 2005, **23**, 630.
125. S. Meister, U. Schubert, K. Neubert, K. Herrmann, R. Burger, M. Gramatzki, S. Hahn, S. Schreiber, S. Wilhelm, M. Herrmann, H. M. Jack and R. E. Voll, *Cancer Res.*, 2007, **67**, 1783.
126. K. Neubert, S. Meister, K. Moser, F. Weisel, D. Maseda, K. Amann, C. Wiethe, T. H. Winkler, J. R. Kalden, R. A. Manz and R. E. Voll, *Nat. Med.*, 2008, **14**, 748.
127. Y. Ma and L. M. Hendershot, *Nat. Rev. Cancer*, 2004, **4**, 966.
128. D. T. Rutkowski and R. J. Kaufman, *Trends Biochem. Sci.*, 2007, **32**, 469.
129. M. Moenner, O. Pluquet, M. Bouchecareilh and E. Chevet, *Cancer Res.*, 2007, **67**, 10631.
130. R. Mathew, V. Karantza-Wadsworth and E. White, *Nat. Rev. Cancer*, 2007, **7**, 961.
131. C. V. Dang, J.-w. Kim, P. Gao and J. Yustein, *Nat. Rev. Cancer*, 2008, **8**, 51.
132. R. Geller, M. Vignuzzi, R. Andino and J. Frydman, *Genes Dev.*, 2007, **21**, 195.
133. C. Dai, L. Whitesell, A. B. Rogers and S. Lindquist, *Cell*, 2007, **130**, 1005.
134. J. Adams, *Nat. Rev. Cancer*, 2004, **4**, 349.
135. P. G. Richardson, P. Sonneveld, M. W. Schuster, D. Irwin, E. A. Stadtmauer, T. Facon, J.-L. Harousseau, D. Ben-Yehuda, S. Lonial, H. Goldschmidt, D. Reece, J. F. San-Miguel, J. Blade, M. Boccadoro, J. Cavenagh, W. S. Dalton, A. L. Boral, D. L. Esseltine, J. B. Porter, D. Schenkein and K. C. Anderson, *N. Engl. J. Med.*, 2005, **352**, 2487.
136. A. Rodriguez-Gonzalez, T. Lin, A. K. Ikeda, T. Simms-Waldrip, C. Fu and K. M. Sakamoto, *Cancer Res.*, 2008, **68**, 2557.
137. T.-W. Mu, D. S. T. Ong, Y.-J. Wang, W. E. Balch, J. R. Yates, L. Segatori and J. W. Kelly, *Cell*, 2008, **134**, 769.

138. D. S. T. Ong, T.-W. Mu, A. E. Palmer and J. W. Kelly, *Nat. Chem. Biol.*, 2010, **6**, 424.
139. X. Wang, J. Venable, P. LaPointe, D. M. Hutt, A. V. Koulov, J. Coppinger, C. Gurkan, W. Kellner, J. Matteson, H. Plutner, J. R. Riordan, J. W. Kelly, J. R. Yates and W. E. Balch, *Cell*, 2006, **127**, 803.
140. T.-W. Mu, D. M. Fowler and J. W. Kelly, *PLoS Biol.*, 2008, **6**, e26.
141. J. F. Morley, H. R. Brignull, J. J. Weyers and R. I. Morimoto, *Proc. Natl. Acad. Sci. USA*, 2002, **99**, 10417.
142. A. Trott, J. D. West, L. Klaic, S. D. Westerheide, R. B. Silverman, R. I. Morimoto and K. A. Morano, *Mol. Biol. Cell*, 2008, **19**, 1104.
143. S. D. Westerheide, J. D. Bosman, B. N. A. Mbadugha, T. L. A. Kawahara, G. Matsumoto, S. Kim, W. Gu, J. P. Devlin, R. B. Silverman and R. I. Morimoto, *J. Biol. Chem.*, 2004, **279**, 56053.
144. K.-I. Tanaka, W. Tomisato, T. Hoshino, T. Ishihara, T. Namba, M. Aburaya, T. Katsu, K. Suzuki, S. Tsutsumi and T. Mizushima, *J. Biol. Chem.*, 2005, **280**, 31059.
145. S. Tsutsumi, T. Namba, K. I. Tanaka, Y. Arai, T. Ishihara, M. Aburaya, S. Mima, T. Hoshino and T. Mizushima, *Oncogene*, 2006, **25**, 1018.
146. R. Rajaiah and K. D. Moudgil, *Autoimmun. Rev.*, 2009, **8**, 388.

CHAPTER 4

Amyloid Channel Modulation by Metal Ions

B. L. KAGAN, M.D., Ph.D.

Professor of Psychiatry, Dept. of Psychiatry & Biobehavioral Sciences, Semel Institute for Neuroscience and Human Behavior, David Geffen School of Medicine at UCLA, Los Angeles, CA 90095-1759, USA

4.1 Amyloid Disease

4.1.1 Amyloid Disease and Neurodegeneration

At least two dozen distinct clinical syndromes involving amyloid proteins and peptides have been described (see Table 4.1). Many of these syndromes, including Alzheimer's disease (AD), Down's syndrome (trisomy 21), Parkinson's disease, Huntington's disease (HD), amyotrophic lateral sclerosis (ALS), prion diseases (Creutzfeldt–Jakob, fatal familial insomnia, bovine spongiform encephalopathy (mad cow), Gerstmann–Straussler–Scheinker, and scrapie), and familial amyloid neuropathy involve neurodegeneration. Although the level of evidence varies amongst these diseases, there is considerable consensus that amyloid plays a critical pathophysiologic role. For example, in AD and Down's syndrome, evidence implicating amyloid includes the fact that amyloid deposits occur nearly universally in the brains of affected patients. Patients with Down's syndrome, who have an extra dose of the amyloid precursor protein gene on chromosome 21, develop amyloid deposits and dementia at an early age. In familial AD affected patients frequently have mutations in the amyloid precursor protein, and these mutations often occur near the cleavage sites

RSC Drug Discovery Series No. 7
Neurodegeneration: Metallostasis and Proteostasis
Edited by Danilo Milardi and Enrico Rizzarelli
© Royal Society of Chemistry 2011
Published by the Royal Society of Chemistry, www.rsc.org

Table 4.1 Amyloid diseases and proteins.

Disease	Protein	Abbreviation
Alzheimer's disease	Amyloid precursor protein	APP
Down's Syndrome (Trisomy 21)	(Abeta 1–42)	(Abeta 1–42)
Heredity cerebral angiopathy (Dutch)		
Kuru	Prion protein	PrP^c/PrP^{sc}
Gerstmann-Straussler- Scheinker Syndrome (GSS)		
Creutzfeldt-Jakob Disease		
Scrapie (sheep)		
Bovine spongiform encephalopathy ("mad cow")		

Type II diabetes mellitus
  (adult onset)
	Islet amyloid polypeptide
	  (amylin)
		IAPP

Dialysis-associated amyloidosis	Beta-2-microglobulin	B2M
Senile cardiac amyloidosis	Atrial natriuretic factor	ANF
Familial amyloid polyneuropathy	Transthyretin	TTR
Reactive amyloidosis	Serum amyloid A	SAA
Familial Mediterranean fever		
Familial amyloid polyneuropathy (Finnish)	Gelsolin	Agel
Macroglobulinemia	Gamma-1 heavy chain	AH
Primary systemic amyloidoses	Ig - lambda, Ig - kappa	AL
Familial polyneuropathy – Iowa (Irish)	Apolipoprotein A1	ApoA1
Hereditary cerebral myopathy (Iceland)	Cystatin C	Acys
Nonneuropathic hereditary amyloid with renal disease	Fibrinogen Alpha	AFibA
Nonneuropathic hereditary amyloid with renal disease	Lysozyme	Alys
Familial British dementia	FBDP	A Bri
Familial Danish dementia	FDDP	A Dan

which ultimately generate the amyloid peptide found in deposits. The Alzheimer's beta-amyloid peptide (Abeta) has been shown to be cytotoxic and neurotoxic in culture, to inhibit long-term potentiation, to induce mitochondrial depolarization and dysfunction, and to induce cellular membrane depolarization and calcium dysregulation.[1] The amyloid precursor protein or the amyloid peptide itself can be transgenically inserted into mice to create an Alzheimer-like disease characterized by memory dysfunction that onsets with age and amyloid deposits in the brain. Similar evidence, though less extensive, exists for PD, where alpha-synuclein is the precursor protein of the amyloid peptide, and prion diseases where the cellular prion protein (PrP^c) is converted into a toxic or scrapie-like form (PrP^{sc}) which then is highly prone to forming amyloid deposits. In familial amyloid polyneuropathy, which is characterized by amyloid deposits of the protein transthyretin, a number of known mutations are associated with the occurrence of this disease in various families. Most recently, evidence has accrued for ALS where the enzyme superoxide dismutase appears to be induced to form amyloid by the presence of a variety of mutations. Huntington's

Table 4.2 Other protein misfolding diseases.

Disease	*Protein*	*Abbreviation*
Diffuse Lewy body disease Parkinson's disease	Alpha-synuclein	AS
Fronto-temporal dementia	tau	tau
Amyotrophic lateral sclerosis	Superoxide dismutase-1	SoD-1
Triplet-repeat diseases:	**Polyglutamine tracts in the following proteins:**	**PG**
(Huntington's, Spinocerebellar ataxias, *etc.*)	Huntingtin	
Spinal and bulbar muscular atrophy	Androgen receptor	
Spinocerebellar ataxias	Ataxins	
Spinocerebellar ataxia 17	TATA box-binding protein	

disease, a triplet-repeat disease, characterized by an expansion of repeats of the trinucleotide CAG resulting in extended tracts of polyglutamine in the protein *huntingtin,* is a non-classic amyloid disease, but it shows some curiously similar properties (Table 4.2).

4.1.2 Amyloid Structure

Amyloid was first described in 1854 by Rudolf Virchow. He described "starch-like" deposits in tissues that stained with iodine. Subsequent investigation determined that these deposits were largely proteinaceous and stained intensely with dyes such as Congo red. The characteristic apple-green birefringence exhibited by these deposits led to amyloid becoming a term used for all such deposits found in pathologic specimens. Later studies show that these deposits consisted of fibrils 80 to 100 angstroms in width and of indeterminate length. (See Sipe and Cohen, 2000 for review.[2]) Although protein is the major component of these deposits, they also contain the pentraxin amyloid P and glycosaminoglycans. Eventually, it became clear that the staining with Congo red and the characteristic appearance of amyloid in microscopic specimens reflected the *beta-sheet* nature of the involved proteins. Although these proteins differed in their functions, amino acid sequences, and tertiary conformations, it is now well established that in amyloid deposits all of these proteins have regions that have assumed a beta-sheet nonnative conformation. Many factors can induce native proteins to adopt a beta-sheet amyloidogenic confirmation. Factors that have been identified so far include high protein concentration, pH, amino acid mutations, protein cleavage, aging, and the presence of membranes.[3]

Although much work had been devoted to the investigation of amyloid deposits and fibrils, recent evidence has suggested that while these fibrils are characteristic of amyloid diseases, they may, in fact, not be pathogenic molecules. Indeed, these fibrils seem not to be toxic in typical cellular or *in vitro* assays. Moreover, it has also been discovered that the proteins and peptides which lead to amyloid formation are generally not toxic in their monomeric state. It appears that it is predominantly oligomeric forms of amyloid peptides that possess cytotoxic properties.[4]

4.1.3   Channel Hypothesis of Amyloid Disease

The mechanism by which amyloid proteins and peptides cause disease remains a subject of controversy and intense investigation. Although many hypotheses of amyloid toxicity have been proposed, none has acquired a persuasive evidence base. No enzymatic or highly specific receptor-binding activity has yet been identified for amyloid peptides. The channel hypothesis of amyloid disease was proposed in 1993 when Arispe *et al.*[5] discovered that Abeta could form ion-permeable channels in artificial lipid bilayer membranes. These channels were large, heterodisperse, nonspecifically cation-selective, permeable to calcium, and sensitive to blockade by aluminium and zinc.[5] The authors went on to propose that channel formation could be a pathogenic mechanism leading to membrane depolarization and calcium influx. Indeed, they calculated that a single Abeta channel could actually cause a significant leak in the plasma membrane of a target neuron. Their work has been repeated several times in different laboratories and has been extended to a number of other amyloid peptides, including those involved in PD, prion diseases, FAP (Familial Amyloid Polyneuropathy), ALS, and Huntington's disease (see Kagan *et al.*, 2004 for review).[6]

The detection of channels by electrophysiologic methods has also been complemented by the detection of pore-like structures using atomic force microscopy (AFM) and electron microscopy.[7] Furthermore, subsequent investigations have shown that amyloid peptide channels can induce cell and mitochondrial depolarization and calcium influx.[1,8]

Molecular dynamic simulations have suggested that a 24-mer of Abeta can form a channel structure with dimensions similar to those observed by AFM. The subunits predicted by these simulations are highly mobile and rearrange-able, which might account for the heterogeneity of channels observed in electrophysiologic studies.[9]

Channels have also been observed using electron microscopy in pathologic specimens from Alzheimer-diseased brain.[10] All of these lines of evidence converge to suggest that channel formation could be the pathogenic mechanism of amyloid in amyloid diseases. The rest of this chapter will be devoted to describing the modulation of amyloid channel function by metal ions. Very few modulators or blockers of these amyloid channels have been described so far, and metals, including zinc, have been the most frequently studied blockers. Because blocking of amyloid channels can inhibit toxicity and might lead to therapeutic drugs that could ameliorate the course of disease, it is vital to understand the modulation and blockade of amyloid channels by metal ions.

4.2 Amyloid Peptide Channels

4.2.1 Channel Size

Amyloid peptide channels exhibit a number of physiologic properties that are striking because they seem to be in common across this class of very diverse peptides and proteins. The first channel property of note is the size of the single channel conductance. The single channel conductances of all amyloid peptide channels are large compared to the conductances of typical physiologic channels, such as the voltage-gated sodium or potassium channels of nerves.

The relatively large conductance of amyloid channels would cause a significant leak in neuronal membranes and, in effect, would short-circuit the current generated by sodium or potassium channels. The large single channel conductances are also consistent with what appears to be a large physical size observed by AFM and electron microscopy. In these studies, size estimates of approximately 10–12 angstroms for the inner diameter of the channel have been consistently observed.[7] This would support the large single channel conductances recorded in physiologic studies. Although one can never infer specific physical size from single channel conductance, there is often a correlation between the two. The large conductance and physical size are also in accordance with molecular dynamic simulations of possible channel conformations.[11]

4.2.2 Hetereogeneity

Amyloid channels are generally found to be heterodisperse, *i.e.*, a number of different single channel conductances are usually observed. This has been the case for amyloid channels involved with neurodegeneration, such as the beta amyloid peptide (Abeta), the prion peptide, transthyretin, and alpha synuclein 65-95. Multiple single channel conductances are consistent with the idea that aggregates of amyloid peptides are the channel forming entities. The aggregates would likely incorporate a variable number of monomers, which would lead to different pore sizes and different single channel conductances. This hypothesis is consistent with the observation that single channel conductance seems to grow with time in the membrane or with aging of the peptide in solution which tends to produce larger aggregates.[12]

In contrast, the islet amyloid polypeptide (IAPP), or amylin, which is found in the islets of Langerhans in the pancreas of patients with type 2 diabetes, was observed to have a homogeneous single channel conductance suggesting a unique molecular organization and confirmation for this peptide oligomer.[13]

4.2.3 Aggregation

Another common property of amyloid peptide channels is that channel formation is promoted by treatments such as aging of the peptide in solution or acidic pH. Both of these treatments tend to enhance peptide aggregation and oligomer formation. These treatments also enhance interaction of the peptides with lipid membranes.[14] Of the peptides so far observed to form channels, virtually all show an enhancement of channel formation at acidic pH.[6] This is particularly vigorous in the presence of biological membranes – there seems to be a synergistic effect of lipid membranes and low pH on the aggregation and insertion process of beta-sheet peptides.[15]

4.2.4   Ionic Selectivity

Although some of the amyloid peptides characterized so far exhibit a limited degree of ionic selectivity, for the most part they are relatively nonselective for common physiologic ions (see Table 4.3). This is in contrast to most biological channels, such as sodium, potassium or calcium channels, which tend to show a relatively high degree of selectivity for a specific ion. This ion selectivity is necessary to their physiologic function and for electrical signaling in nerve cells. The relative lack of ion selectivity of amyloid peptide channels and, in particular, their permeability to calcium, tends to make these channels into a "leak" in the neuronal or mitochondrial membrane. Leakage pathways degrade the physiologic functioning of the cell, requiring enhanced energy expenditure on ion pumps such as the sodium/potassium ATPase, or thereby enhancing energy expenditure in the form of increased electron transport through mitochondria to supply the required ATP. In either case, leakage pathways use up energy that would otherwise go to cellular maintenance and functioning. Since the energy demands on neurons for electrical signaling are already quite high, the leaks imposed by amyloid peptides could pose a metabolic threat to neurons in particular compared to other cells. Neurons are uniquely vulnerable to such leaks because excitable cells maintain large ion gradients across their plasma membranes in order to carry out electrical signaling.

In addition to their effects on the plasma membranes of cells, amyloid channels have been suggested to insert into mitochondrial membranes and degrade the ability of mitochondria to maintain the H^+ gradient necessary for ATP synthesis. Such a degradation in the efficiency of mitochondrial ATP production induces a particularly vicious cycle – lowered ATP levels compromise the ability of the Na and Ca pumps to extrude the additional Na^+ and Ca^{2+} leaking in across the plasma membrane, and elevated cytoplasmic Ca^{2+} causes a switch in mitochondrial function from ATP generation to Ca^{2+} sequestration,[16] which causes a further decrease in ATP production. Maintenance of low intracellular Ca^{2+} is of paramount importance for the health of nerve cells, since elevated intracellular Ca^{2+} levels trigger massive exocytotic release of neurotransmitters, derangement of second messenger systems, and activation of intracellular caspases. Elevated intracellular Ca^{2+} levels are considered to be a final common pathway for apoptotic cell death.

4.2.5   Irreversibility/Cytotoxicity

All the amyloid peptide channels thus far described are irreversible, *i.e.*, once inserted into the membrane, they do not leave the membrane and seem to be

Table 4.3 Metal blockade of amyloid peptides.

Peptide	Single channel conductance	Ion selectivity (Permeability ratio)	Blockade by zinc	Blockade by copper	Blockade by aluminum	Blockade by cadmium	Inhibition by Congo red	Reference
Aβ25-35	10-400 pS	Cation ($P_K/P_{Cl} = 1.6$)	+				+	85, 86
Aβ1-40	10-2000 pS	Cation ($P_K/P_{Cl} = 1.8$)	+	+				37
Aβ1-40	50-4000 pS	Cation ($P_K/P_{Cl} = 11.1$)	+					87, 88
Aβ1-42	10-2000 pS	Cation ($P_K/P_{Cl} = 1.8$)	+				+	37
Aβ1-42		Cation	+	+				31, 32, 33
CT105 (C-terminal fragment of amyloid precursor protein (APP))	120 pS	Cation	+				+	90
Islet amyloid poly-peptide (Amylin)	7.5 pS	Cation ($P_K/P_{Cl} = 1.9$)	+				+	51
PrP106-126	10-400 pS	Cation ($P_K/P_{Cl} = 2.5$)	+				+	60
PrP106-126	140,900, 1444 pS	Cation ($P_K/P_{Cl} > 10$)	+	+				91
PrP 82-146		Cation (variable)		+				12, 63
PrP 106-126 (deamidated)								

Peptide	Single channel conductance	Ion selectivity (Permeability ratio)	Blockade	Inhibition by Congo red	Reference
Serum Amyloid A	10-1000 pS	Cation ($P_K/P_{Cl} = 2.9$)	+	+	9
C-type natriuretic peptide	21,63 pS	Cation ($P_K/P_{Cl} > 10$)	+	+	93, 94
Beta2-microglobulin	0.5-120 pS	Non-selective	+	+	3
Transthyretin	Variable	Cation (Variable)	+	+	7
Polyglutamine (AVG MW = 6000)	19-220 pS	Non-selective	−	−	6
NAC (Alpha-synuclein 65-95)	10-300 pS	Variable	+	+	8, 31
ABri	Variable	N.D.	+	+	13
ADan	Variable	N.D.	+	+	13

open most or all of the time. This property makes them particularly devastating to the gradients of ions and metabolites normally maintained across cell membranes, and to physiologic homeostasis. Because it is technically difficult to identify amyloid peptide channels *in vivo* due to their relative lack of selectivity, it has been hard to determine where their primary mode of action is. *In vitro* cellular experiments indicate that amyloid peptide channels can kill cells through their action at the plasma membrane as demonstrated by the observation that blockade of amyloid peptide channels with zinc or other agents can rescue cells from cytotoxicity.[17] At least in this cytotoxicity model, Abeta channels appear to remain localized in the plasma membrane and to conduct their toxic effects from that location.

Evidence with other amyloid peptide channels, such as those associated with Parkinson's and Huntington's diseases, suggests that damage to mitochondrial function is also an important feature of amyloid peptide action. Amyloid peptides can be observed on the surfaces of mitochondria, and mitochondrial function in these diseases is severely degraded in terms of energy production, release of calcium from mitochondria, and depolarization of the mitochondrial membrane potential.[18] These effects suggest that Abeta is able to traverse the cell's plasma membrane and reach intracellular membrane compartments, such as the mitochondria. In light of this observation, it is reasonable to hypothesize that Abeta channels also affect other internal membranes that are more difficult to observe but may be just as important in pathogenesis, including endoplasmic reticulum, lysosomal membranes, and nuclear membranes.

4.2.6   Calcium Dysregulation

In a number of *in vitro* and *in vivo* models, calcium dysregulation has been implicated in the pathogenesis of neurodegenerative diseases caused by amyloid peptides. Channel formation is a logical proximate cause of calcium dysregulation, both directly through collapsing calcium gradients and indirectly by compromising mitochondrial function (see above). While it is possible for other kinds of physiologic processes to impair the homeostatic regulation of calcium levels in the cytoplasm, the channel model gives a direct and obvious explanation for this near universal finding in amyloid diseases.

4.3   Zinc Blockade

4.3.1   Abeta 1-40

The interaction of zinc ions and the Alzheimer Abeta channels was first described by Arispe *et al.* in 1996.[5] Previous work had shown that the zinc ion could bind to Abeta 1-40 with high affinity and had been implicated in the formation of amyloid plaques.[19] Using the planar lipid bilayer *in vitro* system, Arispe *et al.* incorporated Abeta 1-40 channels into the membrane in an asymmetric one-sided manner. They found that both channel gating and conductance could be modulated by the addition of zinc ion to one side of the channel. The presence of zinc appeared to induce an increased rate of transitions from the maximal, fully open channel state to lower conductance states of the channel. For example, a channel with an open maximal conductance of 110 pS was induced by the presence of 250 μM zinc to more frequently transition to levels of 82, 44, and 19 pS. These effects were reversible by removing zinc from the medium. Intriguingly, the presence of zinc only on one side of the membrane induced this channel-blocking behavior. The presence of zinc on the other side did not, indicating that zinc itself could not permeate the channel.

Inhibition by zinc was observed both in channels of low conductance (less than 400 pS) and in the so-called giant channels (conductance greater than 400 pS). In the giant conductance channels, the presence of zinc ion also seemed to induce a transition to lower conductance states. The blockade induced by zinc was not affected by the transmembrane voltage. This indicates that the zinc binding site does not see any significant fraction of the voltage drop across the membrane. This places the zinc binding site outside the hydrocarbon core of the lipid bilayer. The zinc concentrations used in these experiments varied from 50–1500 μM.

4.3.2 Implications for Pore Structure

The authors interpreted their results in terms of molecular models previously proposed for the beta amyloid peptide channel.[20] They specifically noted that a predicted beta hairpin region of Abeta contains three histidines, at positions 6, 13 and 14, and that there are two nearby negatively charged residues, aspartate 7 and glutamate 11. They note that histidines and negatively charged residues are essential for the association of zinc with metalloproteases, but that the exact arrangement of the histidines and negative charges is not critical. In one of their predicted oligomeric channel structures, there are rings of histidine 6's and aspartate 7's, contributed by multiple subunits, encircling one pore entrance.

In these authors' model, the other entrance is enclosed by rings of glutamate 11's, histidine 13's and histidine 14's. They also note that the location of these sites is consistent with the one-sided nature of their zinc effect and the lack of voltage dependence of the blockade induced by zinc. Five critical features of the zinc effect they describe have subsequently been reproduced in a number of other amyloid peptides. These features are: A. Dose dependence; B. Reversibility; C. One-sidedness; D. Incompleteness of the block; E. Voltage independence.

4.3.3 Shorter Abeta Peptides

Lin *et al.* (2002)[21] reported that a shorter fragment of the beta amyloid peptide, Abeta 25-35, consisting of only 11 amino acids, could also form functional channels which could be blocked by zinc. In addition, they reported that cadmium and copper ions at micromolar concentrations could induce blockade of this channel. They also noted that they had previously reported zinc and cadmium blockade of the PrP 106-126 channel at millimolar concentrations, thus suggesting that the prion amyloid channel might be less sensitive than the Abeta channel to divalent ion blockade.[12] However, the Abeta 25-35 fragment does not occur anywhere in nature and may not be the most appropriate model for looking at metal ion blockade of channels.

4.3.4   Other Amyloids

Zinc blockade was subsequently demonstrated for the Abeta 1-42 channel by Hirakura *et al.*, 2000.[22] The blockade exhibited similar properties to that of the Abeta 1-40 channel. Zinc blockade was also demonstrated for the CT 105 fragment, that is, the C-terminal fragment of the amyloid precursor protein which contains Abeta 1-42 within it.[23]

Zinc blockade has been demonstrated for numerous amyloid peptides including IAPP,[12,24] PrP106-126,[12] serum amyloid A,[25] C-type natriuretic peptide,[26] beta-2 microglobulin,[27] transthyretin,[27] and alpha synuclein 65-95.[28]

All of these investigations repeated the essential findings of Arispe *et al.*[5] that zinc blockade occurred at micromolar concentrations in a reversible, asymmetric, incomplete and voltage-independent manner. This suggested that the nature of the blockade was similar in the various amyloid peptides in spite of the fact that all these peptides have widely diverging primary sequences. This suggested that zinc blockade reflected a common feature of amyloid channel structure rather than any specific amino acid sequence.

Interestingly, the non-amyloid channel-forming polyglutamine peptide found in the mutant huntingtin protein was found not to be blocked by zinc. Since this illness is not a true amyloid disease, but is characterized by amyloid-like inclusion bodies, this result is not entirely unexpected.[29,30]

4.4   Copper Blockade

4.4.1   Prion Channels

Lin *et al.* (1997)[12] demonstrated that a neurotoxic fragment of the prion protein, PrP (106-126), formed relatively nonselective ion permeable channels in planar lipid bilayers, and characterized these channels in terms of their concentration and lipid dependence. They showed that channel formation was inhibited by premixing PrP with Congo red, and channel function was blocked by addition of zinc to one side after channel formation. Kourie *et al.* (2001)[31,32] refined our knowledge of PrP channels by describing three isoforms of the PrP 106-126 channel that could be formed in lipid bilayers. Although these channels exhibited distinct and characteristic biophysical properties, they shared a common response to the addition of copper ion (Cu^{2+}). PrP 106-126 channels showed an overall reduction of activity and conductance due to the addition of copper. More specifically, copper shifted the kinetics of the channel from the open state to a bursting state. In this bursting state, channel activity would occur with rapid opening and closing in small bursts separated by long periods of inactivity. Copper's action on the open channel activity was both time dependent and voltage dependent.

Copper induced these changes in channel kinetics but did not change the conductance of open channels, strongly suggesting that copper was binding at the mouth of the channel rather than within the pore itself where it would be subject to the drop in voltage. Kourie *et al.* (2003)[32] went on to show that copper had a similar effect on the channel activity of the deamidated isoforms of this prion protein fragment. These isoforms involved replacing the asparagine at 108 with aspartate or isoaspartate. These isoforms showed biophysical properties and a response to copper ion similar to the wild type protein. Taken together, these effects of copper on PrP 106-126 channels strongly imply that the copper binding site might be located at methionine 109 and histidine 111 of the prion fragment.

4.4.2 Prion Core Mutants

Kourie *et al.* (2003)[33] extended their study of copper blockade of prion channels to the study of PrP 106-126 mutant peptides in order to investigate the role of the hydrophobic core, AGAAAAGA. They replaced this core with a less hydrophobic VVAASSS. This mutant has reduced hydrophobicity of the core due to the hydrophilic serine residues. This resulted in a significant decrease in channel activity, most likely stemming from the decrease in beta-sheet structure which was confirmed by CD-spectroscopy. The biophysical properties of the channel were generally similar to other amyloid channel properties. Copper caused a shift in the kinetics of the channel from the open or burst state into a spiking mode. Copper reduced both the probability of channel opening and the mean open time. Copper also increased channel opening frequency and the mean closed time. Copper did not induce any changes in open channel conductance suggesting that copper was binding at the mouth of the pore via a fast channel block mechanism rather than binding to the hydrophobic core. Again, this may point to methionine 109 and histidine 111 as a potential binding site for copper. While it is clear from the mutant channel data that the hydrophobic core sequence plays a key role in PrP 106-126 channel formation, it appears not to be a binding site for copper since the effects of copper on these mutant channels are nearly identical to the effects on the wild-type PrP 106-126 channels.

4.4.3 Abeta

Bahadi *et al.* (2003)[34] studied the effects of copper on Abeta 1-42 channels. They found that these channels were inhibited by copper at micromolar concentrations in a voltage- and concentration-dependent manner. This channel blockade was fully reversible at concentrations between 50 and 100 µM but only partially reversible at 250 µM. At the lower concentration of copper, channel blockade could be reversed by the addition of clioquinol, a copper chelating agent and antibiotic, at micromolar concentrations. However, the effects of 250 μM copper on burst and intraburst kinetic parameters of the Abeta 1-42 channels were *not* fully reversed by either washout of the copper or application of clioquinol.

Kinetic analysis suggested that the copper-induced inhibition was most likely mediated by both desensitization of the channel and an open channel block mechanism. It was proposed that copper was binding to histidine residues located at the mouth of the pore. This interpretation is similar to that proposed for the PrP 106-126 channels. Copper interactions with Abeta and PrP had previously been characterized by Curtain *et al.*[35] in work that reported that Abeta was strongly aggregated by copper in acidic conditions. They also identified an attomolar copper affinity binding site on Abeta 1-42 and it was suggested that this copper binding generated an ordered membrane-penetrating structure containing superoxide dismutase-like subunits.[35] These authors went on to note that Abeta channels had a higher affinity for copper than for zinc, and that copper was inducing a fast open channel block as opposed to the slow block induced by zinc. Since copper is present at micromolar concentrations at synapses, they also suggested that copper binding to Abeta 1-42 might have an effect on Abeta-induced toxicity, perhaps by forming membrane-penetrating structures. However, the effect of copper on the already formed Abeta channels appears to down-regulate their activity via a blocking mechanism.

4.5   Aluminium and Cadmium

Because aluminium was found in the amyloid deposits that characterize Alzheimer's disease, many scientists once thought it played an etiologic role in the illness. Many of them even gave up their aluminium cookware. Subsequent investigation revealed that aluminium was a latecomer to the amyloid deposits. Amyloid peptides were capable of strong aggregation and fibrilization in the absence of aluminium. Arispe *et al.*[36] showed that Al^{3+} could irreversibly block Abeta 1-40 channels in lipid membranes. However, this type of block by trivalent cations is common and likely to be due to strong electrostatic interactions. Thus it is not a specific blocking mechanism of aluminium that is being observed.

Cadmium (Cd^{2+}) was reported to reversibly block Abeta 25-35 channels by Lin *et al.* (2002).[37] The biological significance of this is unknown. Cadmium is not associated with Alzheimer's disease, and the Abeta 25-35 fragment is not found in AD tissue. Still this work suggests that we may not have fully investigated all the potential effects of metals on amyloid channels.

Acknowledgements

This work was supported in part by grants from the Alzheimer's Association and NIH. We are grateful to Erik Schweitzer, PhD, and Doris Finck for editorial assistance.

References

1. I. Bezprozvanny and M. P. Mattson, *Trends Neurosci.*, 2008, **31**, 454–463.
2. J. D. Sipe and A. S. Cohen, *J. Struct. Biol.*, 2000, **130**, 88–98.
3. F. Chiti and C. M. Dobson, *Annu. Rev. Biochem.*, 2006, **75**, 333–366.
4. F. Rahimi, A. Shanmugam and G. Bitan, *Curr. Alzheimer. Res.*, 2008, **5**, 319–341.
5. N. Arispe, H. B. Pollard and E. Rojas, *Proc. Natl. Acad. Sci. USA*, 1996, **93**, 1710–1715.
6. B. L. Kagan, R. Azimov and R. Azimova, *J. Membr. Biol.*, 2004, **202**, 1–10.
7. A. Quist, I. Doudevski, H. Lin, R. Azimova, D. Ng, B. Frangione, B. Kagan, J. Ghiso and R. Lal, *Proc. Natl. Acad. Sci. USA*, 2005, **102**, 10427–10432.
8. N. Dragicevic, M. Mamcarz, Y. Zhu, R. Buzzeo, J. Tan, G. W. Arendash and P. C. Bradshaw, *J. Alzheimers Dis.*, 2010, **20**, S535–S550.
9. H. Jang, F. T. Arce, S. Ramachandran, R. Capone, R. Azimova, B. L. Kagan, R. Nussinov and R. La, *Proc. Natl. Acad. Sci. USA*, 2010, **107**, 6538–6543.
10. S. Inoue, *Amyloid*, 2008, **15**, 223–233.
11. Y. Miller, B. Ma and R. Nussinov, *Proc. Natl. Acad. Sci. USA*, 2010, **107**, 9490–9495.
12. M. Lin, T. Mirzabekov and B. L. Kagan, *J. Biol. Chem.*, 1997, **272**, 44–47.
13. T. Mirzabekov, M. Lin and B. L. Kagan, *J. Biol. Chem.*, 1996, **271**, 1988–1992.
14. J. McLaurin and A. Chakrabartty, *Eur. J. Biochem.*, 1997, **245**, 355–363.
15. B. L. Kagan and J. Thundimadathil, in *Proteins: Membrane Binding & Pore Formation*, ed. G. Anderluh and J. Lakey, Landis Biosciences, Austin, TX, 2010, pp. 178–192.
16. J. X. Chen and S. S. Yan, *J. Alzheimers Dis.*, 2010, **20**, S569–S5678.
17. N. Arispe, J. C. Diaz and M. Flora, *Biophys. J.*, 2008, **95**, 4879–4889.
18. I. L. Ferreira, R. Resende, E. Ferreiro, A. C. Rego and C. F. Pereira, *Curr. Drug Targets*, 2010, **11**, 1193–1206.
19. A. I. Bush, W. H. Pettingell, G. Multhaup, M. D. Paradis, J. P. Vonsattel, J. F. Gusella, K. Beyreuther, C. L. Masters and R. E. Tanzi, *Science*, 1994, **265**, 1464–1467.
20. S. R. Durell, H. R. Guy, N. Arispe, E. Rojas and H. B. Pollard, *Biophys. J.*, 1994, **67**, 2137–2145.
21. M.-C. Lin, T. Mirzabekov and B. L. Kagan, *Peptides*, 2002, **23**, 1215–1228.
22. Y. Hirakura and W. W. Yiu, *Amyloid*, 2000, **7**, 194–199.
23. H. J. Kim, Y. H. Suh, M. H. Lee and P. D. Ryu, *Neuroreport*, 1999, **10**, 1427–1431.
24. T. Mirzabekov, M. C. Lin, W. L. Yuan, P. J. Marshall, M. Carman, K. Tomaselli, I. Lieberburg and B. L. Kagan, *Biochem. Biophys. Res. Commun.*, 1994, **202**, 1142–1148.
25. Y. Hirakura, I. Carreres, J. D. Sipe and B. L. Kagan, *Amyloid*, 2002, **9**, 13–23.

26. J. I. Kourie, *FEBS Lett.*, 1999, **445**, 57–62.
27. Y. Hirakura and B. L. Kagan, *Amyloid*, 2001, **8**, 94–100.
28. A. Quist, I. Doudevski, H. Lin, R. Azimova, D. Ng, B. Frangione, B. Kagan, J. Ghiso and R. Lal, *Proc. Natl. Acad. Sci. USA*, 2005, **102**, 10427–10432.
29. Y. Hirakura, R. Azimov, R. Azimova and B. L. Kagan, *J. Neurosci. Res.*, 2000, **60**, 490–494.
30. Y. Hirakura, R. Azimov, R. Azimova, F. Javier, E. S. Schweitzer and B. L. Kagan, *Amyloid: J. Prot. Fold. Dis.*, 2010 (in press).
31. J. I. Kourie, E. A. Hanna and C. L. Henry, *Can. J. Physiol. Pharmacol.*, 2001, **79**, 654–664.
32. J. I. Kourie, P. V. Farrelly and C. L. Henry, *J. Neurosci. Res.*, 2001, **66**, 214–220.
33. J. I. Kourie, B. L. Kenna, D. Tew, M. F. Jobling, C. C. Curtain, C. L. Masters, K. J. Barnham and R. Cappai, *J. Membr. Biol.*, 2003, **193**, 35–45.
34. R. Bahadi, P. V. Farrelly, B. L. Kenna, J. I. Kourie, F. Tagliavini, G. Forloni and M. Salmona, *Am. J. Physiol. Cell Physiol.*, 2003, **285**, C862–C872.
35. C. C. Curtain, F. Ali F, I. Volitakis, R. A. Cherny, R. S. Norton, K. Beyreuther, C. J. Barrow, C. L. Masters, A. I. Bush and K. Barnham, *J. Biol. Chem.*, 2001, **276**, 20466–20473.
36. N. Arispe, E. Rojas and H. B. Pollard, *Proc. Natl. Acad. Sci. USA*, 1993, **90**, 567–571.
37. M.-C. Lin, T. Mirzabekov and B. L. Kagan, *Peptides*, 2002, **23**, 1215–1228.

Metal Ions and the Clearance of Misfolded Proteins

G. GRASSO,[a] D. LA MENDOLA[b] AND D. MILARDI[b]*

[a] Dipartimento di Scienze Chimiche, Università degli Studi di Catania, Viale A. Doria 6, 95125 Catania, Italy; [b] Istituto di Biostrutture e Bioimmagini, Consiglio Nazionale delle Ricerche, Unità Operativa e di Supporto di Catania, Viale A. Doria 6, 95125 Catania, Italy

5.1 Regulation and Quality Control: The Two Issues of Proteome Maintenance

Like all components of the cell, the proteome is in a dynamic equilibrium of synthesis and degradation. Extracellular proteins such as the blood coagulation factors, immunoglobulins, albumin, cargo-carrying proteins and peptide hormones are taken up and carried via a series of vesicles (endosomes) that fuse with intracellular organelles called lysosomes, where they are degraded. During this process the extracellular proteins are never exposed to the intracellular (cytosolic) environment and remain "extracellular" throughout. Degradation of proteins in lysosomes is not specific, and all proteins exposed to lysosomal proteases are degraded at the same rate.[1]

The lysosomes have a pH of 4–5, and contain a wide spectrum of hydrolytic enzymes, which play a major role in the intracellular degradation not only of the proteins, but also of other biomolecules such as polysaccharides and phospholipids. In particular, lysosomal proteases (or cathepsins) are the most important group of these enzymes: they belong to the aspartic, cysteine or

RSC Drug Discovery Series No. 7
Neurodegeneration: Metallostasis and Proteostasis
Edited by Danilo Milardi and Enrico Rizzarelli
© Royal Society of Chemistry 2011
Published by the Royal Society of Chemistry, www.rsc.org

serine protease families of hydrolytic enzymes and are expressed ubiquitously. Besides the gross degradation of proteins internalized through endocytosis, they have many other different functions such as antigen processing within early endosomes, protein processing at secretory vesicles, and degradation of matrix constituents in the extracellular space. Most recently, lysosomal proteases have been proposed to contribute to the initiation of apoptotic processes within the cytosol.[2] Lysosomes degrade intracellular proteins by quite different mechanisms, including endocytosis, crinophagy and the various autophagies.[3] It is interesting to note that the lysosomal compartments are rich in metal ions such as iron and copper, owing to the degradation of many metalloproteins. The contemporary presence of reducing agents can allow also the presence of the reduced forms of these metal ions, able to generate reactive oxygen species. As a result, lysosomes are very sensitive to oxidative stress and their membranes can be damaged by peroxidation, leading to the release of lytic enzymes. If the lysosome leak is moderate, the cell may survive, while the apoptotic machinery is activated if the leak is more substantial.[4,5]

On the other hand, many observations have demonstrated that degradation of intracellular proteins by other means is carried out by another completely distinct mechanism. In fact, intracellular proteolysis is highly specific, and different proteins may have half-life times that vary from a few minutes (*e.g.*, the tumor suppressor p53) to several days (*e.g.*, the muscle proteins actin and myosin) and up to a few years as crystalline.[1]

A unifying theme in eukaryotic intracellular protein degradation is found in the tagging and destruction mechanism called the ubiquitin proteasome system (UPS): a particular protein is targeted for degradation by the recursive covalent addition of the small protein ubiquitin (Ub), which leads to recognition of the resulting multiubiquitin chain by the 26S proteasome.[6] Initially, the Ub-activating enzyme E1 activates Ub in an ATP-requiring reaction to generate a high-energy thiol ester intermediate, E1–S–Ub. E2 enzymes transfer the activated Ub moiety from E1, via an additional high-energy thiol ester intermediate, E2–S–Ub, to the substrate that is specifically bound to a member of the Ub-protein ligase family named E3. E3 catalyzes the last step in the conjugation process: covalent attachment of Ub to the substrate. The Ub molecule is generally transferred to a NH_2 group of an internal lysine (Lys) residue in the substrate to generate a covalent isopeptide bond. In some cases, however, Ub is conjugated to the NH_2-terminal amino group of the substrate. The next step is the synthesis of a polyUb chain by successive addition of activated Ub moieties to internal Lys residues on the previously conjugated Ub molecule. The chain is recognized by the downstream 26S proteasome complex (Figure 5.1).

The 26S proteasome complex is a multifunctional, 2 500 000 Da proteolytic molecular machine, in which several enzymatic (proteolytic, ATPase, de-ubiquitinating) activities function together, with the ultimate goal of protein degradation.[7] In eukaryotes, 26S proteasomes are composed of the cylinder-shaped multimeric protein complex referred to as the 20S proteasome core particle, capped at each end by the regulatory component termed the 19S

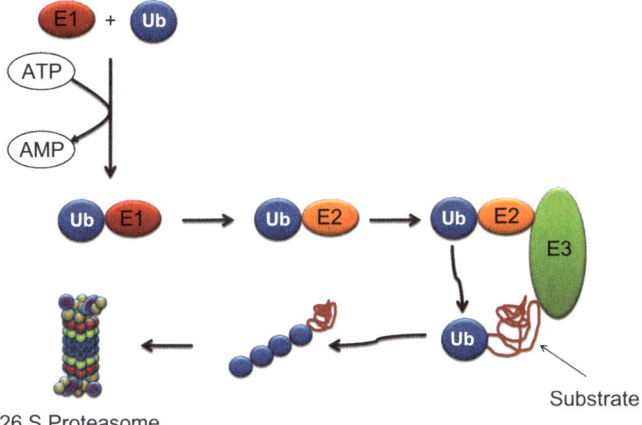

Figure 5.1 The ubiquitin–proteasome system. A cascade of enzymatic reactions leads to ubiquitination of lysine residues of the substrate.

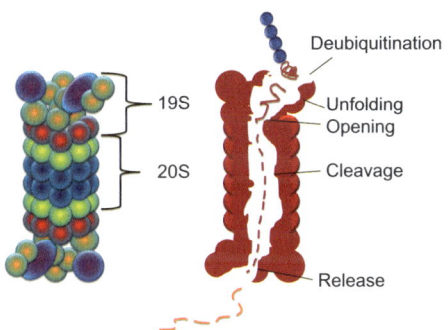

Figure 5.2 A schematic representation of the different parts of the 26S proteasome.

complex (regulatory particle or PA700).[7,8] The substrates are processed at the active sites located within the inner cavity of the 20S proteasome, whereas the 19S regulatory particle is responsible for recognition, unfolding, and translocation of the selected substrates into the lumen of the 20S proteasome.[9] The 20S proteasome (Figure 5.2) is a large, cylinder-shaped protease with a molecular weight of about 700 000 Da.[10] The complex is formed by 28 protein subunits, which are arranged in four stacked rings, each comprising seven subunits.[11]

The mammalian proteasome shows at least five distinct peptidase activities, classified as: chymotrypsin-like activity (ChT-L), trypsin-like activity (T-L), peptidylglutamyl-peptide hydrolyzing (PGPH), branched chain amino acid preferring (BrAAP), and small neutral amino acid preferring (SNAAP), which cleave bonds on the carboxyl side of hydrophobic, basic, acidic, branched chain, and small neutral amino acids, respectively.[12] It is interesting to note that lysosomes may also degrade oxidized proteins, which are usually thought to be degraded by the proteasome.[13] Lysosomes and proteasomes seem to be able to substitute for each other, and misfolded, aggregated and oxidized proteins are handled by both systems.[14] Despite the wide range of uses and recognition mechanisms for UPS substrates, there are two general themes in protein degradation, referred to here as regulation and quality control. In regulation, Ub-mediated degradation is used to exert changes in a specific protein's levels and activity. Examples include the regulated degradation of p53,[15] temporally programmed destruction of cyclins and other cell cycle regulators,[16] and the selective degradation of glucose-synthesizing enzymes after feeding.[17] A distinct second theme in intracellular proteolysis is protein degradation for purposes of quality control (QC) or, in other words, the selective intracellular degradation of misfolded or inappropriately assembled proteins. Protein quality control is implicated in maintaining acceptably low levels of aberrant proteins and is thought to be an important component in the management of cellular stress.[18–21] It has been observed that UPS-mediated degradative quality control is, like the regulation of individual proteins, highly specific and selective. Because of our limited understanding of the degradative QC machinery it can only be speculated that these pathways may similarly work to affect the turnover of many proteins, ensuring cellular proteostasis. It is conceivable that the QC pathways are equipped to promote the turnover of normal proteins, recognizing the extreme members of the ensemble of structures that occur during a normal protein's "breathing" within its normally folded state. According to this speculative model, QC mechanisms would promote the proteolysis of the naturally accessed conformations during their dynamic motions, and the more extreme cases of truly misfolded proteins caused by mutations or adverse external conditions.[6]

This "crossing" of the conceptual boundary between protein regulation and quality control has important biological implications. The idea of employing quality control pathways through selective misfolding of target proteins has been proven to be broadly used in biology: the degradation of Hmg2p, one of the two isoenzyems of HMG-CoA reductase implicated in the sterol pathway, is a well known example of "Regulated Protein Quality Control".[22–24] Regulated quality control may also have applications in drug discovery: the idea that a specific protein can be regulated by the QC pathways of the cell implies that there could be a new class of drugs that would work by specifically driving a protein target down a quality control pathway to achieve its clinically desirable elimination or diminution. Such "degradation modulators" could be as selective as enzyme inhibitors or receptor antagonists: they would work by selectively misfolding only the desired target. In a sense, this idea is the opposite of the "pharmaceutical chaperones".[25]

In fact, a pharmaceutical chaperone binds to a particular protein by virtue of a specific binding site on that protein and promotes its folding or stabilization. On the other hand, an example of a "degradation-modulator" would be the opposite case: the specific binding of a molecule would cause selective misfolding rather than improved folding. Whether such compounds can be found and exactly how to find them is still an open question but hopefully one of sufficient interest to promote attempts to explore these ideas.

5.2 The Failure of Intracellular Protein Degradation in Neurodegenerative Diseases

The correct maintenance of the proteome refers to controlling the levels, conformation, binding interactions, and trafficking of individual proteins in different biological compartments. The competition between cellular protein folding and degradation is one of the numerous processes influencing the cellular homeostasis of proteins and recent experiments reveal that the proteome maintenance capacity of the cell can be exceeded when adverse environmental conditions occur and misfolding-prone proteins appear.[26] Several disorders, including the prevalent dementias and encephalopathies, are now believed to arise from the same mechanism. In each, there is abnormal accumulation of insoluble aggregates that usually consist of fibers containing misfolded protein with a beta-sheet conformation, termed amyloid. The gradual accumulation of these aggregates and the acceleration of their formation by stressful environmental factors explain the characteristic late or episodic onset of the clinical symptoms. The understanding of these processes at the molecular level is opening prospects of more rational approaches to investigation and therapy.[27] The age-associated decline in maintaining the delicate equilibrium between protecting "on-pathway" folding intermediates and the efficient clearance of misfolded species acts often in concert with an increase in protein oxidation and modification that exacerbates aggregation phenomena.[28–31]

Environmental factors that might catalyze protein misfolding include changes in concentrations of metal ions, pathological chaperone proteins, pH or oxidative stress, macromolecular crowding and increases in the concentration of the misfolding protein. Many of these alterations are associated with ageing, consistent with the late onset of neurodegenerative diseases.[32] An increasing number of observations indicate that transition metals are capable of accelerating the aggregation process of various pathologic proteins, *e.g.*, α-synuclein (α-syn), the amyloid β peptide (Aβ), β2-microglobulin (β2-m) and fragments of the prion protein (PrP).[33,38] In particular, there are two generic reactions of relevance to these diseases. Firstly, a metal–protein association may lead to protein aggregation; this reaction may involve redox-inert metal ions such as Al(III), Zn(II), or redox-active metal ions such as Cu(II), Fe(II), Fe(III) and Mn(II).[35–38] Secondly, metal-catalyzed protein oxidation may lead to protein damage and denaturation; this reaction involves a redox-active metal ion.[37–40] Growing evidence indicates that failure to eliminate misfolded proteins can lead to the formation of potentially toxic aggregates, inactivation of functional proteins, and ultimately cell death. The number of disease states linked to aberrant protein conformations underscores the importance of effective quality control for cell survival.[41] Because the accumulation of Ub protein conjugates is a diagnostic hallmark of many neurodegenerative disorders,[42] it has been suggested that neurodegeneration results from a failure of the UPS.[43] Remarkably, a variety of oxidative conditions are known to functionally inactivate the active catalytic centers in the barrel of the proteasome,[44] and there are likely to be neuronal conditions that increase the oxidative nature of the cytoplasm.[45] In addition, proteasome activity has been shown to decrease with age.[46,47] However, defects in the UPS leading to intracellular amyloidogenesis do not have to depend only on a direct alteration of the proteasome catalytic sites. It is also possible that accumulation of ubiquitinated protein could be due to a misregulation of some other component in the pathway: for instance, the successful targeting of a substrate for ubiquitination, or its delivery to the proteasome could be impaired.

5.2.1 Parkinson's Disease (PD)

The first indication that altered protein turnover in the cell could be a crucial factor in the pathogenesis of PD is the presence of proteinaceous aggregates known as Lewy bodies, within the remaining dopamine cells in the substantia nigra compacta (SNc).[48,49] Lewy bodies accumulate a wide range of free and ubiquitinated proteins, which might be normal or abnormal.[49] These include Ub,[48] neurofilaments, torsin-A,[50] parkin,[51] Ub carboxyl-terminal hydrolase,[52] proteasomal elements,[53] protein adducts of 3-nitrotyrosine,[54] and α-synuclein, which can be extensively nitrated.[55,56] Although Lewy bodies have been recognized for many decades to be a characteristic feature of PD neuropathology, the mechanism by which these protein aggregates are formed is unclear. Misfolded, denatured and oxidatively damaged proteins that accumulate tend to aggregate and form insoluble inclusions.[57,58] Indeed, misfolding and/or oxidative modification of proteins leads to the exposure of hydrophobic regions, which crosslink extensively with other damaged proteins to form insoluble aggregates.[59] Such aggregated proteins are relatively refractory to degradation by normal proteolytic mechanisms, so they are transported to perinuclear microtubule-organizing centres (centrosomes). Here, they become associated with components of the UPS, and are encapsulated by intermediate filaments to form large structures called aggresomes.[57,58,60,61] Aggresomes seem to be sites of enhanced proteolysis, and their formation might serve to protect the nucleus and other organelles from exposure to the cytotoxic effects of abnormal proteins.[58] The presence of ubiquitinating and proteolytic enzymes in Lewy bodies, as well as tubulin and other cytoskeletal elements, indicates that these inclusions could be specialized aggresome-related structures formed in dopamine neurons as a mean of controlling excessive levels of abnormal proteins. However, defects in the 26/20S proteasome or in the Ub conjugating cascade, if coupled with the relentless production of abnormal proteins, could exceed the degradation capacity of the UPS and cause poorly degraded proteins to aggregate extensively, promoting the formation of insoluble Lewy body inclusions in the dopamine neurons of PD patients. In support of these hypotheses, impairment of the UPS was shown to be associated with neuro-degeneration and the formation of inclusion bodies in cultured dopamine neurons or Lewy-body-like inclusions in animal models of parkinsonism.[62,63] On the whole, these observations raise the possibility that Lewy body formation might be a cytoprotective event in which dopamine neurons attempt to sequester and compartmentalize poorly degraded proteins into insoluble aggregates and thereby protect against protein-mediated neurotoxicity. This might relate to a failure of protein ubiquitination necessary for crosslinking and polymerization of proteins, and the formation of insoluble aggregates or inclusion bodies.[57,64,65]

5.2.2 Alzheimer's Disease (AD)

The aberrant and misprocessed proteins that accumulate in AD brain constitute the neuropathological hallmarks of AD. The two most pronounced proteinaceous deposits in AD are neurofibrillary tangles (NFT), formed by intracellular accumulations of the hyperphosphorylated protein tau, and plaques, which are extracellular deposits of the 40–42-amino acid amyloid peptide (Aβ), processed from the amyloid precursor protein (APP).[66] In 1987, Mori *et al.* and Perry *et al.* were the first to describe the presence of Ub in paired helical filaments (PHFs), the major components of the tangles in AD brains.[67,68] The presence of Ub and the finding of ubiquitinated proteins in AD brain were the initial clues suggesting that the UPS was involved in the pathogenesis of AD. Later, an aberrant form of Ub (UBB+1) was discovered, which also accumulates in the AD brain.[69] Direct evidence for involvement of the UPS in AD is impressive (see Table 5.1): (i) ubiquitinated proteins accumulate in AD brain, (ii) proteasome subunit immunoreactivity is detected in disease-related areas, (iii) proteasome activity is decreased in AD brain, and (iv) different UPS-related mRNA expression profiles were observed in studies with AD brain tissues.

The substantial role that UPS plays in AD pathology is increasingly recognized. For example, proteasome activity was found to be lower in AD brains than in age-matched controls.[82,83] In addition, high levels of Ub were detected in brain homogenates and cerebrospinal fluid samples (both lumbar puncture samples and postmortem) of AD patients[84] and protein inclusions in AD brains generally contain ubiquitinated proteins.[85] These characteristics are not specific for AD, and are detected in other neurodegenerative diseases as well. Intriguingly, tau and Aβ, the two major players in AD pathology, as well as the mutant form of Ub, UBB+1, were found to alter proteasome activity. These findings strongly support the relevance of altered proteasomal degradation in AD. The tau protein normally exists as an unfolded protein and was suggested to be degraded by the 20S proteasome *in vitro*, both from the N to C and from the C to N terminus.[86] This implies that tau, just like other unfolded proteins, can be degraded by the 20S proteasome in an Ub-independent manner. In PHFs, however, tau was reported to be monoubiquitinated,[87] but to

Table 5.1 Proteins implicated in UPS impairment in AD.

Protein	Function	Implications	References
E1	Ubiquitinating enzyme	Decreased levels and activity in AD	83
E2-25K	Ub-conjugating enzyme	Mediates Aβ toxicity *in vitro*	96
E2 enzymes	Ub-conjugating enzyme	Down-regulated in AD and ageing	70,71
CHIP	Ub ligase	Serves as E3 enzyme for phosporylated tau	88
UCH-L1	Deubiquitinating enzyme	Down-regulated in AD brain. Accumulates in tangles. Oxidatively modified in AD brain	52,72,73
UBB+1	Unknown	Accumulates in AD. Both substrate and inhibitor of the proteasome	69,74
20S β subunits	Proteolysis	All its proteolytic activities are decreased in AD affected brain areas	82
20S α5 subunit	Confining proteolytic chambers	Down-regulated in AD brain	70
S6b	19S ATPase	Immunoreactivity in neurofibrillary tangles	85
S1	19S non-ATPase	Down-regulated in AD brain	70
Aβ amyloid	Unknown, product of Aβ processing	Aβ1-42 accumulates in plaques but also intraneuronally. Inhibits proteasomal activity *in vitro*	94
APP	Unknown, membrane-spanning protein	Degraded by proteasome	75
Presenilin	Component of γ-secretase complex	C terminus of APP is degraded by proteasome	76,77
Pen-2	Component of γ-secretase complex	Degraded by proteasome through ERAD	78,79
Tau	Microtubule-associated protein	Proteasome substrate. Accumulates in tangles, is monoubiquitinated	86,90
ApoE ε4	Lipid transport and cholesterol homeostasis	Associated with decreased Aβ clearance and oxidative stress in AD. Degraded by proteasome	80
LPR receptor	Receptor for apoE and mediates Aβ clearance	Cytosolic fragment processed by proteasome	81

the best of our knowledge these results were not confirmed by other studies, and the *in vivo* ubiquitination of normal tau was not conclusively demonstrated. The monoubiquitinated form of tau could hypothetically reflect a deubiquitinated state of polyubiquitinated tau. Phosphorylated tau extracted from AD brain was recently found to be ubiquitinated *in vitro* by the E2 enzyme UbcH5B and a CHIP–Hsc70 complex as the E3 ligase, the latter being immunodetected in tau aggregates.[88,89] In addition, a positive correlation was found between the amount of proteasome-bound tau and the extent of proteasome inhibition. PHF–tau isolated from AD brain also significantly inhibited proteasomal activity *in vitro*.[90] This inhibition was caused by the aggregation rather than the phosphorylation state of tau. Other aggregated proteins, such as polyglutamine protein aggregates, were also reported to inhibit the proteasome.[91] It is not clear whether this phenomenon is simply due to clogging of the proteasome or if other mechanisms are involved. *In vitro* studies have demonstrated that the C-terminal part of APP can also be processed by the 20S proteasome, which decreased γ-secretase processing.[92] Together, these findings support the view that an AD-associated decline in proteasome activity would lead to increased γ-secretase APP processing, which would result in elevated Aβ levels. The origin and mechanism of Aβ-mediated toxicity remain elusive. Both extracellular and intracellular Aβ have been widely discussed as mediators of neurotoxicity. Intraneuronal accumulation of Aβ was also detected in a triple transgenic mouse model of AD expressing human tau, APP and presenilin.[93] The clearance of tau pathology in these mice was mediated by the proteasome, which is a first indication of UPS involvement in transgenic AD mouse models. It was reported earlier that Aβ can bind to the 20S core of the proteasome and inhibit its activity in a 20–200 mM range *in vitro*.[94,95] A study that presents a possible mechanism for indirect proteasome inhibition by Aβ demonstrates that the toxicity of extracellular Aβ in neuronal cell lines is mediated by the E2 Ub-conjugating enzyme E2-25K/Hip2.[96] E2-25K functions both as an E2 Ub-conjugating enzyme and as an unusual Ub ligase to produce Ub–Ub and unanchored poly-Ub chains, without further requirement for other E3 ligases.[97] Intriguingly, E2-25K is also capable of ubiquitinating UBB+1,[98] which accumulates in AD brains. UBB+1 is a mutant Ub resulting from molecular misreading of the Ub-B gene.[69] This mutant Ub accumulates in the neuritic plaques and tangles in AD patients and in non-demented elderly controls with initial AD pathology. UBB+1 lacks the C-terminal glycine (Gly) of wild-type Ub and instead has a 19-amino acid extension. This mutant Ub can be ubiquitinated but not covalently attached to other proteins.[99,100] However, UBB+1 is also a potent and specific inhibitor of the proteasome.[98,101] Proteasome inhibition by UBB+1 requires a certain threshold concentration to be reached, which implies that other pathogenic mechanisms that interfere with proteasomal degradation precede the accumulation of UBB+1. As mentioned earlier, proteasome inhibition by Aβ may be mediated by its up-regulation of E2-25K.[96] An increase in E2-25K levels would lead to a rise in ubiquitinated UBB+1, which in turn would inhibit the proteasome and lead to neurodegeneration. Therefore, due to its dual substrate/inhibitor nature, UBB+1 seems to be an endogenous marker for UPS dysfunction, not only in AD but also in other neurodegenerative diseases.[102,103] Therefore UBB+1 may be an important determinant of neurotoxicity, contributing to an environment that favors the accumulation of misfolded proteins.

5.3   Metal Ions and the Derangement of the Ubiquitin Proteasome System

Neurodegenerative disorders are known to include a number of different pathological conditions, which share similar critical metabolic processes, such as protein aggregation and oxidative stress, both of which are associated with the involvement of metal ions. This evidence, coupled with the demonstration that the failure of the UPS is involved in many neurodegenerative disorders, prompted studies aimed to examine whether metal ions may contribute to a general impairment of the proteolytic machinery. In fact, it is known that the age-dependent rise in the brain of metal ions, in particular Cu(II) and Zn(II), might contribute to hypermetallate many proteins, thus triggering oxidative stress, misfolding and precipitation.[37] Oxidatively induced accumulation of ubiquitinated proteins in mouse neuronal cells has been demonstrated and suggests that the UPS pathway is closely involved in the cell response to metal-induced oxidative stress.[104] Furthermore, recent studies concerning the effect of oxidative stress induced by neurotoxic metal ions on the properties of the 20S proteasome have demonstrated that exposure of the 20S proteasome to increasing amounts of Fe(III), Fe(II), Cu(II) or Zn(II) affects its main hydrolytic activities: trypsin-like (T-L), chymotrypsin-like (ChT-L), peptidylglutamyl-peptide hydrolase (PGPH), branched-chain amino acid preferring (BrAAP), and caseinolytic activities.[105] As reported above, the presence of Ub-positive protein aggregates is a biomarker of neurodegeneration,[106] but the molecular mechanism underlying their accumulation is unknown. Protein aggregation is believed to be favored by metal ions, such as Cu(II) and Zn(II), whose levels are increased in the brains of patients with PD and AD.[37,107]

So far, Ub has been widely used as a model for protein stability, folding and structural studies,[108,109] and carefully characterized both in solution[110,111] and in solid state.[112,113] However, despite this plethora of structural investigations,

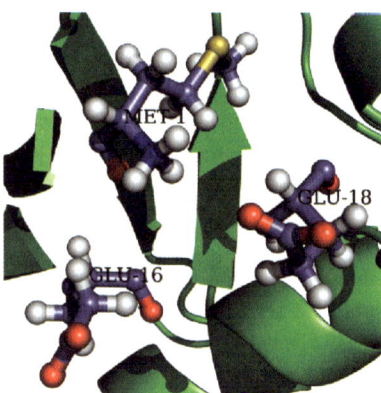

Figure 5.3   A view of the putative binding pocket of Cu(II) in Ub. The residues involved in Cu(II) binding are represented as sticks and balls.

only few studies concerning the interaction of Ub with metal ions are available. To fill this gap, some of us have recently undertaken a description of metal binding to Ub.[114] Ub was shown to bind Cu(II) in a well-defined region at the N-terminus of the protein in a tetragonal N_1O_3 (type II) site involving the NH_2- moiety of Met1, the CO of Val17, together with the CO and the carboxylate side chain of Glu18 (Figure 5.3). Since conditions that destabilize the native state of a protein render the macromolecular system more prone to aggregation, the Cu(II)-induced alterations in Ub structure and stability were investigated. It was demonstrated that Cu(II) binding at specific Ub sites coupled with a moderately low dielectric medium guides the protein through well-identified aggregation pathways.[115]

Differently from Cu(II), whose presence in the intracellular milieu is open to discussion,[116] free Zn(II) ions are stored at up to 300 μM in neurons.[117] In particular, the recent development of novel tools, including zinc-sensitive fluorescent probes, selective chelators and genetically modified animal models, has brought a deeper understanding of the roles of this cation as a crucial intra- and intercellular signaling ion of the central nervous system, thus suggesting a major role for Zn(II) in neurodegeneration, as it is a key component of the amyloid plaques observed in AD.[118–121] It is also known that brain trauma such as ischemia and seizure[122–125] can lead to an abnormal accumulation of cytoplasmic Zn(II) in neurons. Ischemia-induced neurodegeneration can be prevented by Zn(II)-chelating agents,[126,127] suggesting that the accumulation of cytoplasmic Zn(II) is an important trigger for ischemic neuronal death. Furthermore, exposure of mature cortical cell cultures to micromolar concentrations of Zn(II) induces neuronal death.[126] These findings indicate that Zn(II) is a determining factor in post-ischemic neuronal stress and couples neurodegeneration with UPS impairment. Notwithstanding these results would imply a deepening in the Zn(II) binding features of all the components of UPS, to the best of our knowledge only one structural investigation has been performed on Zn(II)–Ub adducts in the solid state revealing as a preferential anchoring site the His68. At higher Zn(II) concentration additional sites, close to the N-terminus of the protein, are increasingly populated; the stereochemistry of Ub assemblies appears to depend on the clustering of deshielded backbone hydrogen-bond patches, and Zn(II) ions were shown to foster this process.[128]

5.4 The Role of Metal Ions in the Functioning of Proteases involved in Amyloid Clearance

Many general perturbations of the proteolytic degradation machinery in neurons have been implicated in the pathogenesis of AD. However, it is currently believed that the two major landmarks in AD research are the so-called "amyloid cascade hypothesis" and the abnormal tau phosphorylation in neuronal tangles. Although missense mutations in the tau gene are sufficient to cause dementia independently of the presence of Aβ,[129] amyloid-centered research still remains the main focus in the search for a cure for AD. Several studies have suggested that Aβ or Aβ aggregates can induce tau phosphor-ylation and tangle formation or interfere with synaptic function, and Aβ has been shown to kill neurons in cell culture and in the brain *in vivo*.[130] Never-theless, in AD mouse models and in the normal ageing human brain, large amounts of amyloid plaques have been observed with minor neuronal alterations, indicating that the relationship between Aβ accumulation and Aβ toxicity is not straightforward.[131] Emerging evidence also suggests that the steady-state levels of Aβ are determined by the balance between its production and degradation.[132–134] Moreover, while in the past it was generally believed that extracellular accumulation of Aβ following secretion of soluble Aβ into the extracellular space was responsible for the AD pathology, it is now acknowl-edged that accumulation of Aβ may occur both intra- and extracellularly.[135] Thus, intracellular accumulation of Aβ may also be relevant to the pathogenesis of AD.

The proteolytic processes leading to Aβ formation have been extensively studied[130,136–138] and they will not be further discussed here. On the contrary, less is known about the proteases that degrade Aβ intra- and extracellularly and the environmental factors that might alter their activity.[139,140] Proteases acting at the site of Aβ generation and/or within the secretory pathway may degrade the peptide intracellularly, thus limiting the amount of the peptide available for secretion. The concentration of secreted Aβ may be further regulated by direct degradation by extracellular proteases and by receptor-mediated endocytosis or phagocytosis followed by lysosomal degradation. Catabolism of Aβ peptides at each of these steps would limit the accumulation of extracellular Aβ, and dis-ruption of this catabolism may be a risk factor for AD. Additionally, the identification of enzymes that degrade Aβ intracellularly and extracellularly may lead to development of novel therapeutics aimed at reducing Aβ concentration by enhancing its removal. Therefore, enzymes such as NEP (neprilysin), IDE (insulin-degrading enzyme), ECE-1 (endothelin converting enzyme 1), ECE-2, ACE (angiotensin-converting enzyme), MMPs (matrix metalloproteinases), PreP (presequence peptidase), and plasmin[141] have recently been widely studied because besides carrying out their different specific functions, they have also been demonstrated to be able to degrade Aβ peptides. However, with the data available at this time, it is impossible to determine which of the identified proteases contributes most to Aβ degradation in the brain. Plasmin, like trypsin, belongs to the family of serine proteases and is formed by cleavage of the peptide bond between Arg-560 and Val-561 of plasminogen by tissue plasminogen activator (tPA), urokinase plasminogen activator (uPA), and factor XII (Hageman factor). On the contrary, plasmin is inactivated by alpha 2-antiplasmin, a serine protease inhibitor (serpin). Although the main function of plasmin is to dissolve fibrin blood clots, it was also found that purified plasmin is able to degrade both monomeric Aβ, at multiple sites with physiolo-gically relevant efficiency, and Aβ fibrils.[142] Moreover, exogenously added plasmin blocks Aβ neurotoxicity, suggesting that the plasmin pathway is induced by aggregated Aβ, which can lead to Aβ degradation and inhibition of Aβ actions.[143] However, it was also suggested that plasmin does not regulate steady-state Aβ levels in non-pathologic conditions, and that plasminogen deficiency does not result in an increase of Aβ in the brain or in the plasma of adult mice.[144]

PreP is postulated to degrade mitochondrial targeting peptides that are cleaved off by mitochondrial processing peptidases following import. Proteolysis is crucial for maintaining mitochondrial morphology and function because it removes misfolded or damaged proteins and peptides and protects organelles from the toxicity of potentially harmful peptides. Recently, in addition to this previously identified function, PreP was also found to be responsible for the degradation of both Aβ1–40 and Aβ1–42.[145] Such a result is highly significant since Aβ is found in mitochondria, and Aβ-induced mitochondrial toxicity is associated with AD.[146] In contrast to IDE, which is a functional analogue of PreP, the latter does not degrade insulin but does degrade insulin B-chain.

MMPs are a family of Zn-dependent endo-peptidases known for their ability to cleave several components of the extracellular matrix, but which can also cleave many non-matrix proteins. There is a large amount of evidence that MMPs are involved in physiologic and pathologic processes, and a huge effort has been put into the development of possible inhibitors that could reduce the activity of MMPs, as it is clear that the ability to monitor and control such activity plays a pivotal role in the search for potential drugs aimed at finding a cure for several diseases such as pulmonary emphysema, rheumatoid arthritis, fibrotic disorders and cancer.[147] Many studies provide evidence that MMP-9 is also capable of degrading fibrillar Aβ1–42, and this property is not shared by other Aβ-degrading proteases.[148] Moreover, significant increases in the steady-state levels of Aβ have been found in the brains of MMP-2 and -9 knockout mice compared with wild-type controls, while pharmacologic inhibition of the MMPs with specific inhibitors increased brain interstitial fluid Aβ levels in mice, suggesting that MMP-2 and -9 may contribute to extracellular brain Aβ clearance by promoting Aβ catabolism.[149,150]

ACE, NEP, and ECE can be categorized as vasopeptidases based on their ability to generate or inactivate vasoactive peptides.[151] ACE is a circulating zinc metallopeptidase and a dipeptidyl carboxypeptidase that cleaves 2 amino acids from the C terminus of angiotensin (Ang) I and converts Ang I to the vasoactive and aldosterone-stimulating peptide Ang II.[152] In addition, it has been reported that ACE is able to convert Aβ1–42 to Aβ1–40 and to degrade Aβs under physiological conditions.[153] However, in another study, no evidence was found that ACE directly regulates endogenous steady-state Aβ levels *in vivo*, as mice lacking ACE expression in the brain showed no alterations in Aβ concentration, and treatment with ACE inhibitors did not cause a detectable increase in Aβ concentration in brain or in plasma even when ACE activity was substantially inhibited.[154] These conflicting results represent a good example of the different conclusions that sometimes are drawn based on studies carried out under different experimental conditions (for example results obtained *in vitro* rather than *in vivo*). In contrast to ACE, there is a growing and unambiguous body of evidence that ECE[155] not only generates endothelin and potentially degrades bradykinin, but that its activity is also critical for limiting Aβ accumulation in the brain.[156] For this reason, concerns regarding the use of ECE inhibitors as anti-hypertension drugs have been raised. Because most ECE inhibitors will inhibit the enzyme at certain concentrations, the use of these drugs may be particularly risky if ECE plays a physiologic role in the degradation of Aβ in the brain, as blocking this activity could lead to the development of AD. NEP is a nonmatrix zinc endoprotease and is associated with the inactivation and degradation of signal peptides including the enkephalins, bradykinin and endothelin.[157] NEP has been shown to regulate the steady-state levels of both Aβ40 and Aβ42,[158] therefore the identification of pharmacologic means to selectively upregulate brain NEP activity has been considered a target for new therapeutic opportunities.[159] Moreover, in order to investigate the subcellular compartments where NEP degrades Aβ most efficiently, NEP-chimeric proteins containing various subcellular compartment-targeting domains in neurons were expressed and their effects on Aβ clearance were examined.[160] The results indicate that the majority of extracellular Aβ degradation by NEP takes place on the cell surface. Evidence from animal studies strongly suggests that NEP[161] and IDE[162,163] are the key Aβ degradation enzymes *in vivo*. Furthermore, as hyperinsulinemia is associated with a high risk of AD,[164] IDE has also been considered to be particularly important for degrading Aβ,[165] playing a critical role in the mechanism associating hyperinsulinemia and type 2 diabetes with AD. For this reason, since the discovery of IDE cleavage action on Aβ,[166] much effort has been put into trying to understand some key questions regarding cleavage sites,[167,168] kinetics of interaction,[169,170] whether IDE is capable of degrading Aβ-bearing pathogenic amino acid substitutions,[171] and specific features of the IDE-produced amyloid fragments.[172,173] Although it is often reported that the common feature shared by IDE substrates is the ability to form, under certain physiologic conditions, amyloid fibrils,[174] the molecular basis by which IDE shows high selectivity but degenerate cleavage sites for a broad range of substrates has so far remained elusive.[175] Furthermore, in the attempt to control the enzyme activity, many studies focus on the investigation of the interaction between IDE and some of its possible modulators.[176,177]

Generally, all the abovementioned Aβ-degrading enzymes have distinct subcellular localizations, and are mediated by different cellular signals. Upregulation of the expression levels of these enzyme or their proteolytic activities, especially by certain pharmaceutical agents, may provide novel and viable therapeutic strategies for AD.[178] For this reason, a search for new analytical procedures that allow rapid *in vitro* screening of the Aβ-degrading enzymes has been carried out.[147,179,180] While IDE is originally described as a cytosolic enzyme, it has also been reported to be present in other subcellular compartments including peroxisomes, endosomes, and the nucleus.[181] An isoform of IDE with an N-terminal mitochondrial targeting sequence generated by translation at an in-frame initiation codon upstream of the canonical translation start has been recently identified, localized to mitochondria in cell lines, and shown to degrade mitochondrial targeting peptides.[182] Moreover, besides playing a critical role in degradation of intracellular Aβ, IDE is also localized at the cell surface. Therefore, although there are many cellular locations proposed for IDE, there is a lack of a definitive study that clearly establishes within which cellular compartment Aβ hydrolysis occurs.[183] In addition to decreased expression of Aβ-degrading proteases, Aβ accumulation in the AD brain may be the result of decreased Aβ sensitivity to protease activity, possibly due to other biochemical factors consistent with conditions in the AD brain. In this scenario, metals play an important role in tuning enzyme activity and the three different pathways by which metal ions can affect the proteolytic system can be summarized as follows (Figure 5.4):

   i. a change of conformation induced by metal ions (discussed in Chapter 6) in the peptides that are substrates of specific degradative enzymes can produce proteolytic resistant species;
  ii. direct binding of metal ions to degradative enzymes alter the activity of the latter;
 iii. the signaling cascades promoted by some metal ions have a strong impact on the production of some metalloproteases assigned to degrade Aβ peptides.

In the current literature the information regarding the three points mentioned above is very often mixed and the result is quite fragmentary. Particularly, the role that metal ions have on regulating some of the enzymes involved in the degradation of Aβ peptides is still under debate,[184,185] and conflicting results have been reported regarding the real effect of the chelating therapies.[186] Indeed, while some authors propose new chelating agents as possible drugs to be used to cure AD,[187,188] others observed a certain toxicity

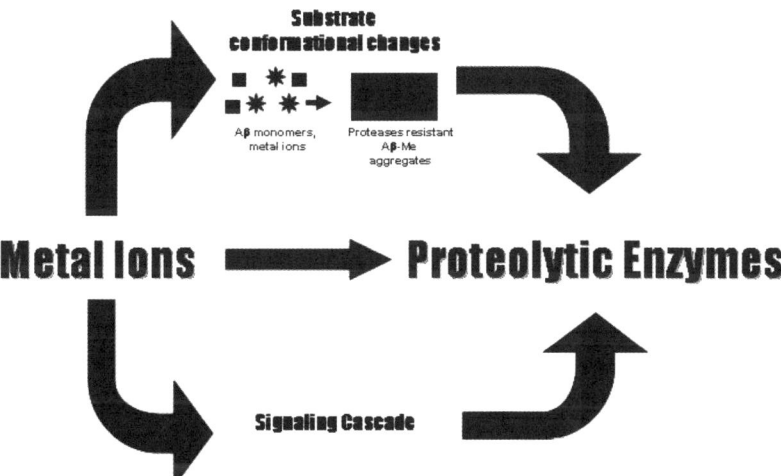

Figure 5.4   The three different pathways through which metal ions can affect pro-teolytic enzyme activity.

for some of the proposed compounds on APP transgenic mice[189] and doubted the clinical benefit for patients with AD.[190,191]

Different studies have shown that some metals induce β-sheet and subsequent amyloid formation in a truncated form of Aβ.[192] In Chapter 6 of this volume the conformational changes that Aβ peptides undergo under metal addition will be discussed in detail. Furthermore, such conformational changes have been also correlated to the alteration of the proteolytic action of some Aβ-degrading enzymes. Indeed, Crouch et al.[193] reported that Zn but not Cu induces the formation of protease resistant (NEP and IDE) Aβ amyloid. These data indicate that external factors, such as high metal concentrations that promote Aβ amyloid formation, may contribute to Aβ accumulation by decreasing the peptide's sensitivity to proteolytic degradation.

On the other hand, metal ions have also long been known to stabilize and activate enzymes.[194,195] For several proteins, binding of metal cofactors, such as calcium, magnesium, manganese, copper, zinc and iron in different oxidation states, has been shown to increase the thermal stabilities of the proteins[196,197] and to substantially influence their structure and activity.[198–200] Many metallonucleases contain two active-site metal ions required for hydrolysis of phosphodiester bonds[201] and they are affected by metal ions, as perturbing an endogenous metal that is vital to enzymatic action can render an enzyme inactive.[202–204] Even in the case of metalloproteases, a metal ion, generally zinc, is present in the catalytic site together with the carboxyl group of a glutamic acid (or aspartic acid) side chain. A very common proteolytic mechanism usually involves the zinc ion that has a tetrahedral coordination with three coordination sites provided by the enzyme and a water molecule being the fourth ligand. When the substrate enters the catalytic chamber of the enzyme the water molecule becomes more nucleophilic, being pushed toward the carboxyl group of the glutamic acid side chain. This favors a general base-type mechanism of peptide bond cleavage with the attack of the water molecule on the carbonyl carbon of the scissile peptide bond of the substrate.[205] Modulation of the proteolytic activity of metalloproteases can occur through different routes such as coordination of exogenous ligands to the catalytic metal, or substitution or removal of the metal ion.[206–213] Moreover, metals can coordinate to active site residues to block substrate interaction or coordinate to residues outside the active site to affect structural integrity of the enzyme.[214,215] The physiopathologic significance of these effects, in particular for neurological disorders such as AD, are very complex as an imbalance of proteolytic as well as antiproteolytic systems appears to be a crucial event for the formation of both neuritic plaques and neurofibrillary tangles, which are the major hallmarks of the disease.

Serum and cerebrospinal fluid (CSF) copper levels have been reported to be elevated in AD patients, while the activity of some copper-dependent as well as copper-nondependent enzymes shows the opposite trend to copper concentration.[216] The metal involvement in the proteolytic system is further complicated as it has been reported that the capacity to increase intracellular metal bioavailability can activate neuroprotective cell signaling pathways pertinent to AD.[185,217] As metal dishomeostasis is a key factor in the etiology of AD,[204,218] in the attempt to up-regulate the Aβ-degrading metalloproteases, much effort has been put toward the synthesis of metal compounds that can shuttle metals (mainly copper and zinc) intracellularly.[219] Several recent studies have highlighted a potential role for modulating metal levels as a basis for AD therapy,[220] and the chelating therapy approach has been scrutinized.[221,222] Clioquinol (CQ) has been shown to have possible beneficial effects in animal models of AD,[223] and preliminary data suggest that CQ has positive effects also on AD patients.[224] A variety of different mechanisms of action have been proposed to account for these effects on disease (copper and zinc chelation[223] or increased level of a metalloprotease[217]) and the most updated view refers to its action in regulating the activity of the enzyme Cyclin-dependant like Kinase 1 (CLK1).[225] Furthermore, it was also demonstrated that in human M17 cells, CQ can protect against oxidative stress by activating the PI3K-dependent survival pathway and blocking p53-mediated cell death.[226] These findings have important implications for the development of protective metal ligand-based therapies for treatment of disorders involving oxidative stress.

In summary, it seems that restoration of intracellular metal levels (mainly copper) could restore the imbalance encountered in AD and promote a decrease in Aβ accumulation. The mechanisms associated with this are yet not well known and more studies that take into account all of the three different pathways through which metal ions act, as outlined above, are very much needed.

5.5 Metal Binding Compounds: Therapeutic Perspectives

All these sources of evidence have prompted a number of studies aimed to elucidate the potential therapeutic use of different metal-binding molecules targeting the proteolytic pathways of the cell. For instance, it has been found that treatment with pyrrolidine dithiocarbamate (PDTC) resulted in the accumulation of several proteasome substrates in HeLa cells.[227] PDTC belongs to the dithiocarbamate family, a class of metal chelating and antioxidant compounds, previously used in the treatment of bacterial and fungal interactions. The PDTC effect was due to an extended half-life of these proteins through the mobilization of zinc. PDTC and/or zinc also increased fluorescence intensity of the UbG76V–GFP fusion protein that is degraded rapidly by the UPS. Treatment of cells with zinc induced formation of ubiquitinated inclusions in the centrosome, a histological marker of proteasome inhibition. Western blotting showed zinc-induced increase in laddering bands of poly-ubiquitin-conjugated proteins. *In vitro*, Zn(II) inhibited the Ub-independent proteasomal degradations of a-synuclein.[227] It has been reported that organic copper complexes can potently and selectively inhibit the chymotrypsin-like activity of the proteasome *in vitro* and *in vivo*. In particular, bis-8-hydroxy-quinoline copper(II) [Cu(8-OHQ)$_2$] was able to inhibit the chymotrypsin-like activity of purified 20S proteasome.[228] Furthermore, it has been found that copper-mediated inhibition of purified 20S proteasome cannot be blocked by a reducing agent and that organic copper compounds do not generate hydrogen peroxide in the cells, suggesting that proteasome inhibition and apoptosis induction are not due to copper-mediated oxidative damage to proteins. These results suggest that certain types of organic ligand could form potent proteasome inhibitors in the presence of copper, whose mechanism of action is different from oxidation.[229] In addition, it has been recently reported that CQ and PDTC (Figure 5.5) can interact with copper to form cancer-specific proteasome inhibitors and apoptosis inducers in human breast cancer cells.[230]

Until recently, the effect of these small molecule drugs has been attributed also to their chelating properties. Subsequently, it was demonstrated that their activity was due to the ability to act as "ionophores", transporting the metal ion through cellular membrane, releasing it, and then repeating the process with other metal ions present in the extracellular space. The "ionophore" may diffuse from the extracellular to the intracellular space and in reverse. An "ionophore" is not a simple metal-shuttle and its activity can be amplified by increasing metal ion concentration. The significance of this is that rather than just increasing cellular zinc(II) and/or copper(II) levels, such compounds increase cellular bioavailability of these metal ions, restoring dysregulated metal homeostasis.[231] Notwithstanding these promising results, because of the central role played by the UPS in a broad array of cellular processes, development of metal-based drugs that modulate the activity of the system may be difficult. Straightforward inhibition of enzymes common to the entire pathway, such as the proteasome, may affect many processes nonspecifically with many undesired side-effects. A completely different approach to drug development may be, however, the development of small molecules that bind and inhibit specific enzymes involved in the UPS. Indeed, effective development of these novel drugs implies a detailed knowledge of all the reactions involved in the correct functioning of the UPS. However, the enzymology of ubiquitination, and the mechanisms of polyUb chain formation are currently not clear. For example, we do not yet have unequivocal answers to basic questions such as whether the E3 enzymes conjugate intact Ub chains to the substrate or whether they synthesize the chain *de novo* on the substrate. In the latter case, how does

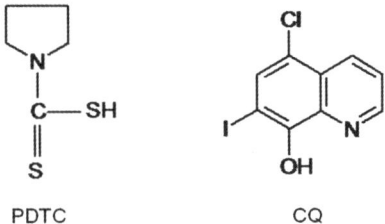

Figure 5.5   Two copper-chelating proteasome inhibitors: pyrrolidine dithiocarbamate (PDTC) and CQ.

the enzyme reach the distal end of the growing chain? How intimately are the ubiquitination and the proteasome machineries linked? How are polyUb-tagged proteins recognized by the proteasome for degradation? And ultimately, what is the role played by metal ions in each of these steps? What we can certainly expect in the near future is the identification of an ever-increasing number of substrates of the Ub system and their specific E2/E3 complexes. It remains to be seen how abnormal environmental factors (*e.g.*, metal dyshomeostasis) may affect each single step of the UPS. Consequently, we can expect to see the development of a new exciting area focused on specific modulators/drugs that can interfere with specific substrate recognition at different levels of the system in a variety of pathologic states. The metal-mediated UPS inhibition hypothesis is elegant and could provide a unifying mechanism for diseases involving protein misfolding and oxidative stress, but this model has yet to be tested in a definitive manner, both *in vitro* and *in vivo*. In conclusion, although there are many tantalizing clues suggesting that the UPS is crucial for neurodegenerative disease pathogenesis, determination of the specific mechanisms and confirmation *in vivo* remain elusive goals. In part, this is because we still lack a clear understanding at a molecular level of how misfolded proteins are targeted for ubiquitination and degradation. Otherwise, we need to define clearly what factors may interfere with the cascade of molecular events involved in the UPS. Understanding the details of these mechanisms will be important for developing novel therapeutic interventions.

Acknowledgements

This work was financially supported by Consiglio Nazionale delle Ricerche (RSTL n° 620), and MiUR (PRIN2008).

References

1. M. H. Glickman and A. Ciechanover, *Physiol. Rev.*, 2002, **82**, 373.
2. J. D. Colbert, S. P. Matthews, G. Miller and C. Watts, *Eur. J. Immunol.*, 2009, **39**, 2955.
3. E. Knecht, C. Aguado, J. Carcel, I. Esteban, J. M. Esteve, G. Ghislat, J. F. Moruno, J. M. Vidal and R. Saez, *Cell. Mol. Life Sci.*, 2009, **66**, 2427.
4. B. Turk, V. Stoka, J. Rozman-Pungercar, T. Cirman, G. Droga-Mazovec, K. Oresic and V. Turk, *Biol. Chem.*, 2002, **383**, 1035.
5. F. Antunes, E. Cadenas and U. T. Brunk, *Biochem. J.*, 2001, **356**, 549.
6. R. Y. Hampton and R. M. Garza, *Chem. Rev.*, 2009, **109**, 1561.
7. R. Hough, G. Pratt and M. Rechsteiner, *J. Biol. Chem.*, 1986, **261**, 2400.
8. E. Eytan, D. Ganoth, T. Armonand and A. Hershko, *Proc. Natl. Acad. Sci. USA*, 1989, **86**, 7751.
9. D. Voges, P. Zwickl and W. Baumeister, *Annu. Rev. Biochem.*, 1999, **68**, 1015.

10. J. Harris, *Biochim. Biophys. Acta*, 1968, **150**, 534.

11. R. Hegerl, G. Pfeifer, G. Puhler, B. Dahlmann and W. Baumeister, *FEBS Lett.*, 1991, **283**, 117.

12. M. Orlowski and S. Wilk, *Arch. Biochem. Biophys.*, 2000, **383**, 1.

13. I. Jariel-Encontre, G. Bossis and M. Piechaczyk, *Biochim. Biophys. Acta*, 2008, **1786**, 153.

14. U. B. Pandey, Z. Nie, Y. Batlevi, B. A. McRay, G. P. Ritson, N. B. Nedelsky, S. L. Schwartz, N. A. Di Prospero, M. A. Knight, O. Schuldiner, R. Padmanabhan, M. Hild, D. L. Berry, D. Garza, C. C. Hubbert, T. P. Yao, E. H. Baehrecke and J. P. Taylor, *Nature*, 2007, **447**, 859.

15. S. Fang, J. P. Jensen, R. L. Ludwig, K. H. Vousden and A. M. Weissman, *J. Biol. Chem.*, 2000, **275**, 8945.

16. J. Pines, *Trends Cell Biol.*, 2006, **16**, 55.

17. J. Regelmann, T. Schule, F. S. Josupeit, J. Horak, M. Rose, K. D. Entian, M. Thumm and D. H. Wolf, *Mol. Biol. Cell*, 2003, **14**, 1652.

18. R. G. Gardner, Z. W. Nelson and D. E. Gottschling, *Cell*, 2005, **120**, 803.

19. A. J. McClellan, M. D. Scott and J. Frydman, *Cell*, 2005, **121**, 739.

20. R. Hengge and B. Bukau, *Mol. Microbiol.*, 2003, **49**, 1451.

21. A. Mogk, T. Haslberger, P. Tessarz and B. Bukau, *Biochem. Soc. Trans.*, 2008, **36**, 120.

22. R. G. Gardner and R. Y. Hampton, *J. Biol. Chem.*, 1999, **274**, 31671.

23. R. G. Gardner and R. Y. Hampton, *EMBO J.*, 1999, **18**, 5994.

24. R. Y. Hampton, R. G. Gardner and J. Rine, *Mol. Biol. Cell*, 1996, **7**, 2029.

25. V. Bernier, M. Lagace, D. G. Bichet and M. Bouvier, *Trends Endocrinol. Metab.*, 2004, **15**, 222.

26. T. Gidalevitz, A. Ben-Zvi, K. H. Ho, H. R. Brignull and R. I. Morimoto, *Science*, 2006, **311**, 1471.

27. R. W. Carrell and D. A. Lomas, *Lancet*, 1997, **350**, 134.

28. Q. Zhang, E. T. Powers, J. Nieva, M. E. Huff, M. A. Dendle, J. Bieschke, C. G. Glabe, A. Eschenmoser, P. Wentworth Jr., R. A. Lerner and J. W. Kelly, *Proc. Natl. Acad. Sci. USA*, 2004, **101**, 4752.

29. A. C. Massey, R. Kiffin and A. M. Cuervo, *Cell Cycle*, 2006, **5**, 1292.

30. R. R. Erickson, L. M. Dunning, J. L. Holtzman and J. Geron, *A: Biol. Sci. Med. Sci.*, 2006, **61A**, 435.

31. B. K. Derham and J. J. Harding, *Biochem. J.*, 1997, **328**, 763.

32. C. Soto, *Nat. Rev. Neurosci.*, 2003, **4**, 49.

33. M. Hashimoto, L. J. Hsu, Y. Xia, A. M. Takeda, M. Sundsmo and E. Masliah, *Neuroreport*, 1999, **10**, 717.

34. S. R. Paik, H. J. Shin, J. H. Lee, C. S. Chang and J. Kim, *Biochem. J.*, 1999, **340**, 821.

35. A. I. Bush, R. D. Moir, K. M. Rosenkranz and R. E. Tanzi, *Science*, 1995, **268**, 1921.

36. C. J. Morgan, M. Gelfand, C. Atreya and A. D. Miranker, *J. Mol. Biol.*, 2001, **309**, 339.

37. A. I. Bush, *Trends Neurosci.*, 2003, **26**, 207.

38. C. J. Frederickson, J.-Y. Koh and A. I. Bush, *Nat. Rev. Neurosci.*, 2005, **6**, 449.

39. J. K. Andersen, *Nat. Med.*, 2004, **10**, S18.

40. A. I. Bush, *Neurobiol. Aging*, 2002, **23**, 1031.

41. A. J. McClellan, S. Tam, D. Kaganovich and J. Frydman, *Nat. Cell Biol.*, 2005, **7**, 736.

42. R. J. Mayer, *Drug News Perspect.*, 2003, **16**, 103.

43. K. S. McNaught, C. W. Olanow, B. Halliwell, O. Isacson and P. Jenner, *Nat. Rev. Neurosci.*, 2001, **2**, 589.

44. A. L. Bulteau, K. C. Lundberg, K. M. Humphries, H. A. Sadek, P. A. Szweda, B. Friguet and L. I. Szweda, *J. Biol. Chem.*, 2001, **276**, 30057.

45. N. Jha, M. J. Kumar, R. Boonplueang and J. K. Andersen, *J. Neurochem.*, 2002, **80**, 555.

46. G. Carrard, A. L. Bulteau, I. Petropoulos and B. Friguet, *Int. J. Biochem. Cell. Biol.*, 2002, **34**, 1461.

47. B. Friguet, A. L. Bulteau, N. Chondrogianni, M. Conconi and I. Petropoulos, *Ann. N. Y. Acad. Sci.*, 2000, **908**, 143.

48. L. S. Forno, *J. Neuropathol. Exp. Neurol.*, 1996, **55**, 259.

49. M. S. Pollanen, D. W. Dickson and C. Bergeron, *J. Neuropathol. Exp. Neurol.*, 1993, **52**, 183.

50. P. Shashidharan, P. F. Good, A. Hsu, D. P. Perl, M. F. Brin and C. W. Olanow, *Brain Res.*, 2000, **877**, 379.

51. H. Shimura, N. Hattori, S. Kubo, M. Yoshikawa, T. Kitada, H. Matsumine, S. Asakawa, S. Minoshima, Y. Yamamura, N. Shimizu and Y. Mizuno, *Ann. Neurol.*, 1999, **45**, 668.

52. J. Lowe, H. McDermott, M. Landon, R. J. Mayer and K. D. Wilkinson, *J. Pathol.*, 1990, **161**, 153.

53. K. Ii, H. Ito, K. Tanaka and A. Hirano, *J. Neuropathol. Exp. Neurol.*, 1997, **56**, 125.

54. P. F. Good, A. Hsu, P. Werner, D. P. Perl and C. W. Olanow, *J. Neuropathol. Exp. Neurol.*, 1998, **57**, 338.

55. M. G. Spillantini, R. A. Crowther, R. Jakes, M. Hasegawa and M. Goedert, *Proc. Natl. Acad. Sci. USA*, 1998, **95**, 6469.

56. B. I. Giasson, J. E. Duda, I. V. J. Murray, Q. Chen, J. M. Souza, H. I. Hurtig, H. Ischiropoulos, J. Q. Trojanowski and V. M.-Y. Lee, *Science*, 2000, **290**, 985.

57. J. A. Johnston, C. L. Ward and R. R. Kopito, *J. Cell. Biol.*, 1998, **143**, 1883.

58. R. R. Kopito, *Trends Cell Biol.*, 2000, **10**, 524.

59. K. J. Davies, *Biochimie*, 2001, **83**, 301.

60. M. Y. Sherman and A. L. Goldberg, *Neuron*, 2001, **29**, 15.

61. W. C. Wigley, R. P. Fabunmi, M. G. Lee, C. R. Marino, S. Muallem, G. N. DeMartino and P. J. Thomas, *J. Cell. Biol.*, 1999, **145**, 481.

62. E. Masliah, E. Rockenstein, I. Veinbergs, M. Mallory, M. Hashimoto, A. Takeda, Y. Sagara, A. Sisk and L. Mucke, *Science*, 2000, **287**, 1265.

63. M. B. Feany and W. W. Bender, *Nature*, 2000, **404**, 394.
64. H. Shimura, M. G. Schlossmacher, N. Hattori, M. P. Frosch, A. Trockenbacher, R. Schneider, Y. Mizuno, K. S. Kosik and D. J. Selkoe, *Science*, 2001, **293**, 263.
65. C. J. Cummings, *Neuron*, 1999, **24**, 879.
66. H. Braak and K. Del Tredici, *Neurobiol. Aging*, 2004, **25**, 19.
67. H. Mori, J. Kondo and Y. Ihara, *Science*, 1987, **235**, 1641.
68. G. Perry, R. Friedman, G. Shaw and V. Chau, *Proc. Natl. Acad. Sci. USA*, 1987, **84**, 3033.
69. F. W. Van Leeuwen, D. P. de Kleijn, H. H. van den Hurk, A. Neubauer, M. A. Sonnemans, J. A. Sluijs, S. Koycu, R. D. Ramdjielal, A. Salehi, G. J. Martens, F. G. Grosveld, J. Peter, H. Burbach and E. M. Hol, *Science*, 1998, **279**, 242.
70. J. F. Loring, X. Wen, J. M. Lee, J. Seilhamer and R. Somogyi, *DNA Cell Biol.*, 2001, **20**, 683.
71. T. Lu, Y. Pan, S.-Y. Kao, C. Li, I. Kohane, J. Chan and B. A. Yankner, *Nature*, 2004, **429**, 883.
72. G. M. Pasinetti, *J. Neurosci. Res.*, 2001, **65**, 471.
73. A. Castegna, M. Aksenov, V. Thongboonkerd, J. B. Klein, W. M. Pierce, R. Booze, W. R. Markesbery and D. A. Butterfield, *Free Radic. Biol. Med.*, 2002, **33**, 562.
74. K. Lindsten, V. Menendez-Benito, M. G. Masucci and N. P. Dantuma, *Nat. Biotechnol.*, 2003, **21**, 897.
75. J. Nunan, N. A. Williamson, A. F. Hill, M. F. Sernee, C. L. Masters and D. H. Small, *J. Neurosci. Res.*, 2003, **74**, 378.
76. P. E. Fraser, G. Levesque, G. Yu, L. R. Mills, J. Thirlwell, M. Frantseva, S. E. Gandy, M. Seeger, P. L. Carlen and P. St. George-Hyslop, *Neurobiol. Aging*, 1998, **19**, S19.
77. T. W. Kim, W. H. Pettingell, O. G. Hallmark, R. D. Moir, W. Wasco and R. E. Tanzi, *J. Biol. Chem.*, 1997, **272**, 11006.
78. A. Bergman, E. Hansson, S. E. Pursglove, M. R. Farmery, L. Lannfelt, U. Lendahl, J. Lundkvist and J. Naslund, *J. Biol. Chem.*, 2004, **279**, 16744.
79. A. S. Crystal, V. A. Morais, R. R. Fortna, D. Carlin, T. C. Pierson, C. A. Wilson, V. M. Lee and R. W. Doms, *Biochemistry*, 2004, **43**, 3555.
80. C. Wenner, S. Lorkowski, T. Engel and P. Cullen, *Biochem. Biophys. Res. Commun.*, 2001, **282**, 608.
81. P. May, Y. K. Reddy and J. Herz, *J. Biol. Chem.*, 2002, **277**, 18736.
82. J. N. Keller, K. B. Hanni and W. R. Markesbery, *J. Neurochem.*, 2000, **75**, 436.
83. M. Lopez Salon, L. Morelli, E. M. Castano, E. F. Soto and J. M. Pasquini, *J. Neurosci. Res.*, 2000, **62**, 302.
84. T. Kudo, K. Iqbal, R. Ravid, D. F. Swaab and I. Grundke-Iqbal, *Brain Res.*, 1994, **639**, 1.
85. J. Fergusson, M. Landon, J. Lowe, S. P. Dawson, R. Layfield, D. P. Hanger and R. J. Mayer, *Neurosci. Lett.*, 1996, **219**, 167.

86. D. C. David, R. Layfield, L. Serpell, Y. Narain, M. Goedert and M. G. Spillantini, *J. Neurochem.*, 2002, **83**, 176.
87. M. Morishima-Kawashima, M. Hasegawa, K. Takio, M. Suzuki, K. Titani and Y. Ihara, *Neuron*, 1993, **10**, 1151.
88. L. Petrucelli, D. Dickson, K. Kehoe, J. Taylor, H. Snyder, A. Grover, M. De Lucia, E. McGowan, J. Lewis, G. Prihar, J. Kim, W. H. Dillmann, S. E. Browne, A. Hall, R. Voellmy, Y. Tsuboi, T. M. Dawson, B. Wolozin, J. Hardy and M. Hutton, *Hum. Mol. Genet.*, 2004, **13**, 703.
89. H. Shimura, D. Schwartz, S. P. Gygi and K. S. Kosik, *J. Biol. Chem.*, 2004, **279**, 4869.
90. S. Keck, R. Nitsch, T. Grune and O. Ullrich, *J. Neurochem.*, 2003, **85**, 115.
91. N. F. Bence, R. M. Sampat and R. R. Kopito, *Science*, 2001, **292**, 1552.
92. J. Nunan, N. A. Williamson, A. F. Hill, M. F. Sernee, C. L. Masters and D. H. Small, *J. Neurosci. Res.*, 2003, **74**, 378.
93. S. Oddo, A. Caccamo, J. D. Shepherd, M. P. Murphy, T. E. Golde, R. Kayed, R. Metherate, M. P. Mattson, Y. Akbari and F. M. LaFerla, *Neuron*, 2003, **39**, 409.
94. L. Gregori, C. Fuchs, M. E. Figueiredo-Pereira, W. E. Van Nostrand and D. Goldgaber, *J. Biol. Chem.*, 1995, **270**, 19702.
95. L. Gregori, J. F. Hainfeld, M. N. Simon and D. Goldgaber, *J. Biol. Chem.*, 1997, **272**, 58.
96. S. Song, S. Y. Kim, Y. M. Hong, D. G. Jo, J. Y. Lee, S. M. Shim, C. W. Chung, S. J. Seo, Y. J. Yoo, J. Y. Koh, M. C. Lee, A. J. Yates, H. Ichijo and Y. K. Jung, *Mol. Cell*, 2003, **12**, 553.
97. Z. Chen and C. M. Pickart, *J. Biol. Chem.*, 1990, **265**, 21835.
98. Y. A. Lam, C. M. Pickart, A. Alban, M. Landon, C. Jamieson, R. Ramage, R. J. Mayer and R. Layfield, *Proc. Natl. Acad. Sci. USA*, 2000, **97**, 9902.
99. F. M. De Vrij, J. A. Sluijs, L. Gregori, D. F. Fischer, W. T. Hermens, D. Goldgaber, J. Verhaagen, F. W. Van Leeuwen and E. M. Hol, *FASEB J.*, 2001, **15**, 2680.
100. Y. A. Lam, W. Xu, G. N. De Martino and R. E. Cohen, *Nature*, 1997, **385**, 737.
101. K. Lindsten, F. M. S. de Vrij, L. G. G. C. Verhoef, D. F. Fischer, F. W. Van Leeuwen, E. M. Hol, M. G. Masucci and N. P. Dantuma, *J. Cell Biol.*, 2002, **157**, 417.
102. R. de Pril, D. F. Fischer, M. L. Maat-Schieman, B. Hobo, R. A. de Vos and E. R. Brunt, *Hum. Mol. Genet.*, 2004, **13**, 1803.
103. D. F. Fischer, R. A. De Vos, R. Van Dijk, F. M. De Vrij, E. A. Proper, M. A. Sonnemans, M. C. Verhage, J. A. Sluijs, B. Hobo, M. Zouambia, E. N. Steur, W. Kamphorst, E. M. Hol and F. W. Van Leeuwen, *FASEB J.*, 2003, **17**, 2014.
104. M. E. Figueiredo-Pereira and G. Cohen, *Mol. Biol. Rep.*, 1999, **26**, 65.
105. M. Amici, K. Forti, C. Nobili, G. Lupidi, M. Angeletti, E. Fioretti and A. Eleuteri, *J. Biol. Inorg. Chem.*, 2002, **7**, 750.

106. A. Ciechanover and P. Brundin, *Neuron*, 2003, **40**, 427.

107. K. J. Barnham and A. I. Bush, *Curr. Opin. Chem. Biol.*, 2008, **12**, 222.

108. S. E. Jackson, *Org. Biomol. Chem.*, 2006, **4**, 1845.

109. K. L. Schweiker, V. W. Fitz and G. I. Makhatadze, *Biochemistry*, 2009, **48**, 10846.

110. D. L. Di Stefano and A. J. Wand, *Biochemistry*, 1987, **26**, 7272.

111. D. Nash and J. Jonas, *Biochem. Biophys. Res. Commun.*, 1997, **238**, 289.

112. S. Vijay-Kumar, C. E. Bugg and W. J. Cook, *J. Mol. Biol.*, 1987, **194**, 531.

113. K. J. Walters, A. M. Goh, Q. Wang, G. Wagner and P. M. Howley, *BBA-Mol. Cell Res.*, 2004, **1695**, 73.

114. D. Milardi, F. Arnesano, G. Grasso, A. Magrì, G. Tabbì, S. Scintilla, G. Natile and E. Rizzarelli, *Angew. Chem. Int. Ed.*, 2007, **46**, 7993.

115. F. Arnesano, S. Scintilla, V. Calò, E. Bonfrate, C. Ingrosso, M. Losacco, T. Pellegrino, E. Rizzarelli and G. Natile, *PLoS One*, 2009, **4**, e7052.

116. A. K. Boal and A. C. Rosenzweig, *Chem. Rev.*, 2009, **109**, 4760.

117. H. Shen, F. Wang, Y. Zhang, J. Xu, J. Long, H. Qin, F. Liu and J. Guo, *Biol. Trace Elem. Res.*, 2007, **119**, 166.

118. G. Danscher, K. B. Jensen, C. J. Frederickson, K. Kemp, A. Andreasen, S. Juhl, M. Stoltenberg and R. Ravid, *J. Neurosci. Meth.*, 1997, **76**, 53.

119. M. A. Lovell, J. D. Robertson, W. J. Teesdale, J. L. Campbell and W. R. Markesbery, *J. Neurol. Sci.*, 1998, **158**, 47.

120. C. Opazo, S. Luza, V. L. Villemagne, I. Volitakis, C. Rowe, K. J. Barnham, D. Strozyk, C. L. Masters, R. A. Cherny and A. I. Bush, *Aging Cell*, 2006, **5**, 69.

121. D. Religa, D. Strozyk, R. A. Cherny, I. Volitakis, V. Haroutunian, B. Winblad, J. Naslund and A. I. Bush, *Neurology*, 2006, **67**, 69.

122. J.-Y. Koh, S. W. Suh, B. J. Gwag, Y. Y. He, C. Y. Hsu and D. W. Choi, *Science*, 1996, **272**, 1013.

123. S. W. Suh, J. W. Chen, M. Motamedi, B. Bell, K. Listiak, N. F. Pons, G. Danscher and C. J. Frederickson, *Brain Res.*, 2000, **852**, 268.

124. C. J. Frederickson, M. D. Hernandez and J. F. McGinty, *Brain Res.*, 1989, **480**, 317.

125. À. Riba-Bosch and J. Pérez-Clausell, *Neuroscience*, 2004, **125**, 803.

126. J.-Y. Koh and D. W. Choi, *Neuroscience*, 1994, **60**, 1049.

127. A. Calderone, T. Jover, T. Mashiko, K. Noh, H. Tanaka, M. V. L. Bennett and R. S. Zukin, *J. Neurosci.*, 2004, **24**, 9903.

128. G. Falini, S. Fermani, G. Tosi, F. Arnesano and G. Natile, *Chem. Commun.*, 2008, **7**, 5960.

129. M. Goedert, R. A. Crowther and M. G. Spillantini, *Neuron*, 1998, **21**, 955.

130. B. De Strooper, *Physiol. Rev.*, 2010, **90**, 465.

131. H. J. Aizenstein, R. D. Nebes, J. A. Saxton, J. C. Price, C. A. Mathis, N. D. Tsopelas, S. K. Ziolko, J. A. James, B. E. Snitz, P. R. Houck, W. Bi, A. D. Cohen, B. J. Lopresti, S. T. DeKosky, E. M. Halligan and W. E. Klunk, *Arch. Neurol.*, 2008, **65**, 1509.

132. J. Tian, J. Shi, L. Zhang, J. Yin, Q. Hu, Y. Xu, S. Sheng, P. Wang, Y. Ren, R. Wang and Y. Wang, *Curr. Alzheimer Res.*, 2009, **6**, 118.

133. X. Wang, B. Su, S. L. Siedlak, P. I. Moreira, H. Fujioka, Y. Wang, G. Casadesus and X. Zhu, *Proc. Natl. Acad. Sci. USA*, 2008, **105**, 19318.

134. M. A. Leissring, *J. Biol. Chem.*, 2008, **283**, 29645.

135. S. Sudoh, M. P. Frosch and B. A. Wolf, *Biochemistry*, 2002, **41**, 1091.

136. S. Eggert, K. Paliga, P. Soba, G. Evin, C. L. Masters, A. Weidemann and K. Beyreuther, *J. Biol. Chem.*, 2004, **279**, 18146.

137. W. Xia, *Curr. Alzheimer Res.*, 2008, **5**, 172.

138. H.-J. Park, S.-S. Kim, Y.-M. Seong, K.-H. Kim, H. G. Goo, E. J. Yoon, D. S. Min, S. Kang and H. Rhim, *J. Biol. Chem.*, 2006, **281**, 34277.

139. J. S. Miners, S. Baig, J. Palmer, L. E. Palmer, P. G. Kehoe and S. Love, *Brain Pathol.*, 2008, **18**, 240.

140. L. B. Hersh, *Curr. Pharm. Design*, 2003, **9**, 449.

141. N. N. Nalivaeva, L. R. Fisk, N. D. Belyaev and A. J. Turner, *Curr. Alzheimer Res.*, 2008, **5**, 212.

142. H. M. Tucker, M. Kihiko, J. N. Caldwell, S. Wright, T. Kawarabayashi, D. Price, D. Walker, S. Scheff, J. P. McGillis, R. E. Rydel and S. Estus, *J. Neurosci.*, 2000, **20**, 3937.

143. J. P. Melchor, R. Pawlak and S. Strickland, *J. Neurosci.*, 2003, **23**, 8867.

144. H. M. Tucker, J. Simpson, M. Kihiko-Ehmann, L. H. Younkin, J. P. McGillis, S. G. Younkin, J. L. Degen and S. Estus, *Neurosci. Lett.*, 2004, **368**, 285.

145. A. Falkevall, N. Alikhani, S. Bhushan, P. F. Pavlov, K. Busch, K. A. Johnson, T. Eneqvist, L. Tjernberg, M. Ankarcrona and E. Glaser, *J. Biol. Chem.*, 2006, **281**, 29096.

146. C. Caspersen, N. Wang, J. Yao, A. Sosunov, X. Chen, J. W. Lustbader, H. W. Xu, D. Stern, G. McKhann and S. D. Yan, *FASEB J.*, 2005, **19**, 2040.

147. G. Grasso, R. D'Agata, E. Rizzarelli, G. Spoto, L. D'Andrea, C. Pedone, A. Picardi, A. Romanelli, M. Fragai and K. J. Yeo, *J. Mass Spectrom.*, 2005, **40**, 1565.

148. P. Yan, X. Hu, H. Song, K. Yin, R. J. Bateman, J. R. Cirrito, Q. Xiao, F. F. Hsu, J. W. Turk, J. Xu, C. Y. Hsu, D. M. Holtzman and J.-M. Lee, *J. Biol. Chem.*, 2006, **281**, 24566.

149. K.-J. Yin, J. R. Cirrito, P. Yan, X. Hu, Q. Xiao, X. Pan, R. Bateman, H. Song, F.-F. Hsu, J. Turk, J. Xu, C. Y. Hsu, J. C. Mills, D. M. Holtzman and J.-M. Lee, *J. Neurosci.*, 2006, **26**, 10939.

150. J. R. Backstrom, G. P. Lim, M. J. Cullen and Z. A. Tökés, *J. Neurosci.*, 1996, **16**, 7910.

151. G. Molinaro, J. L. Rouleau and A. Adam, *Curr. Opin. Pharmacol.*, 2002, **2**, 131.

152. P. Corvol, T. A. Williams and F. Soubrier, *Method Enzymol.*, 1995, **248**, 283.

153. K. Zou, H. Yamaguchi, H. Akatsu, T. Sakamoto, M. Ko, K. Mizoguchi, J.-S. Gong, W. Yu, T. Yamamoto, K. Kosaka, K. Yanagisawa and M. Michikawa, *J. Neurosci.*, 2007, **27**, 8628.

154. E. A. Eckman, S. K. Adams, F. J. Troendle, B. A. Stodola, M. A. Kahn, A. H. Fauq, H. D. Xiao, K. E. Bernstein and C. B. Eckman, *J. Biol. Chem.*, 2006, **281**, 30471.

155. K. Barnes, C. Brown and A. J. Turner, *Hypertension*, 1998, **31**, 3.
156. E. A. Eckman, D. K. Reed and C. B. Eckman, *J. Biol. Chem.*, 2001, **276**, 24540.
157. C. Oefner, A. D'Arcy, M. Hennig, F. K. Winkler and G. E. Dale, *J. Mol. Biol.*, 2000, **296**, 341.
158. N. Iwata, S. Tsubuki, Y. Takaki, K. Shirotani, B. Lu, N. P. Gerard, C. Gerard, E. Hama, H. J. Lee and T. C. Saido, *Science*, 2001, **292**, 1550.
159. T. Saito, N. Iwata, S. Tsubuki, Y. Takaki, J. Takano, S.-M. Huang, T. Suemoto, M. Higuchi and T. C. Saido, *Nat. Med.*, 2005, **11**, 434.
160. E. Hama, K. Shirotani, N. Iwata and T. C. Saido, *J. Biol. Chem.*, 2004, **279**, 30259.
161. L. B. Hersh and D. W. Rodgers, *Curr. Alzheimer Res.*, 2008, **5**, 225.
162. J. R. McDermott and A. M. Gibson, *Neurochem. Res.*, 1997, **22**, 49.
163. M. A. Leissring, W. Farris, A. Y. Chang, D. M. Walsh, X. Wu, X. Sun, M. P. Frosch and D. J. Selkoe, *Neuron*, 2003, **40**, 1087.
164. J. A. Luchsinger, M. X. Tang, S. Shea and R. Mayeux, *Neurology*, 2004, **63**, 1187.
165. W. Q. Qiu and M. F. Folstein, *Neurobiol. Aging*, 2006, **27**, 190.
166. I. V. Kurochkin and S. Goto, *FEBS Lett.*, 1994, **345**, 33.
167. A. Mukherjee, E. Song, M. Kihiko-Ehmann, J. P. Goodman Jr, J. S. Pyrek, S. Estus and L. B. Hersh, *J. Neurosci.*, 2000, **20**, 8745.
168. G. Grasso, E. Rizzarelli and G. Spoto, *J. Mass Spectrom.*, 2007, **42**, 1590.
169. G. Grasso, A. I. Bush, R. D'Agata, E. Rizzarelli and G. Spoto, *Eur. Biophys. J.*, 2009, **38**, 407.
170. M. A. Leissring, A. Lu, M. M. Condron, D. B. Teplow, R. L. Stein, W. Farris and D. J. Selkoe, *J. Biol. Chem.*, 2003, **278**, 37314.
171. L. Morelli, R. Llovera, S. A. Gonzalez, J. L. Affranchino, F. Prelli, B. Frangine, J. Ghiso and E. M. Castaño, *J. Biol. Chem.*, 2003, **278**, 23221.
172. G. Grasso, P. Mineo, E. Rizzarelli and G. Spoto, *Int. J. Mass Spectrom.*, 2009, **282**, 50.
173. V. Chesneau, K. Vekrellis, M. R. Rosner and D. J. Selkoe, *Biochem. J.*, 2000, **351**, 509.
174. I. V. Kurochkin, *FEBS Lett.*, 1998, **427**, 153.
175. Y. Shen, A. Joachimiak, M. R. Rosner and W.-J. Tang, *Nature*, 2006, **443**, 870.
176. C. Ciaccio, G. F. Tundo, G. Grasso, G. Spoto, D. Marasco, M. Ruvo, S. Marini, E. Rizzarelli and M. Coletta, *J. Mol. Biol.*, 2009, **385**, 1556.
177. G. Grasso, E. Rizzarelli and G. Spoto, *BBA-Proteins Proteom.*, 2008, **1784**, 1122.
178. J. Zhao and G. Pei, *Cell Res.*, 2008, **18**, 803.
179. G. Grasso, M. Fragai, E. Rizzarelli, G. Spoto and K. J. Yeo, *J. Mass Spectrom.*, 2006, **41**, 1561.
180. G. Grasso, M. Fragai, E. Rizzarelli, G. Spoto and K. J. Yeo, *J. Am. Soc. Mass Spectrom.*, 2007, **18**, 961.

181. W. C. Duckworth, R. G. Bennett and F. G. Hamel, *Endocr. Rev.*, 1998, **19**, 608.

182. M. A. Leissring, W. Farris, X. Wu, D. C. Christodoulou, M. C. Haigis, L. Guarente and D. J. Selkoe, *Biochem. J.*, 2004, **383**, 439.

183. L. B. Hersh, *Cell. Mol. Life Sci.*, 2006, **63**, 2432.

184. D. Strozyk, L. J. Launer, P. A. Adlard, R. A. Cherny, A. Tsatsanis, I. Volitakis, K. Blennow, H. Petrovitch, L. R. White and A. I. Bush, *Neurobiol. Aging*, 2009, **30**, 1069.

185. P. S. Donnelly, A. Caragounis, T. Du, K. M. Laughton, I. Volitakis, R. A. Cherny, R. A. Sharples, A. F. Hill, Q.-X. Li, C. L. Masters, K. J. Barnham and A. R. White, *J. Biol. Chem.*, 2008, **283**, 4568.

186. J. L. Domingo, *J. Alzheimers Dis.*, 2006, **10**, 331.

187. C. Rodriguez-Rodriguez, N. Sanchez de Groot, A. Rimola, A. Alvarez-Larena, V. Lloveras, J. Vidal-Gancedo, S. Ventura, J. Vendrell, M. Sodupe and P. Gonzalez-Duarte, *J. Am. Chem. Soc.*, 2009, **131**, 1436.

188. P. A. Adlard, R. A. Cherny, D. I. Finkelstein, E. Gautier, E. Robb, M. Cortes, I. Volitakis, X. Liu, J. P. Smith, K. Perez, K. Laughton, Q.-X. Li, S. A. Charman, J. A. Nicolazzo, S. Wilkins, K. Deleva, T. Lynch, G. Kok, C. W. Ritchie, R. E. Tanzi, R. Cappai, C. L. Masters, K. J. Barnham and A. I. Bush, *Neuron*, 2008, **59**, 43.

189. S. Shäfer, F. G. Pajonk, G. Multhaup and T. A. Bayer, *J. Mol. Med.*, 2007, **85**, 405.

190. E. Sampson, L. Jenagaratnam and R. McShane, *Cochrane Data Base Syst. Rev.*, 2008, **23**, CD005380.

191. S. Bolognin, P. Zatta, D. Drago, P. P. Parnigotto, F. Ricchelli and G. Tognon, *Neuromol. Med.*, 2008, **10**, 322.

192. D. S. Yang, J. McLaurin, K. Qin, D. Westaway and P. E. Fraser, *Eur. J. Biochem.*, 2000, **267**, 6692.

193. P. J. Crouch, D. J. Tew, T. Du, D. N. Nguyen, A. Caragounis, G. Filiz, R. E. Blake, I. A. Trounce, C. P. W. Soon, K. Laughton, K. A. Perez, Q.-X. Li, R. A. Cherny, C. L. Masters, K. J. Barnham and A. R. White, *J. Neurochem.*, 2009, **108**, 1198.

194. K. L. Epting, C. Vieille, J. G. Zeikus and R. M. Kelly, *FEBS J.*, 2005, **272**, 1454.

195. W. F. Li, X. X. Zhou and P. Lu, *Biotechnol. Adv.*, 2005, **23**, 271.

196. C. Vieille and G. J. Zeikus, *Microbiol. Mol. Biol. Rev.*, 2001, **65**, 1.

197. A. L. Pey and A. Martinez, *J. Biol. Inorg. Chem.*, 2009, **14**, 521.

198. S. Z. Potter, H. Zhu, B. F. Shaw, J. A. Rodriguez, P. A. Doucette, S. H. Sohn, A. Durazo, K. F. Faull, E. B. Gralla, A. M. Nersissian and J. S. Valentine, *J. Am. Chem. Soc.*, 2007, **29**, 4575.

199. Q. Han, Y. Fu, H. Zhou, Y. He and Y. Luo, *FEBS Lett.*, 2007, **581**, 3027.

200. J. M. Hadden, A.-C. Declais, S. E. V. Phillips and D. M. Lilley, *EMBO J.*, 2002, **21**, 3505.

201. A. Pingoud, M. Fuxreiter, V. Pingoud and W. Wende, *Cell. Mol. Life Sci.*, 2005, **62**, 685.

202. J. Aaqvist and A. Warshel, *J. Am. Chem. Soc.*, 1990, **112**, 2860.

203. F. Xie and C. M. Dupureur, *Arch. Biochem. Biophys.*, 2009, **483**, 1.
204. A. Y. Louie and T. J. Meade, *Chem. Rev.*, 1999, **99**, 2711.
205. R. V. Kukreja, S. Sharma, S. Cai and B. R. Singh, *BBA-Proteins Proteom.*, 2007, **1774**, 213.
206. N. Selevsek, S. Rival, A. Tholey, E. Heinzle, U. Heinz, L. Hemmingsen and H. W. Adolph, *J. Biol. Chem.*, 2009, **284**, 16419.
207. S. Maric, S. M. Donnelly, M. W. Robinson, T. Skinner-Adams, K. R. Trenholme, D. L. Gardiner, J. P. Dalton, C. M. Stack and J. Lowther, *Biochemistry*, 2009, **48**, 5435.
208. M. F. Souliere, J.-P. Perreault and M. Bisaillon, *Biochem. J.*, 2009, **420**, 27.
209. J. Arima, Y. Uesugi and T. Hatanaka, *Biochimie*, 2009, **91**, 568.
210. B. Lai, Y. Li, A. Cao and L. Lai, *Biochemistry*, 2003, **42**, 785.
211. B. Krajewska, *J. Enzym. Inhib. Med. Ch.*, 2008, **23**, 535.
212. P. A. Benkovic, C. A. Caperelli, M. de Maine and S. J. Benkovic, *Proc. Natl. Acad. Sci. USA*, 1978, **75**, 2185.
213. S. C. Graham, C. S. Bond, H. C. Freeman and J. M. Guss, *Biochemistry*, 2005, **44**, 13820.
214. G. Falkous, J. B. Harris and D. Mantle, *Clin. Chim. Acta*, 1995, **238**, 125.
215. G. Lupidi, M. Angeletti, A. M. Eleuteri, E. Fioretti, S. Marini, M. Gioia and M. Coletta, *Chem. Rev.*, 2002, **228**, 263.
216. M. Li, M. Sun, Y. Liu, J. Yu, H. Yang, D. Fan and D. Chui, *J. Alzheimers Dis.*, 2010, **19**, 161.
217. A. R. White, T. Du, K. M. Laughton, I. Volitakis, R. A. Sharples, M. E. Xilinas, D. E. Hoke, R. M. D. Holsinger, G. Evin, R. A. Cherny, A. F. Hill, K. J. Barnham, Q.-X. Li, A. I. Bush and C. L. Masters, *J. Biol. Chem.*, 2006, **281**, 17670.
218. P. S. Donnelly, Z. Xiao and A. G. Wedd, *Curr. Opin. Chem. Biol.*, 2007, **11**, 128.
219. P. J. Crouch, L. W. Hung, P. A. Adlard, M. Cortes, V. Lal, G. Filiz, K. A. Perez, M. Nurjono, A. Caragounis, T. Du, K. Laughton, I. Volitakis, A. I. Bush, Q. X. Li, C. L. Masters, R. Cappai, R. A. Cherny, P. S. Donnelly, A. R. White and K. J. Barnham, *Proc. Natl. Acad. Sci. USA*, 2009, **106**, 381.
220. K. A. Price, G. Filiz, A. Caragounis, T. Du, K. M. Laughton, C. L. Masters, R. A. Sharples, A. F. Hill, Q.-X. Li, P. S. Donnelly, K. J. Barnham, P. J. Crouch and A. R. White, *Int. J. Biochem. Cell B.*, 2008, **40**, 1901.
221. H. Zheng, M. B. H. Youdim and M. Fridkin, *J. Med. Chem.*, 2009, **52**, 4095.
222. G. Liu, P. Men, W. Kudo, G. Perry and M. A. Smith, *Neurosci. Lett.*, 2009, **455**, 187.
223. R. A. Cherny, C. S. Atwood, M. E. Xilinas, D. N. Gray, W. D. Jones, C. A. McLean, K. J. Barnham, I. Volitakis, F. W. Fraser, Y. Kim, X. Huang, L. E. Goldstein, R. D. Moir, J. T. Lim, K. Beyreuther, H. Zheng, R. E. Tanzi, C. L. Masters and A. I. Bush, *Neuron*, 2001, **30**, 665.

224. B. Ibach, E. Haen, J. Marienhagen and G. Hajak, *Pharmacopsychiatry*, 2005, **38**, 178.
225. Y. Wang, R. Branicky, Z. Stepanyan, M. Carroll, M.-P. Guimond, A. Hihi, S. Hayes, K. McBride and S. Hekimi, *J. Biol. Chem.*, 2009, **284**, 314.
226. G. Filiz, A. Caragounis, L. Bica, T. Du, C. L. Masters, P. J. Crouch and A. R. White, *Int. J. Biochem. Cell. Biol.*, 2008, **40**, 1030.
227. I. Kim, C. H. Kim, J. H. Kim, J. Lee, J. J. Choi, Z. A. Chen, M. G. Lee, K. C. Chung, C. Y. Hsu and Y. S. Ahna, *Exp. Cell Res.*, 2004, **298**, 229.
228. S. Zhai, L. Yang, Q. C. Cui, Y. Sun, Q. P. Dou and B. Yan, *J. Biol. Inorg. Chem.*, 2010, **15**, 259.
229. K. G. Daniel, P. Gupta, R. H. Harbach, W. C. Guida and Q. P. Dou, *Biochem. Pharmacol.*, 2004, **67**, 1139.
230. K. G. Daniel, D. Chen, B. Yan and Q. Ping Dou, *Front. Biosci.*, 2007, **12**, 135.
231. W. Q. Ding and S. E. Lind, *IUBMB Life*, 2009, **61**, 1013.

CHAPTER 6

The Inorganic Side of Alzheimer's Disease

G. PAPPALARDO,[a] D. MILARDI,[a] E. RIZZARELLI[a,b*] AND I. SOVAGO[c*]

[a] Istituto di Biostrutture e Bioimmagini, Consiglio Nazionale delle Ricerche, Unità Operativa e di Supporto di Catania, Catania, Italy; [b] Dipartimento di Scienze Chimiche, Università degli Studi di Catania, Catania, Italy; [c] Department of Analytical and Inorganic Chemistry, University of Debrecen, Debrecen, Hungary

6.1 Introduction

Alzheimer's disease (AD), the most common form of senile dementia, represents an enormous social problem that still requires an improvement of diagnostic tools and therapies.[1] The major risk factor known for AD is age; 95% of all AD cases have no clear pattern of inheritance and it is believed that both genetic and environmental factors may contribute to the etiology of AD.[2] There are four main consistent features in AD brains: i) extracellular deposition of amyloid-β (Aβ) peptide plaques;[3] ii) intracellular hyperphosphorylation of the microtubule associated protein, tau;[4] iii) elevated oxidative stress to lipids, proteins and nucleic acids;[5] and, as proposed more recently, iv) a loss of metallostasis (bio-metal homeostasis).[6,7] Aβ is neurotoxic at non-physiological (micromolar) concentrations *in vitro*, but it is also produced in health[8] and, at physiological (nanomolar) concentrations, is neurotrophic in cell cultures.[9–11] Furthermore, synthetic Aβ1–42 is known to rapidly self-aggregate into amyloids in solution.[12,13] The sum of these observations has led to the "Amyloid Cascade

RSC Drug Discovery Series No. 7
Neurodegeneration: Metallostasis and Proteostasis
Edited by Danilo Milardi and Enrico Rizzarelli
© Royal Society of Chemistry 2011
Published by the Royal Society of Chemistry, www.rsc.org

Hypothesis", where the overproduction of Aβ is regarded as the major cause of the disease. But the self-aggregating properties of Aβ are insufficient to explain the association of the peptide with AD pathogenesis. Indeed, there is considerable evidence that the soluble, but not the fibrillar, forms of Aβ correlate with morbidity of AD symptoms.[14-16] However, not all forms of soluble Aβ are toxic, since healthy people normally have soluble Aβ in their brains, and Aβ is a soluble component of all biological fluids. Consequently, the presence of toxic forms of soluble Aβ in AD may be hypothesized. Based upon the amyloid hypothesis, the major approaches for developing therapeutics for AD have been focused either to prevent Aβ production (β or γ-secretase BACE-inhibitors) or to control the correct folding of Aβ. In this light, a very recent review has contributed to highlight the substantial role played by chaperones in the maintenance of the proteostatic network in neurodegeneration.[17] Cellular protein homeostasis (proteostasis) refers to controlling the conformation, concentration, binding interactions and location of individual proteins making up the proteome.[18] This is accomplished by a complex network of molecular interactions that balances protein biosynthesis, folding, trafficking, assembly/disassembly, and protein clearance.[19] The proteostasis network is composed of: i) the ribosome, chaperones, aggregases, and disaggregases that direct folding, as well as pathways that select proteins for degradation; ii) signaling pathways that influence the activity of the cell components; iii) genetic and epigenetic pathways, physiologic stressors, and intracellular metabolites that affect the activities of the cell. Two prominent modulators of protein homeostasis are molecular chaperones and stress-inducible responses. Proteostasis regulators can partially correct proteostatic deficiencies that contribute to a broad range of human diseases, some that present at birth, but most upon aging.[18,19] There is substantial evidence that many other neurochemical reactions apart from Aβ production may contribute to neurodegeneration in AD. Amyloid deposition is an age-dependent phenomenon, and if Aβ production is known to increase with age, other age-related neurochemical changes are believed to play an essential role in the reaction that causes Aβ to accumulate in neurons. The age-dependent changes are closely associated with oxidative damage to neuronal cells, which precedes Aβ deposition and is characterized by the involvement of redox active metal ions.[20-22] Bush and Tanzi have recently proposed the "Metal Hypothesis of Alzheimer's Diseases" which stipulates that the neuropathogenic effects of Aβ in AD are promoted by, and possibly even dependent upon, Aβ–metal interactions.[23] Thus, alterations of copper and zinc levels have been reported to lie at the root of the pathogenic cascade of events leading to neurodegeneration in AD. In analogy to protein homeostasis (proteostasis), cellular metal homeostasis (metallostasis) refers to controlling the concentration, binding interactions and location of individual metal ions making up the "metallome". This is accomplished by a highly complex network of molecular interactions that balances intracellular metal uptake, trafficking, storage, speciation and signaling. Prominent modulators of metal homeostasis are metal chaperones, metal transporters, metalloproteins, small chaperone molecules and metal transcription factors. Metallostasis regulators can partially correct metal ion dys-homeostasis that contributes to a broad range of human diseases, some that present at birth, but most upon aging, as observed in AD. In this light, a novel mode of therapeutic treatment of AD may be based on restoring intracellular metallostasis by ionophores which, in turn, re-establish proteostasis by the activation of the kinase cascade involving PI3K, AkT, GSK3β, MAP, JNK and ERK, with a consequent up-regulation of metalloproteases (MMPs) and degradation of extracellular Aβ.[23]

6.2 Metallostasis in AD Brain and its Compartments

A common misunderstanding in the description of the role of metal ions in AD is that the neurological syndromes in which metal ions are implicated are hypothetically caused by toxicological exposure to Cu, Fe, Zn, Al, Hg and Mn. In other words, ingestion or exposure to the metal ions would cause abnormal protein interactions and, consequently, the disease. This misconception is probably an inheritance of the hypothesis that aluminium exposure can cause AD pathology.[24] In terms of total concentrations, the normal brain has sufficient amounts of metal ions to damage or dysregulate numerous proteins and metabolic systems. For example, the concentration of Zn^{2+} that is released during neurotransmission is ≈ 300 μM, which is enough to be rapidly neurotoxic in neuronal cell culture.[25] Therefore, the brain is supposed to have efficient homeostatic mechanisms and buffers in place to prevent an abnormal dys-compartmentalization of metal ions. Also, the blood–brain barrier (BBB) is known to cope effectively with the fluctuating levels of plasma metal ions. Metal homeostasis may thus be perturbed at three different levels in AD patients with respect to healthy people: i) differences in metal ion concentrations in the brain and in biological fluids; ii) differences in the balance between intra- and extracellular concentrations of metal ions; iii) alterations in metal ion transporters and chaperones.

6.2.1 Differences in Copper and Zinc Levels in AD Brain

As mentioned above, alterations in brain copper levels have been implicated in the pathogenesis of several neurological disorders including Alzheimer's, Parkinson's and prion diseases.[26–28] The serum and CSF levels of copper are significantly higher in patients with AD compared to age matched controls,[29,30] and potentially correlate with ceruloplasmin expression.[31] This increased serum copper was reported to correlate well with higher levels of serum peroxides in AD patients.[32] Notably, copper mediates low-density lipoprotein (LDL) oxidation by homocysteine,[33] and plasma homocysteine levels are a well known risk factor for AD.[34] In AD brains, copper has been shown to be associated with senile plaques, significantly increasing copper content from 79 μM (in the normal age-matched neuropil) to 390 μM within these plaques.[35] Lovell *et al.* have also reported an increase in parenchymal copper levels.[35] However, their analysis was restricted to the amygdala and contradicts a larger collection of studies reporting a decrease in bulk tissue copper levels in AD-affected cortical regions.[36,37] Considering the increased copper concentrations in senile plaques with the overall reduction in copper levels, a complex picture emerges where copper seems to be abnormally redistributed in AD brain regions and collects outside the cell. This abnormal redistribution of copper in the brain may involve a lipid component: in fact rabbits maintained on a diet containing elevated cholesterol and copper demonstrated accelerated amyloid Aβ aggregation and promotion of oxidative events in the brain.[38] This is supported by the evidence that copper in combination with a high fat diet increases the risk for AD.[39]

Zinc is another transition metal present in all tissues. Diverse classes of protein require bound zinc for normal function: zinc metalloenzymes (*e.g.* SOD1), transcription factors containing zinc-binding motifs such as zinc fingers (*e.g.* p53 and GAL4),[40] signaling (*e.g.* protein kinase C)[41] and storage to buffer cytosolic zinc pools (*e.g.* metallothioneins).[42] Of all organs, the brain is thought to contain the highest levels of zinc.[43] In the brain, zinc is highly enriched in the glutamatergic nerve terminals (10–15%), where it is released upon neuronal activation. Recent studies have investigated the mechanisms involved in the control of intra-neuronal zinc concentrations.[44] Although cytosolic free zinc is typically in the picomolar range,[45] this is substantially increased in the synapsis to micromolar concentrations. Squitti *et al.* found that non-ceruloplasmin bound copper(II) is elevated in AD patients, and may generate reactive oxygen species (ROS).[46] Notably, copper(II) homeostasis is abnormal with aging.[23,30,47] In subjects greater than 75 years of age copper has repeatedly been reported as elevated in serum[48] and plasma.[49–53] Moreover, several animal and human studies have demonstrated a rise in levels of brain copper from youth to adulthood. Analyses of normal mice (BL6/SJL) have demonstrated a 46% increase in copper content.[54,55] Interestingly, once in adulthood a marked drop in levels from middle age onwards occurs. All these sources of evidence support the hypothesis that these gross changes in copper levels may be region specific.[56] The concentration of zinc ions in both plasma and CSF has also been reported to decrease in AD compared to age-matched control patients.[57–59] While nutritional deficiency is common in advanced age, this further decrease in AD zinc levels is supposed to enhance amyloid pathology. In fact, as previously described with copper, zinc is highly enriched within AD plaques (1055 μM) compared to normal age-matched neuropil (350 μM).[35] Histochemically reactive zinc deposits are also found specifically localized to cerebral amyloid angiopathy deposits and neurofibrillary tangle (NFT)-bearing neurons.[60,61] Dyshomeostasis of zinc in the AD brain may arise from inhibition of zinc export by 4-hydroxynonenal, a peroxidation product of Aβ : copper redox activity that is elevated in AD tissue.[62] In contrast to copper, zinc levels in rodent and human plasma have been reported to be highest at birth and steadily decrease with age.[63–67] One report on healthy men aged 8–89 years indicated that plasma zinc levels had a tendency to remain constant throughout life until the age of 75 where zinc levels suddenly decreased.[48] While these changes have not reflected an age-related alteration in the global zinc levels within the brain,[55] specific areas known to have high concentrations of zinc, such as the hippocampus, are known to exhibit decreased zinc levels with age.[68]

6.2.2 Balance between Intra- and Extracellular Levels of Copper and Zinc in AD

In vivo, Rajendran *et al.* imaged and identified amyloid plaques in brain sections from transgenic mice and simultaneously quantified the trace Cu(II) within them.[69] They found increased Cu(II) concentrations (250 μM) within the amyloid plaques compared to the surrounding tissue (80 μM), while the extracellular concentration Cu(II) is 0.2–1.7 μM in normal brains.[70] Increasing extracellular copper and decreasing intracellular copper reduces secretion of molecules involved in the protection of neurons against oxidative stress, such as cyclophilin A (CypA), or of molecules capable of shifting neuronal cells towards a pro-inflammatory state, such as interleukin (IL)-1a, IL-12, Rantes, neutrophil gelatinase-associated lipocalin (NGAL) and secreted protein acidic and rich in cysteine (SPARC).[71] These alterations in copper balance have been linked to changes in senile plaque deposition in AD.[72,73] Whether copper supplementation is beneficial or harmful for AD pathology is not clear. Sparks *et al.* showed that trace amounts of copper given in drinking water to rabbits on a high cholesterol diet produces AD pathology,[38] while Bayer *et al.* proposed the concept of copper deficiency in AD.[74] If, on one hand, recent clinical trials produced no benefit of copper supplementation to AD patients,[75] on the other hand, PBT2, a transition metal ionophore,[76] appears to confer beneficial effects. Furthermore, other researchers found that supplement of Cu(II) could restore the levels of Cu(II) in the brain and reduce Aβ deposition. For example, Crouch *et al.* reported that increasing Cu bioavailability inhibits Aβ oligomers and tau phosphorylation, *in vitro* and *in vivo*.[77] This apparent contradiction may be explained if one takes into account intra- and extracellular Cu(II) dyshomeostasis and the altered balance between Cu(II) influx and efflux mediated by specific Cu(II) chaperones as well as metallothioneins (MTs). In this light, it has been observed that the intracellular homeostatic levels of Zn(II) are maintained by the activity of MTs, and oxidative stress has been found to be a potent catalyst for the release of this metal ion from these cytosolic proteins.[45,73] Thus the intracellular increase of Zn(II) concentration, as a consequence of ions release from MT or entry from the extracellular space, can induce potent and irreversible disruption of mitochondrial function, a mechanism that can also boost the oxidative stress and neuronal apoptosis.[44] Such an occurrence contributes to lay the groundwork for a vicious loop to be established. In fact, Aβ itself triggers oxidative stress[6] that can cause mobilization of MT-bound Zn(II); Zn(II), even at low nanomolar concentrations (*i.e.*, well in the range of what is estimated to be released from MTs), can trigger additional reactive oxygen species (ROS) generation, leading to release of more Zn(II) from MTs and increased aggregation of Aβ. By the way, the past few years have witnessed noticeable developments in deeper understanding of the roles of zinc as a crucial intra- and intercellular signaling ion of the CNS, and recent studies have clarified the mechanisms involved in maintaining intra-neuronal zinc(II) concentrations, and a multitude of zinc transporters, zinc importing proteins, and an array of buffering proteins have been identified to date.[44] These results, in conjunction with data about copper, also underline the neurophysiological importance of zinc-dependent pathways and the injurious effects of zinc dyshomeostasis.

6.2.3 Alterations of Copper and Zinc Transporters and Chaperones in AD

The delicate balance of intra- and extracellular levels of metal ions and alterations in metal ion transporters represent a paradigmatic link between metallostasis and proteostasis; in addition, metal ions are themselves signals activating the transcription of the proper trafficking agent. In fact, keeping metal ions in homeostasis is an important process for health, but how to keep a balance of metal ions in living organisms? Our brains have mechanisms to tightly control the levels of metals entering the brain, and these mechanisms may become corrupted with aging or by diseases. Although copper(II) is an essential micronutrient that participates in several processes crucial for life, it may be toxic to cell membranes, DNA and protein when accumulated in excess. Therefore, its homeostasis is carefully regulated through a system of protein transporters and chaperones. The biochemically functional metal ion content of the brain is stringently regulated and there is no passive flux of metals from the circulation to the brain. Copper(II) and zinc(II) are increasingly implicated in interactions with the major protein components of neurodegenerative diseases, this is not merely due to increased (*e.g.*, toxicological) exposure to these metals, but rather it is ascribable to a breakdown in the homeostatic mechanisms that compartmentalize and regulate metal ions in neurons. The reactivity of copper and its ability to bind at sites for other metals have resulted in the evolution of homeostatic systems for copper.[78–87] Copper pathways typically involve transporters, which allow the metal to cross membranes, and the diffusible cytosolic metallochaperone Atx1 that binds and delivers copper.[78–82,86–89] The ability of copper ions to exchange electrons is thus tightly regulated via a variety of transporters as also described in Chapter 8 of this book.[90] CCS (Copper Chaperone for Superoxide Dismutase) is a copper-binding protein that delivers copper to several proteins including the antioxidant enzyme Cu/Zn superoxide dismutase (SOD1), X-linked inhibitor of apoptosis protein (XIAP), and possibly BACE1.[91–93] A large body of evidence suggests that CCS may impact on AβPP processing and Aβ production in Alzheimer's disease. Firstly, CCS binds to the intracellular domain of BACE1 and may deliver copper to BACE1.[91] BACE1 is a key enzyme required for the processing of AβPP to produce Aβ.[94] Secondly, CCS binds to the neuronal adaptor protein X11α (also known as munc-18 interacting protein-1).[95] X11α also interacts directly with AβPP, and overexpression of X11α inhibits Aβ production in AβPP transgenic mice.[96–99] Finally, modulating SOD1 levels alters Aβ production.[99] However, whether CCS affects Aβ production is not known. With reference to zinc, a multitude of zinc transporters (ZnTs), zinc-importing proteins (ZIPs) and buffering proteins such as the metallothioneins bind cytosolic zinc to prevent free zinc becoming toxic.[44,68,100–105] Estrogen can also modulate levels of ZnT3 and thus synaptic zinc;[106] this is of particular significance as gender is another major risk factor for AD. Consistent with the metal ions in plaques playing a primary role in Aβ aggregation, experiments in ZnT3 knockout mice have established that presynaptic Zn release causes amyloid formation in AβPP transgenic mice. ZnT3 knockout mice have about 15% less Zn in their cortex, but are otherwise phenotypically subtle.[107] These mice were crossed with Tg2576AβPP transgenic mice, and the progeny characterized. These experiments showed that ZnT3 genetic ablation markedly inhibits amyloid pathology,[108,109] increasing the concentration of soluble Aβ. This suggests that soluble Aβ and soluble Zn exist in a dissociable equilibrium with insoluble plaque Aβ.[110] The increased amyloid deposition in women and female AβPP transgenic mice also may be explained by an estrogen-dependent increase in ZnT3 expression.[111] In accordance, ZnT3 is also decreased with age.[68,112] However, alterations of zinc transporters in AD are summarized in Chapter 8 of this book and will not be further discussed here.

6.3   APP and Aβ Levels are Regulated by Metal Ions

APP is a transmembrane metalloprotein, which is expressed in all tissue and is believed to be involved in metallostasis. The processing of APP involves a number of activities by α-, β- and γ-secretase.[113–115] A direct influence of metals on the secretases and, in turn, on APP processing has been reported. The zinc binding in the Aβ region of the APP sequence spans the α-secretase cleavage site and may, therefore, tune the cleavage of Aβ from APP and also protect Aβ from proteolytic degradation.[116] APP can modulate copper transport, presumably by its extracellular Cu binding domain (CuBD). The binding of Cu(II) to the CuBD is related to the generation of Aβ, intracellular copper deficiency and an increase in extracellular Cu(II), marked by elevated Aβ secretion due to either increasing the expression of APP or enhancing the activity of β-secretase. Intracellular copper deficiency in human neuroblastoma cells significantly increases the activity of β-secretase and Aβ secretion.[117] PC12 cells, if exposed to extracellular copper(II), enhance the expression of both APP and β-secretase in a time and concentration dependent pattern, thus increasing Aβ levels.[118] Copper levels influence APP mRNA expression,[119,120] as reported in a human study showing that a low copper diet was associated with a significant decrease in APP expression in platelets.[121] Conversely, in APP knockout mice a significant increase in copper levels has been found in brain and liver,[122,123] while the overexpression of APP in various transgenic mouse lines carrying a mutated APP gene decreases copper levels[55,75,124,125] and at the same time lowers the SOD-1 activity.[74] In the transgenic mice, dietary copper(II) administration increases bioavailable brain copper(II) levels, restores SOD-1 activity, prevents premature death and decreases Aβ levels.[126] Relative to Aβ, the increase in bioavailable copper in these mice results in a lowering of soluble and insoluble Aβ.[74] Dietary zinc supplementation to two transgenic models of AD, the Tg2576 (overexpressing human APP with the AD-related Swedish mutation) and TgCRND8 (overexpressing human APP with the

AD-related Swedish and Indiana mutations) decreases amyloid-β plaque deposits, although it increases spatial memory impairments.[127] Correspondingly, lowering dietary zinc in another transgenic model of AD causes a significant increase in plaque volume.[128] Interesting results were described with the TgCRND8 mice when crossed with a transgenic model containing the "toxic-milk" mutation in the gene encoding the copper transport protein, ATP7b.[125] The mutation in ATP7b results in an accumulation of intracellular copper that decreases amyloid plaque content as well as the soluble and insoluble Aβ levels, indicating that the elevated intracellular copper level reduces Aβ aggregation, thus suggesting that a tight control of metal complexation to Aβ may provide promising hints in AD therapy.[129]

6.4 Copper and Zinc Interactions with Aβ

The formation of Aβ aggregates and the production of ROS are features of AD.[130] Copper and zinc ions form different kinds of aggregate species characterized by different morphologies while Cu^{2+} coordinated by Aβ is involved in the catalytic formation of ROS.[6] In both cases, the processes are governed by the binding modes of these metal ions and hence determination of the speciation, stability and coordination features of copper(II) and zinc(II) ions is essential to understand their roles in AD. However, the stoichiometry, the metal binding affinities and the unambiguous identification of metal ion binding in copper(II) and zinc(II) complexes with Aβ has remained difficult and no real agreement has emerged in the literature.[131] The amino acid sequence of amyloid-β reveals the high abundance of the effective metal binding sites. The terminal amino groups of oligopeptides are generally considered as the most common anchors for metal binding but hundreds of publications and several recent reviews[132,133] show the primary ligating role of the imidazole side chains of histidyl residues, too. Moreover, amyloid-β peptide contains several side chain carboxylate functions from the aspartyl and/or glutamyl residues and their extra charge has a significant contribution to the overall stability of peptide complexes. As a consequence, the coordination chemistry of amyloid-β peptide and its N- or C-terminally shortened fragments is rather complicated and the clarification of the metal ion speciation in these systems requires the combined application of different experimental techniques with a special emphasis on the appropriate use of the thermodynamic and spectroscopic methods. Huge numbers of studies have been performed to clarify the major metal binding sites of these peptides, focusing mainly on the interactions with copper(II) and zinc(II) ions and the results have already been reviewed by several authors.[134–136] The previous studies include the evaluation of potentiometric and solution spectroscopic studies on the copper(II) complexes of various fragments of amyloid-β.[137–142] These papers describe the major binding modes of the peptides but the generalization of data is hardly possible because only the equimolar samples have been investigated in most cases. NMR and mass spectroscopic techniques were also frequently used to elucidate the coordination environment of zinc(II) in the complexes with amyloid-β peptide.[143–146]

It is now well established from the previous studies that in the amyloid-β peptide, the metal binding sites are located at the N-terminal hydrophilic region encompassing the amino acid residues 1–16, whereas the C-terminal region, which contains hydrophobic amino acid residues, is not believed to be associated with any direct interactions with metal ions. Other studies have highlighted the fact that amyloid-β can host more than one copper(II)[147–149] or zinc(II)[148,150] ion per peptide molecule, but the precipitate formation did not allow the complete characterization of the solution equilibria and structural properties of these polynuclear systems.

It is known that conjugation with the polyethylene glycol (PEG) moiety enhances the water solubility of hydrophobic peptides. Thus, a new Aβ(1-16) conjugate, bearing a PEG moiety at the C-terminus, Aβ(1-16)PEG, was synthesized in our laboratories, allowing the exact determination of the metal complex speciation. In this way, both potentiometric and spectroscopic (Ultraviolet-Visible Spectroscopy (UV-Vis), Circular Dichroism (CD), Electron Paramagnetic Resonance (EPR), Nuclear Magnetic Resonances (NMR)) studies were carried out in aqueous solution at different metal to ligand ratios. Furthermore, in order to elucidate the structure of the metal ion complexes formed in the Aβ(1-16) region at various metal to ligand ratios, we resorted to a comparative study by investigating a series of shorter and/or single point mutated peptide analogues. Thus, the wild-type peptide Aβ(1-16) and its mutant Aβ(1-16)Y10A, the C- and N-terminally shortened fragments Aβ(1-4), Aβ(1-6), Ac-Aβ(1-6) and Ac-Aβ(8-16)Y10A, respectively, were also studied with the metal ions copper(II),[151] zinc(II)[152] and nickel(II).[153] It was one of the major conclusions of these studies that all these metal ions can form di-, tri- or tetra-nuclear complexes with amyloid-β which can result in the formation of mixed metal polynuclear complexes under biological conditions. In the next two subsections the most important results and conclusions obtained for the binary and ternary systems will be briefly summarized and compared to previous literature findings.

6.4.1 Solution Equilibria and Structural Characterization of the Major Species formed in the Copper(II)–, Nickel(II)– and Zinc(II)–Amyloid-β systems

A series of different experimental methods including potentiometry, UV-Vis, CD, EPR and ESI-MS (Electrospray Ionization-Mass Spectrometry) techniques have been used to characterize the metal ion speciation of the copper(II)–Aβ(1-16)PEG system.[151] The potentiometric and ESI-MS data revealed that Aβ(1-16)PEG can keep four copper(II) ions in solution even under alkaline conditions. The high complexity of the speciation curves and the large number of different coordination isomers did not make it possible to structurally characterize all the metal complexes, but the binding modes of the major species were elucidated by means of the various spectroscopic techniques and they are shown by Scheme 6.1.

All data unambiguously prove that the N-terminus of the peptide is the major metal binding site, starting with the involvement of terminal amino and

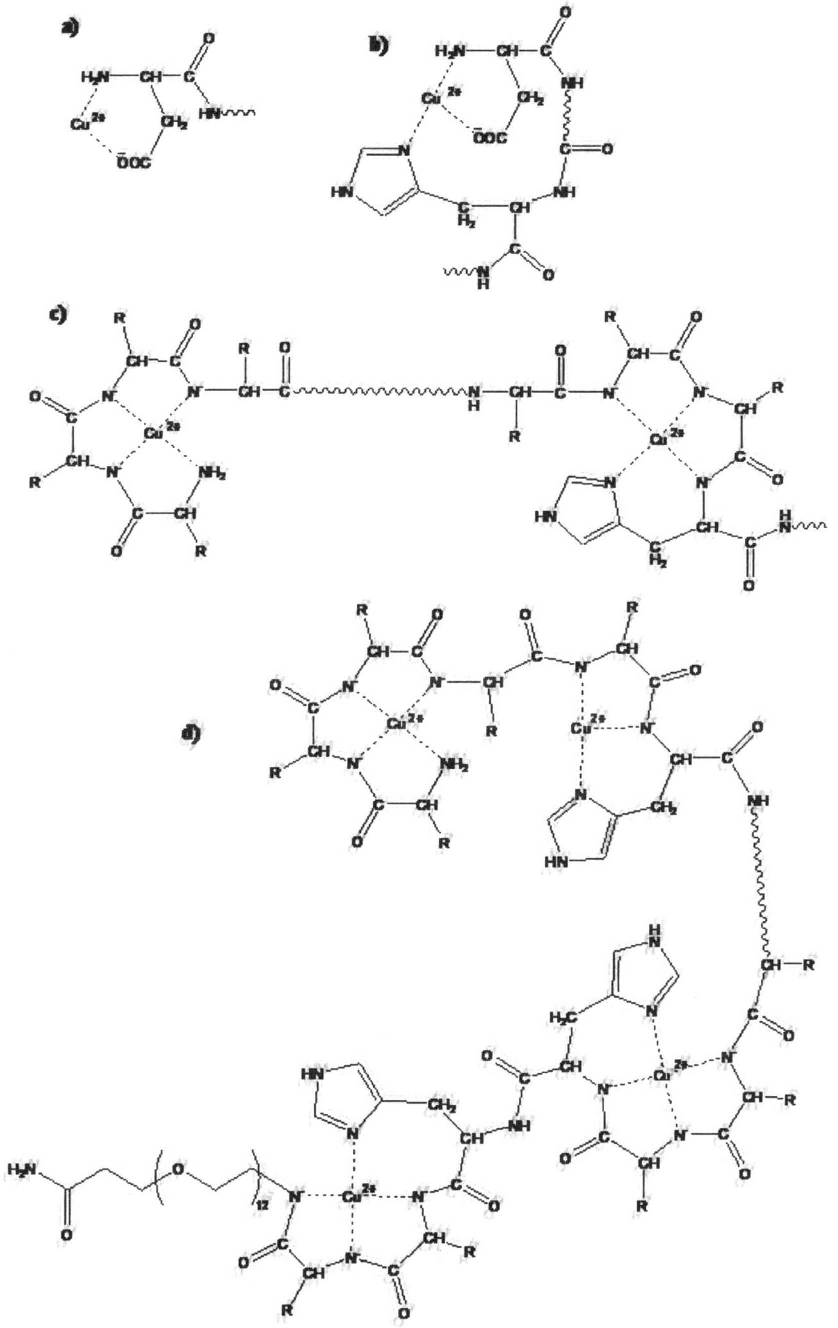

Scheme 6.1

carboxylate functions of aspartyl residues in coordination (Scheme 6.1.a). The copper(II) complexes with the mutated Aβ(1-16)Y10A show similar stability constant values to those of the analogous copper(II) complexes with the wild type peptide. In addition, the same spectroscopic features have been found for the metal complexes both with Aβ(1-16)Y10A and those with Aβ(1-16)PEG in the whole pH range investigated. Thus, we can conclude that the tyrosine residue is not involved in the copper(II) binding of the N-terminal fragment of the peptide. The deprotonation of the side chain imidazole results in the formation of macrochelates, of which several isomers can exist, because one or more histidyl residues can occupy the remaining coordination sites (Scheme 6.1.b). Further increase of pH results in the deprotonation and metal ion coordination of the amide functions subsequent to the amino group. The final species is a 4N complex with (NH_2, N^-, N^-, N^-) coordination mode, which is a single species with the small tetrapeptide fragment Aβ(1-4), but it is formed in overlapping processes with the metal binding of the histidyl sites in the case of Aβ(1-6) and especially Aβ(1-16)PEG. A series of the coordination isomers of the mononuclear complexes can exist under these conditions, but the CD spectra unambiguously prove the preference for the coordination at the N-terminus. Scheme 6.1.c illustrates the binding modes of a dinuclear species when the amino terminus and His13 are the metal binding sites. However, it is important to emphasize that several coordination isomers of this dinuclear species can exist because any of the three histidyl residues can be the anchoring site. The exclusive binding of histidyl residues is also possible, although a preference for the coordination via the N-terminus is found. Scheme 6.1.d represents the binding mode of the major species formed at a 4 : 1 metal to ligand ratio in slightly alkaline solution. It is obvious from Scheme 6.1.d that all four copper(II) ions are coordinated by nitrogen donor atoms, but the direct coordination environments of the four metal ions are slightly different.

Zinc(II) complexes of the same PEG-ylated hexadecapeptide (Aβ(1-16)PEG) have been studied under the same conditions by the combined application of potentiometric, NMR and ESI-MS techniques.[152] It is clear from these studies that the peptide can bind three equivalent zinc(II) ions in solution, in agreement with the assumption of some earlier studies.[154] The speciation of the equimolar system is shown by Figure 6.1, indicating the high ratio of dinuclear complexes at any pH value. It is also obvious from the speciation diagram that in the physiological pH range zinc(II) ions are almost completely bound by the peptide, reflecting the high zinc binding affinity of amyloid-β peptide. The comparison of the data obtained for Aβ(1-16)PEG with those reported for the short fragments, and the analysis of the NMR spectra, made it possible to assess the ratios of coordination isomers. A clear preference for binding at the internal histidyl sites (His13 and His14) was obtained from this comparison and this is just the opposite of the conclusions reported for the corresponding copper(II) complexes. Such an apparent difference in the coordination behavior of the two metal ions might have significant biological consequences in terms of amyloid-β aggregation and toxicity.

Nickel(II) ions are also frequent constituents of many living systems although the direct biological role of this metal ion has not yet been justified in the human

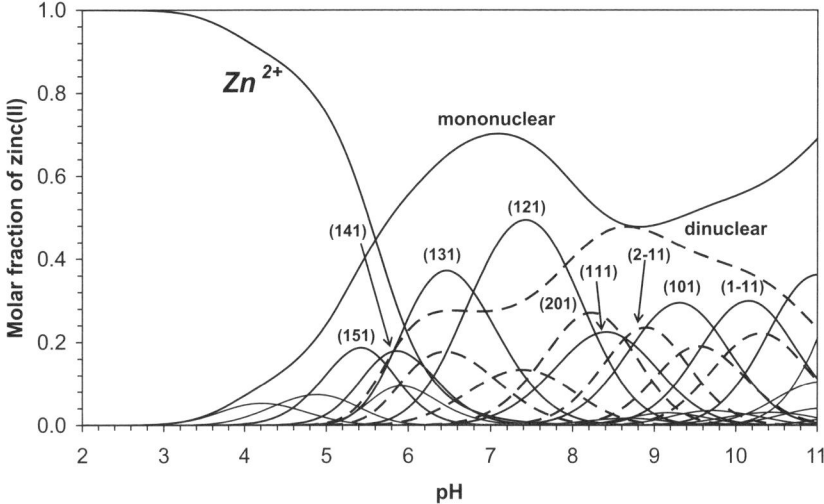

Figure 6.1   Metal ion speciation in the Zn(II)–Aβ(1-16)PEG system in equimolar samples ($c_{Zn} = c_L = 2$ mmol dm^{-3}).

body. At the same time, the coordination chemistry of the nickel(II) peptide systems is quite similar to that of the corresponding copper(II) complexes and, thus, nickel(II) can be a promising model to understand the binding modes of copper(II) complexes. Previous studies on the complexation between nickel(II) ions and amyloid peptides are scarce, but a recent paper suggests a strong interaction of nickel(II) ions with Aβ(1-40).[146] Nickel(II) complexes of Aβ(1-16)Y10A and its smaller fragments, including Aβ(1-4), Aβ(1-6), Ac-Aβ(1-6) and Ac-Aβ(8-16)Y10A, have been studied by potentiometric, UV-Vis and CD spectroscopic measurements in our laboratories.[153] It was found that the hexadecapeptide and its fragments are effective nickel(II) binding ligands and the complex formation processes of nickel(II) ions are quite similar to those of copper(II). Formation of mono- and di-nuclear complexes was detected in the nickel(II)–Aβ(1-16)Y10A system, suggesting the existence of two separated metal binding motifs: the N-terminus and internal histidyl residues. The preference for the coordination at the N-terminus was supported by the spectroscopic measurements but in equilibrium with the metal binding at the internal histidyl sites. The evaluation of CD spectra allowed the assessment of the ratio of coordination isomers, and a percentage of 78% and 22% of the metal ions was obtained for the binding at the N-terminus and the internal histidyl sites, respectively.

6.4.2   Formation of Mixed Metal Complexes of Amyloid-β Peptides

The studies on the copper(II), nickel(II) and zinc(II) complexes of amyloid-β peptides revealed that these peptides can bind more than one metal ion. It is also clear from these studies that the distribution of the different metal ions among the possible binding sites depends on the nature of the metal ions. The preference for binding at the N-terminal site was characteristic of copper(II) (both terminal amino and His6 residues) and nickel(II) (only the terminal amino group), while zinc(II) preferred the coordination at internal histidyl sites (His13 and 14) (Figure 6.2). The different preferences of the metal ions justify the studies on the mixed metal complexes of various peptide fragments.

Mixed metal systems containing nickel(II) and copper(II) ions can be more efficiently monitored by CD spectroscopy, because characteristic differences can be observed in the CD spectra of the various peptide fragments. In this case, both metal ions prefer the same binding sites (the N-terminus of the peptide), but the thermodynamic stabilities of copper(II) complexes are significantly higher than those of nickel(II). As a consequence, nickel(II) ions are forced to partially move to the internal histidyl sites, as demonstrated by Figure 6.3. The pH dependence of CD spectra of the ternary systems unambiguously prove that in the ternary system copper(II) is preferentially bound at the N-terminus while the high majority of nickel(II) ions are bonded at the internal histidyl sites.

The distribution of the metal ions in the 4-component copper(II)–nickel(II)–zinc(II)–Aβ(1-16)Y10A system is very complicated because of the high number of the possible coordination isomers. The most important finding in that respect is linked to the observation that in equimolar solution of the four components the peptide is able to keep all three metal ions in solution, suggesting that both the N-terminus (amino and His6) and the internal histidyl residues (His13 and His14) can work as effective dinuclear binding motifs.

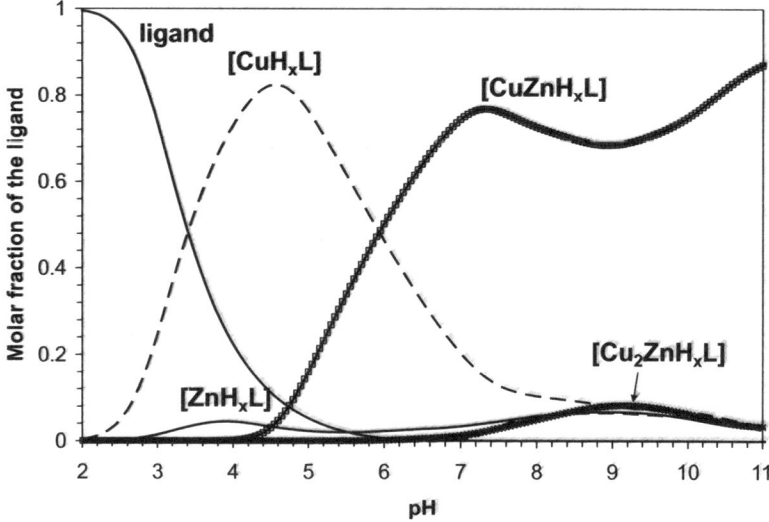

Figure 6.2 Species distribution diagram of the complexes formed in the zinc(II)–copper(II)–Aβ(1-16)PEG system ($c_{Zn} = c_{Cu} = c_L = 2$ mmol dm^{-3}).

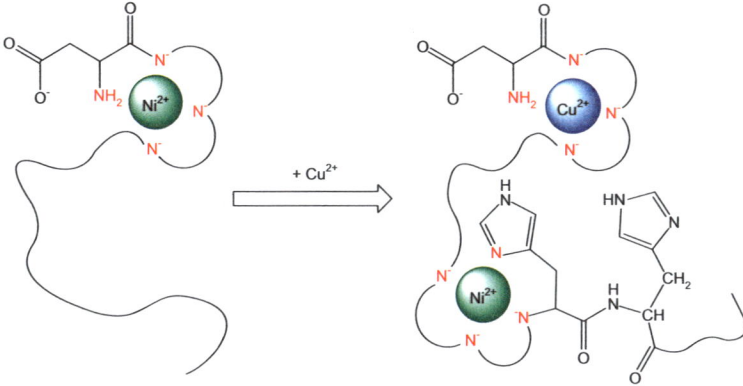

Figure 6.3 The addition of copper(II) to the nickel(II) complex of Aβ(1-16)Y10A at pH 9 shifts nickel(II) to the internal histidyl sites and the N-terminus is occupied by copper(II).

Scheme 6.2 is used to demonstrate the major binding modes of various species predominating in the binary and mixed metal systems. The structures A and B represent the N-terminal, while C and D show the internal histidyl binding motifs. For example, in the binary nickel(II)-Aβ(1-16)Y10A system the single nickel(II) ion of the mononuclear complex can be at either the A or C structural motif, with a preference for A. The dinuclear nickel(II) complex is obtained by the simple addition of these sites (A-C).

As concerns the mixed metal systems, neither zinc(II) nor nickel(II) can substitute for copper(II) in the complexes of Aβ(1-16)Y10A but both metal ions are able to alter the distribution of copper(II) ions among the various binding sites. In Scheme 2 the dinuclear Cu(II)–Ni(II)–Aβ(1-16)Y10A complex can be described as A–C after a simple substitution of Ni(II) with Cu(II) in structure A. Similarly, the dinuclear Cu(II)–Zn(II)–Aβ(1-16)Y10A complex can be constructed as A-D, replacing Ni(II) with Cu(II) in structure A. A series of coordination isomers can exist in the quaternary copper(II)–nickel(II)–zinc(II)–Aβ(1-16)Y10A system. The binding of copper(II) is favored by the N-terminus, while zinc(II) always prefers the binding at the multihistidyl sites. In the quaternary systems, the complex formation with nickel(II) takes place in alkaline samples and this metal ion will occupy the remaining free coordination sites. As a consequence, the major species in the equimolar samples of copper(II)–nickel(II)–zinc(II)–Aβ(1-16)Y10A system can be described as B-D and A-D in Scheme 6.2. Further increase of copper(II) and/or zinc(II) ions, however, liberates free nickel(II) ions which will be hydrolyzed under these alkaline conditions. A series of different experimental techniques was used to study the ternary copper(II)–zinc(II)–Aβ(1-16)-PEG system.[155] The zinc(II) ion is spectroscopically silent, therefore only the small changes observed in the UV-Vis, CD and/or EPR spectra can be used for the evaluation of data. It was shown that zinc(II) cannot completely substitute for copper(II) but can alter its

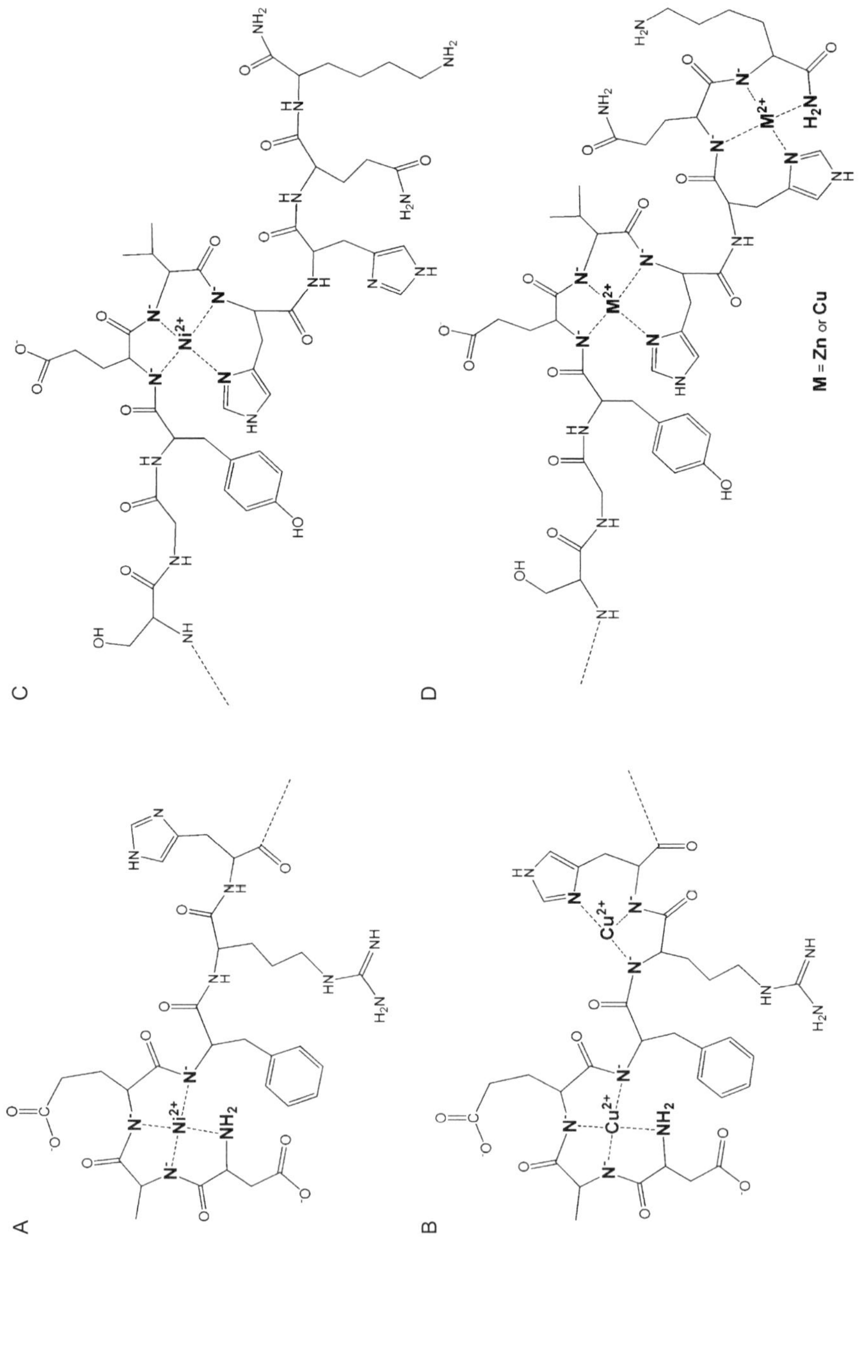

Scheme 6.2

distribution among the available binding sites. Similar observations were reported earlier for the ternary copper(II)–zinc(II)–prion protein system, too.[156] In the case of the copper(II)–zinc(II)–Aβ(1-16)-PEG system it was the major conclusion that zinc(II) prefers the coordination at the histidyl sites and shifts copper(II) towards the N-terminus. The species distribution diagram of the Cu(II)–Zn(II)–Aβ(1-16)PEG system is shown in Figure 6.2, which indicates the greater tendency of copper(II) than zinc(II) to use peptide binding. It is also clear from Figure 6.2 that copper(II) is almost completely bound while zinc(II) is free in the pH range 4–5, while both metal ions are involved in the mixed metal complexes around the physiological pH. A consistent feature of AD is that the affected brain is under a chronic oxidative stress load. Brain regions with higher levels of Aβ show elevated markers of oxidative stress,[157–161] which may involve metal dyshomeostasis.[162,163] *In vitro* studies have shown that Aβ interacts with copper(II), reducing it to copper(I) with H_2O_2 produced as a by-product,[164] but only in the presence of a very large excess of ascorbate,[165] or by electrochemical reduction of copper(II).[166] The H_2O_2 can then react with the copper(I) bound to Aβ, generating •OH radicals,[167] which can cause increased oxidative damage and promote further Aβ aggregation.[168–170] Recently, direct reactivity of Cu^+–Aβ peptide fragments (Aβ6-14, Aβ10-14 and Aβ12-14Val12Phe) with O_2 to produce H_2O_2 from the oxygenated copper(I)-peptide solution has been reported.[171] All three peptides, whether including the third His residue (His6) or not, or including the potentially reactive Tyr10 or not, produce the same amount of H_2O_2 with the same formation rate. The metal ion adopts a two coordinate His13–Cu–His14 environment in the solid state and in aqueous solution, and it is thought to be responsible for Cu–Aβ redox chemistry; a similar linear *bis*-His coordination has also been reported in a Cu^+–Aβ1–16 complex.[172] Zn^{2+} can promote Aβ aggregation and plaque formation, and this activity may be protective since Zn^{2+} has been reported to attenuate Aβ toxicity in cortical cultures.[173] The antioxidant activity includes the ability both to compete with Cu for Aβ binding and, thereby, to inhibit Aβ redox chemistry.[70] Although we explored high zinc(II) to metal ratios to mimic the brain levels of the two metal ions, our results show that the Zn^{2+} is not able to completely substitute for the second ion. More interestingly, the combined potentiometric and spectroscopic data here reported indicate that copper is shifted from binding to the two His residues, failing to stabilize the coordination of His13 and His14, which was previously invoked to explain the redox activity of copper(I)–Aβ. The formation of ternary metal complexes may justify the protective role of zinc(II) in comparison with copper(I), in addition to the recent proposed role of metallothionein-3 (MT3) in reducing Cu–Aβ induced toxicity towards cell cultures by a metal swap, involving the reduction and sequestration of the Cu^+ ion in MT3 and concomitant transfer of a Zn^{2+} ion from MT3 to the Aβ peptide.[174]

6.5 Metal Ion-Induced Aggregation Polymorphism and Neurotoxicity in Aβ

As stressed above, metal ions such as copper(II) and zinc(II) contribute to the neuropathology associated with Aβ fibrils, by affecting the rate of fibril formation,[175] by modifying fibril morphology,[176–178] and by direct chemical reaction with Aβ.[170] Beside this emerging general consensus, controversy still exists in the literature about the effect of Zn^{2+} or Cu^{2+} binding on the aggregation of Aβ, the morphologies of the aggregates as well as the related neurotoxicity.[179] Reports of accelerating and inhibiting effects have been published for both these metals. In some cases metal ion-induced Aβ aggregation was found to depend on pH, where environmental conditions reproducing acidosis (as seen in inflammation) led to a marked metal-induced aggregation of the peptide.[147] Other studies stressed that contamination with trace metals such as copper, zinc or iron can initiate the seeding process and Aβ oligomerization, eventually leading to Aβ amyloid plaques;[38,180] whereas it has been shown by other authors that copper(II) may reduce the Aβ amyloid burden *in vivo*[74,125] and inhibit the aggregation of Aβ1-42 *in vitro*.[181] Further *in vitro* and *in vivo* studies have shown that physiological levels of zinc(II) and copper(II) promote the aggregation process of Aβ and influence the morphology of the aggregates, which were described to be more amorphous, *i.e.* containing no or fewer fibrils.[116,182–184] The morphology of aggregated amyloid-β depends on the concentration of Cu^{2+} ions, as distinct differences in the coordination of Cu^{2+} ions to amyloid-β were observed by electron spin resonance as a function of the increasing metal concentration. The results suggest a correlation between specific Cu^{2+} ion coordination and the overall morphology of aggregates.[185] By using peptide fragments, and related mutants, from the central region of Aβ (*i.e.* Aβ(13-21)), which contains residues His13 and His14 implicated in Aβ metal-ion binding, Lynn and co-workers showed that differences in the kinetics of aggregation, morphology and neurotoxicity of the aggregates can be related to switching between different metal binding modes that in turn differently compromise neuron viability.[186] Other studies reported that copper ions strongly inhibit zinc-induced Aβ aggregation and fibrillogenesis,[187] whereas experimental evidence suggests that the overall morphology of the aggregates also depends on the concentrations (and then on the metal to peptide ratio) of these metal ions, as well as on the coexistence of mixed copper(II) zinc(II) complex species.[177,188,189] Also, the zinc(II) appears to have contradictory effects on Aβ activity: at high concentration it was shown to promote Aβ-induced toxicity both *in vitro*[190] and *in vivo*.[191] However, other experimental work revealed that Zn(II), along with Cu(II), inhibited the β-aggregation (which leads to fibrillogenesis) of the Aβ peptides in a concentration-dependent manner.[192] Secondary structural analysis and microscopic studies revealed that metals induced Aβ to form non-fibrillar aggregates by disrupting β-sheet formation. Interestingly, the ability of Zn(II) and Cu(II) to diminish the formation of β-aggregates resulted in protection of neurons against Aβ-associated cytotoxicity.[192] In agreement with these results, other studies reported that Cu(II) and Zn(II) inhibit Aβ fibrillization and initiate formation of non-fibrillar Aβ aggregates, but in this case the Cu(II)–Aβ aggregates were revealed to be neurotoxic *in vitro* only in the presence of ascorbate, whereas Aβ monomers and Zn(II)–Aβ aggregates were non-toxic.[193] Finally, alternative studies that report increased toxicity at high (1 mM) Zn(II) concentrations also report a neuroprotective effect at lower (<50 μM) concentrations.[194,195] This fact, termed the zinc paradox,[196] has been explained by other authors who consider that Zn(II) at a concentration of a few micromolar, which is too low to affect the precipitation equilibrium of Aβ, can destabilize soluble amyloid-β aggregates by accelerating their precipitation, thus abolishing Aβ toxicity.[197] From the above, it is concluded that transition metals may contribute both directly and indirectly to the pathogenesis of AD, suggesting that neurotoxicity generated by metal ions is more complex than the kinetics of Aβ aggregation. It is now well established that copper(II) and zinc(II) can bind to Aβ in the N-terminal region encompassing amino acid residues 1–16, and the involvement of the histidyl residues in the coordination of metal ions has been strongly suggested.[135,198] In particular, coordination to His13 and His14, and to a lesser extent His6, was found to be necessary for metal-induced Aβ aggregation.[193,199] Interestingly, the structural characterization of Aβ within amyloid fibrils has shown that the N-terminal hydrophilic region of Aβ constitutes the outer wall of the fibrils.[200] Thus it appears to remain accessible for interactions with metal ions, even after amyloid deposition, suggesting that the N-terminal region of Aβ is not involved in the β-sheet network of the amyloid fibril, but may contribute to fibril stability by participating in protofilament packing.[201] This is also in agreement with the results reported by Szalai, who showed that a stoichiometric amount of copper(II) can bind to the same mononuclear coordination environment regardless of the oligomeric state of Aβ.[202–204] On the other hand, Aβ complex formation (and stoichiometry) appears to be strongly dependent on both concentration and metal/peptide ratio.[131] The coordination features of the different complex species change as a function of the metal/peptide stoichiometry.[151,152] Assuming that copper(II) or zinc(II) metal binding sites and coordination modes with the full-length Aβ are identical to those observed with the model peptide Aβ(1-16), it becomes possible to attempt a correlation between the structure of the different complex species present at physiological pH, in the various metal to ligand ratios, the different morphologies of the aggregates and the diverse levels of toxicity of the metal/peptide aggregates according to Scheme 6.3, shown below.

In particular, based on our previous studies demonstrating that the Aβ N-terminus can coordinate up to four copper ions or three zinc ions,[151,152] it can be hypothesized that when an equivalent amount of copper(II) is present, two imidazoles, one from His6 and another one provided by His13 or His14, plus the terminal amino group and carboxyl side chain of Asp1, are around the metal ion in the form of a macrochelate complex. Such a plastic structure

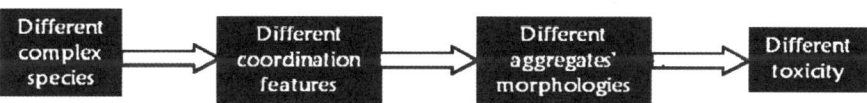

Scheme 6.3

may allow copper(II) to undergo easy redox reactions, thus explaining toxic effects due to ROS production.[205] As the metal to ligand ratio increases the macrochelate is disrupted and copper(II) ions distribute between the amino terminus, His6, and the His13/His14 dyad. Contextually deprotonated peptide nitrogens begin to take part in the coordination sphere of the metals. This new situation renders the metal ions more resistant to redox chemistry and at the same time the structuring effect introduced by the metal ions uploading may affect the morphology of the aggregates, which become no longer prone to fibrillization. In contrast to what was found for the copper(II) ion, which shows a high affinity for the N-terminal part of the Aβ(1-16) sequence, the zinc(II) clearly shows a preference to bind to the His domain toward the C-terminal part of the Aβ(1-16) peptide fragment (*i.e.* His13/His14 residues). Such an apparent difference might have biological consequences, in view of what was reported above about Zn(II)-induced Aβ aggregation morphology and toxicity. Our results establish that the N-terminal region of Aβ can give access also to different Zn(II) coordination environments that change under slightly different experimental conditions.[152] In particular, the stability constant values of the zinc(II)–Aβ(1-16) system show which species form at different metal to ligand ratios and pH values, thereby contributing to answering the question of the formation and survival of different zinc(II) complexes with Aβ in different brain areas, where different concentration values of both Aβ and zinc(II) can occur. Thus, it is very likely that the different metal complex species can lead to profound changes in Aβ self-assembly, morphology, and neurotoxicity. In order to get further insight into this issue, we recently started to investigate the effect of copper and zinc concentration on the aggregation kinetics of the full-length Aβ1-42. We used CD spectroscopy to monitor Aβ1-42 aggregation as a function of copper or zinc concentration. In particular we monitored the decrease of the CD signal at 222 nm over the time of incubation in the presence or in the absence of metals. Our preliminary results suggest that the higher the metal concentration is, the faster the CD signal decreases because of aggregation and precipitation of the peptide. In agreement with the notion that metal complexation induces the formation of more amorphous aggregates, as compared to those formed in the absence of metals, the Th-T fluorescence markedly decreased in the samples containing copper(II) or zinc(II), thus suggesting lower content of amyloid structures. In addition, and in keeping with literature data, it seems that zinc is more effective in precipitating the peptide and produces a greater amount of amorphous aggregates.[206] Overall, the observations described above lead to the conclusion that the inherent polymorphic nature of Aβ assemblies can be associated with different toxic behavior. Consequently, there is a need to delineate the structural variability underlying the formation and stabilities of soluble Aβ oligomers. In this regard the elucidation of the chemistry through which transition-metal ions participate in the assembly and toxicity of Aβ oligomers is important to drug design efforts if inhibition of Aβ-containing bound metal ions becomes a treatment for AD.

Acknowledgments

This work was supported by MIUR, FIRB RBPR05JH2P ITALNANONET, the MTA(Hungary)-CNR(Italy) bilateral program and OTKA 77586 (Hungary) and CNR RSTL 620.

References

1. M. P. Mattson, *Nature*, 2004, **430**, 631–639.
2. L. Migliore and F. Coppede, *Mutat. Res.*, 2009, **667**, 82–97.
3. C. L. Masters, G. Simms, N. A. Weinman, G. Multhaup, B. L. McDonald and K. Beyreuther, *Proc. Nat. Acad. Sci. USA*, 1985, **82**, 4245–4249.
4. J. Avila, *FEBS Lett.*, 2006, **580**, 2922–2927.
5. A. Nunomura, G. Perry, G. Aliev, K. Hirai, A. Takeda, E. K. Balraj, P. K. Jones, H. Ghanbari, T. Wataya, S. Shimohama, S. Chiba, C. S. Atwood, R. B. Petersen and M. A. Smith, *J. Neur. and Exp. Neurol.*, 2001, **60**, 759–767.
6. A. I. Bush, *Trends in Neurosciences*, 2003, **26**, 207–214.
7. C. J. Maynard, A. I. Bush, C. L. Masters, R. Cappai and Q. X. Li, *Int. J. Exp. Path.*, 2005, **86**, 147–159.
8. P. Seubert, C. Vigo-Pelfrey, F. Esch, M. Lee, H. Dovey, D. Davis, S. Sinha, M. Schlossmacher, J. Whaley, C. Swindlehurst, R. McCormack, R. Wolfert, D. Selkoe, I. Lieberburg and D. Schenk, *Nature*, 1992, **359**, 325–327.
9. J. S. Whitson, C. G. Glabe, E. Shintani, A. Abcar and C. W. Cotman, *Neurosci Lett.*, 1990, **110**, 319–324.
10. J. S. Whitson, D. J. Selkoe and C. W. Cotman, *Science*, 1989, **243**, 1488–1490.
11. B. A. Yankner, L. K. Duffy and D. A. Kirschner, *Science*, 1990, **250**, 279–282.
12. C. Hilbich, B. Kisters-Woike, J. Reed, C. L. Masters and K. Beyreuther, *J. Mol. Biol.*, 1991, **218**, 149–163.
13. J. T. Jarrett, E. P. Berger and P. T. Lansbury, *Biochemistry*, 1993, **32**, 4693–4697.
14. L. F. Lue, Y. M. Kuo, A. E. Roher, L. Brachova, Y. Shen, L. Sue, T. Beach, J. H. Kurth, R. E. Rydel and J. Rogers, *Am. J. Pathol.*, 1999, **155**, 853–862.
15. C. McLean, R. Cherny, F. Fraser, S. Fuller, M. Smith, K. Beyreuther, A. Bush and C. Masters, *Ann. Neurol.*, 1999, **46**, 860–866.
16. J. Wang, D. W. Dickson, J. Q. Trojanowski and V. M. Lee, *Exp. Neurol.*, 1999, **158**, 328–337.
17. V. N. Uversky, *Chem. Rev.* 2010; in press, doi: 10.102/cr100186d.
18. E. T. Powers, R. I. Morimoto, A. Dillin, J. W. Kelly and W. E. Balch, *Annu. Rev. Biochem.*, 2009, **78**, 959–991.
19. E. A. Kikis, T. Gidalevitz and R. I. Morimoto, *Adv. Exp. Med. Biol.*, 2010, **694**, 138–159.

20. A. Nunomura, G. Perry, G. Aliev, K. Hirai, A. Takeda, E. K. Balraj, P. K. Jones, H. Ghanbari, T. Wataya, S. Shimohama, S. Chiba, C. S. Atwood, R. B. Petersen and M. A. Smith, *J. Neuropathol. Exp. Neurol.*, 2001, **60**, 759–767.

21. A. Nunomura, G. Perry, M. A. Pappolla, R. P. Friedland, K. Hirai, S. Chiba and M. A. Smith, *J. Neuropathol. Exp. Neurol.*, 2000, **59**, 1011–1017.

22. A. Nunomura, G. Perry, M. A. Pappolla, R. Wade, K. Hirai, S. Chiba and M. A. Smith, *J. Neurosci.*, 1999, **19**, 1959–1964.

23. A. I. Bush and R. E. Tanzi, *Neurotherapeutics*, 2008, **5**, 421–432.

24. T. P. Flaten, *Brain Res. Bull.*, 2001, **55**, 187–196.

25. C. J. Frederickson, *Int. Rev. Neurobiol.*, 1989, **31**, 145–328.

26. D. R. Brown and H. Kozlowski, *Dalton Trans.*, 2004, 1907–1917.

27. J. F. Mercer, *Trends Mol. Med.*, 2001, **7**, 64–69.

28. G. Torsdottir, J. Kristinsson, S. Sveinbjornsdottir, J. Snaedal and T. Johannesson, *Pharmacol. Toxicol.*, 1999, **85**, 239–243.

29. H. Basun, L. G. Forssell, L. Wetterberg and B. Winblad, *J. Neural. Transm. Park. Dis. Dement. Sect.*, 1991, **3**, 231–258.

30. R. Squitti, D. Lupoi, P. Pasqualetti, G. Dal Forno, F. Vernieri, P. Chiovenda, L. Rossi, L. Cortesi, M. Cassetta and P. M. Rossini, *Neurology*, 2002, **59**, 1153–1161.

31. R. Squitti, P. Pasqualetti, G. Dal Forno, F. Moffa, E. Cassetta, D. Lupoi, F. Vernieri, L. Rossi, M. Baldassini and P.M. Rossini, *Neurology*, 2005, **64**, 1040–1046.

32. R. Squitti, P. M. Rossini, E. Cassetta, F. Moffa, P. Pasqualetti, M. Cortesi, A. Colloca, L. Rossi and A. Finazzi-Agro, *Eur. J. Clin. Invest.*, 2002, **32**, 51–59.

33. E. Nakano, M. P. Williamson, N. H. Williams and H. J. Powers, *Biochim. Biophys. Acta*, 2004, **1688**, 33–42.

34. S. Seshadri, A. Beiser, J. Selhub, P. F. Jacques, I. H. Rosenberg, R. B. D'Agostino, P. W. Wilson and P. A. Wolf, *N. Engl. J. Med.*, 2002, **346**, 476–483.

35. M. A. Lovell, J. D. Robertson, W. J. Teesdale, J. L. Campbell and W. R. Markesbery, *J. Neurol. Sci.*, 1998, **158**, 47–52.

36. M. A. Deibel, W. D. Ehmann and W. R. Markesbery, *J. Neurol. Sci.*, 1996, **143**, 137–142.

37. D. A. Loeffler, P. A. LeWitt, P. L. Juneau, A. A. Sima, H. U. Nguyen, A. J. DeMaggio, C. M. Brickman, G. J. Brewer, R. D. Dick, M. D. Troyer and L. Kanaley, *Brain Res.*, 1996, **738**, 265–274.

38. D. L. Sparks and B. G. Schreurs, *Proc. Natl. Acad. Sci. USA*, 2003, **100**, 11065–11069.

39. M. C. Morris, D. A. Evans, C. C. Tangney, J. L. Bienias, J. A. Schneider, R. S. Wilson and P. A. Scherr, *Arch. Neurol.*, 2006, **63**, 1085–1088.

40. B. L. Vallee, J. E. Coleman and D. S. Auld, *Proc. Natl. Acad. Sci. USA*, 1991, **88**, 999–1003.

41. Z. Szallasi, K. Bogi, S. Gohari, T. Biro, P. Acs and P. M. Blumberg, *J. Biol. Chem.*, 1996, **271**, 18299–18301.

42. M. Ebadi, M. A. Elsayed and M. H. Aly, *Biol. Signals*, 1994, **3**, 123–126.
43. C. J. Frederickson, *Int. Rev. Neurobiol.*, 1989, **31**, 145–238.
44. S. Sensi, P. Paoletti, A. I. Bush and I. Sekler, *Nat. Rev. Neurosci.*, 2009, **10**, 780–791.
45. C. J. Frederickson, J. Y. Koh and A. I. Bush, *Nat. Rev. Neurosci.*, 2005, **6**, 449–462.
46. R. Squitti, M. Ventriglia, G. Barbati, E. Cassetta, F. Ferreri, G. Dal Forno, S. Ramires, F. Zappasodi and P. Rossini, *J. Neur. Transm.*, 2007, **114**, 1589–1594.
47. D. B. Milne and P. E. Johnson, *Clin. Chem.*, 1993, **39**, 883–887.
48. A. Madaric, E. Ginter and J. Kadrabova, *Physiol. Res.*, 1994, **43**, 107–111.
49. C. Ekmekcioglu, *Nahrung*, 2001, **45**, 309–316.
50. M. Iskra, J. Patelski and W. Majewski, *J. Trace Elem. Electrolytes Health Dis.*, 1993, **7**, 185–188.
51. D. McMaster, E. McCrum, C. C. Patterson, M. M. Kerr, D. O'Reilly, A. E. Evans and A. H. Love, *Am. J. Clin. Nutr.*, 1992, **56**, 440–446.
52. A. Menditto, G. Morisi, A. Alimonti, S. Caroli, F. Petrucci, A. Spagnolo and A. Menotti, *J. Trace Elem. Electrolytes Health Dis.*, 1993, **7**, 251–253.
53. D. B. Milne and P. E. Johnson, *Clin. Chem.*, 1993, **39**, 883–887.
54. C. J. Maynard, R. Cappai, I. Volitakis, R. A. Cherny, C. L. Masters, Q. X. Li and A. I. Bush, *J. Inorg. Biochem.*, 2006, **100**, 952–962.
55. C. J. Maynard, R. Cappai, I. Volitakis, R. A. Cherny, A. R. White, K. Beyreuther, C. L. Masters, A. I. Bush and Q. X. Li, *J. Biol. Chem.*, 2002, **277**, 44670–44676.
56. M. Wender, J. Szczech, S. Hoffmann and W. Hilczer, *Neuropatol. Pol.*, 1992, **30**, 65–72.
57. H. Basun, L. G. Forssell, L. Wetterberg and B. Winblad, *J. Neural. Transm. Park Dis. Dement. Sect.*, 1991, **3**, 231–258.
58. L. Baum, I. H. Chan, S. K. Cheung, W. B. Goggins, V. Mok, L. Lam, V. Leung, E. Hui, C. Ng, J. Woo, H. F. Chiu, B. C. Zee, W. Cheng, M. H. Chan, S. Szeto, V. Lui, J. Tsoh, A. I. Bush, C. W. Lam and T. Kwok, *Biometals*, 2010, **23**, 173–179.
59. J. A. Molina, F. J. Jimenez-Jimenez, M. V. Aguilar, I. Meseguer, C. J. Mateos-Vega, M. J. Gonzalez-Munoz, F. de Bustos, J. Porta, M. Orti-Pareja, M. Zurdo, E. Barrios and M. C. Martinez-Para, *J. Neural. Transm.*, 1998, **105**, 479–488.
60. A. L. Friedlich, J. Y. Lee, T. van Groen, R. A. Cherny, I. Volitakis, T. B. Cole, R. D. Palmiter, J. Y. Koh and A. I. Bush, *J. Neurosci.*, 2004, **24**, 3453–3459.
61. S. W. Suh, K. B. Jensen, M. S. Jensen, D. S. Silva, P. J. Kesslak, G. Danscher and C. J. Frederickson, *Brain Res.*, 2000, **852**, 274–278.
62. J. L. Smith, S. Xiong and M. A. Lovell, *Neurotoxicology*, 2006, **27**, 1–5.
63. V. W. Bunker, L. J. Hinks, M. F. Stansfield, M. S. Lawson and B. E. Clayton, *Am. J. Clin. Nutr.*, 1987, **46**, 353–359.
64. E. Martinez Lista, J. Sole, L. Arola and A. Mas, *Biol. Neonate*, 1993, **64**, 47–52.

65. A. L. Monget, P. Galan, P. Preziosi, H. Keller, C. Bourgeois, J. Arnaud, A. Favier and S. Hercberg, *Int. J. Vitam. Nutr. Res.*, 1996, **66**, 71–76.

66. H. N. Munro, P. M. Suter and R. M. Russell, *Annu. Rev. Nutr.*, 1987, **7**, 23–49.

67. G. Ravaglia, P. Forti, F. Maioli, B. Nesi, L. Pratelli, L. Savarino, D. Cucinotta and G. Cavalli, *J. Clin. Endocrinol. Metab.*, 2000, **85**, 2260–2265.

68. P. A. Adlard, J. M. Parncutt, D. I. Finkelstein and A. I. Bush, *J. Neurosci.*, 2010, **30**, 1631–1636.

69. R. Rajendran, M. Q. Ren, M. D. Ynsa, G. Casadesus, M. A. Smith, G. Perry, B. Halliwell and F. Watt, *Biochem. Biophys. Res. Comm.*, 2009, **382**, 91–95.

70. D. G. Smith, R. Cappai and K. J. Barnham, *Biochim. Biophys. Acta-Biomembranes*, 2007, **1768**, 1976–1990.

71. E. Spisni, M. C. Valerii, M. Manerba, A. Strillacci, E. Polazzi, T. Mattia, C. Griffoni and V. Tomasi, *NeuroToxicology*, 2009, **30**, 605–612.

72. D. Strozyk, L. J. Launer, P. A. Adlard, R. A. Cherny, A. Tsatsanis, I. Volitakis, K. Blennow, H. Petrovitch, L. R. White and A. I. Bush, *Neurobiology of Aging*, 2009, **30**, 1069–1077.

73. P. A. Adlard and A. I. Bush, *J. Alzheimers Dis.*, 2006, **10**, 145–163.

74. T. A. Bayer, S. Schafer, A. Simons, A. Kemmling, T. Kamer, R. Tepest, A. Eckert, K. Schussel, O. Eikenberg, C. Sturchler-Pierrat, D. Abramowski, M. Staufenbiel and G. Multhaup, *Proc. Natl. Acad. Sci. USA*, 2003, **100**, 14187–14192.

75. H. Kessler, T. A. Bayer, D. Bach, T. Schneider-Axmann, T. Supprian, W. Herrmann, M. Haber, G. Multhaup, P. Falkai and F. G. Pajonk, *J. of Neur. Transmission*, 2008, **115**, 1181–1187.

76. P. A. Adlard, R. A. Cherny, D. I. Finkelstein, E. Gautier, E. Robb, M. Cortes, I. Volitakis, X. Liu, J. P. Smith, K. Perez, K. Laughton, Q. X. Li, S. A. Charman, J. A. Nicolazzo, S. Wilkins, K. Deleva, T. Lynch, G. Kok, C. W. Ritchie, R. E. Tanzi, R. Cappai, C. L. Masters, K. J. Barnham and A. I. Bush, *Neuron*, 2008, **59**, 43–55.

77. P. J. Crouch, L. W. Hung, P. A. Adlard, M. Cortes, V. Lal, G. Filiz, K. A. Perez, M. Nurjono, A. Caragounis, T. Du, K. Laughton, I. Volitakis, A. I. Bush, Q. X. Li, C. L. Masters, R. Cappai, R. A. Cherny, P. S. Donnelly, A. R. White and K. J. Barnham, *Proc. Natl. Acad. Sci. USA*, 2009, **106**, 381–386.

78. A. Odermatt and M. Solioz, *J. Biol. Chem.*, 1995, **270**, 4349–4354.

79. R. A. Pufahl, C. P. Singer, K. L. Peariso, S. J. Lin, P. J. Schmidt, C. J. Fahrni, V. C. Culotta, J. E. Penner-Hahn and T. V. O'Halloran, *Science*, 1997, **278**, 853–856.

80. J. S. Valentine and E. B. Gralla, *Science*, 1997, **278**, 817–818.

81. T. V. O'Halloran and V. C. Culotta, *J. Biol. Chem.*, 2000, **275**, 25057–25060.

82. L. A. Finney and T. V. O'Halloran, *Science*, 2003, **300**, 931–936.

83. H. S. Carr and D. R. Winge, *Acc. Chem. Res.*, 2003, **36**, 309–316.

84. S. Tottey, K. J. Waldron, S. J. Firbank, B. Reale, C. Bessant, K. Sato, T. R. Cheek, J. Gray, M. J. Banfield, C. Dennison and N. J. Robinson, *Nature*, 2008, **455**, 1138–1142.

85. L. Macomber and J. A. Imlay, *Proc. Natl. Acad. Sci. USA*, 2009, **106**, 8344–8349.

86. A. K. Boal and A. C. Rosenzweig, *Chem. Rev.*, 2009, **109**, 4760–4779.

87. N. J. Robinson and D. R. Winge, *Annu. Rev. Biochem.*, 2010, **79**, 537–562.

88. S. Tottey, P. R. Rich, S. A. Rondet and N. J. Robinson, *J. Biol. Chem.*, 2001, **276**, 19999–20004.

89. S. Tottey, S. A. Rondet, G. P. Borrelly, P. J. Robinson, P. R. Rich and N. J. Robinson, *J. Biol. Chem.*, 2002, **277**, 5490–5497.

90. L. Banci, I. Bertini, F. Cantini and S. Ciofi-Baffoni, *Cell Mol. Life Sci.*, 2010, **67**, 2563–2589.

91. B. Angeletti, K. J. Waldron, K. B. Freeman, H. Bawagan, I. Hussain, C. C. Miller, K. F. Lau, M. E. Tennant, C. Dennison, N. J. Robinson and C. Dingwall, *J. Biol. Chem.*, 2005, **280**, 17930–17937.

92. P. C. Wong, D. Waggoner, J. R. Subramaniam, L. Tessarollo, T. B. Bartnikas, V. C. Culotta, D. L. Price, J. Rothstein and J. D. Gitlin, *Proc. Natl. Acad. Sci. USA*, 2000, **97**, 2886–2891.

93. G. F. Brady, S. Galban, X. Liu, V. Basrur, J. D. Gitlin, K. S. Elenitoba-Johnson, T. E. Wilson and C. S. Duckett, *Mol. Cell. Biol.*, 2010, **30**, 1923–1936.

94. R. Vassar, D. M. Kovacs, R. Yan and P. C. Wong, *J. Neurosci.*, 2009, **29**, 12787–12794.

95. D. M. McLoughlin, C. L. Standen, K.-F. Lau, S. Ackerley, T. P. Bartnikas, J. D. Gitlin and C. C. J. Miller, *J. Biol. Chem.*, 2001, **276**, 9303–9307.

96. C. C. Miller, D. M. McLoughlin, K. F. Lau, M. E. Tennant and B. Rogelj, *Trends Neurosci.*, 2006, **29**, 280–285.

97. J. H. Lee, K. F. Lau, M. S. Perkinton, C. L. Standen, S. J. Shemilt, L. Mercken, J. D. Cooper, D. M. McLoughlin and C. C. Miller, *J. Biol. Chem.*, 2003, **278**, 47025–47029.

98. J. C. Mitchell, M. S. Perkinton, D. M. Yates, K. F. Lau, B. Rogelj, C. C. Miller and D. M. McLoughlin, *J. Alzheimers Dis.*, 2010, **20**, 31–36.

99. C. Iadecola, F. Zhang, K. Niwa, C. Eckman, S. K. Turner, E. Fischer, S. Younkin, D. R. Borchelt, K. K. Hsiao and G. A. Carlson, *Nat. Neurosci.*, 1999, **2**, 157–161.

100. T. B. Cole, H. J. Wenzel, K. E. Kafer, P. A. Schwartzkroin and R. D. Palmiter, *Proc. Natl. Acad. Sci. USA*, 1999, **96**, 1716–1721.

101. D. H. Linkous, J. M. Flinn, J. Y. Koh, A. Lanzirotti, P. M. Bertsch, B. F. Jones, L. J. Giblin and C. J. Frederickson, *J. Histochem. Cytochem.*, 2008, **56**, 3–6.

102. R. D. Palmiter, T. B. Cole, C. J. Quaife and S. D. Findley, *Proc. Natl. Acad. Sci. USA*, 1996, **93**, 14934–14939.

103. A. Takeda, A. Minami, S. Takefuta, M. Tochigi and N. Oku, *J. Neurosci. Res.*, 2001, **63**, 447–452.

104. W. Chowanadisai, S. L. Kelleher and B. Lonnerdal, *J. Nutr.*, 2005, **135**, 1002–1007.

105. G. Lyubartseva, J. L. Smith, W. R. Markesbery and M. A. Lovell, *Brain Pathol.*, 2010, **20**, 343–350.

106. J. Y. Lee, J. H. Kim, S. H. Hong, R. A. Cherny, A. I. Bush, R. D. Palmiter and J. Y. Koh, *J. Biol. Chem.*, 2004, **279**, 8602–8607.

107. T. B. Cole, H. J. Wenzel, K. E. Kafer, P. A. Schwartzkroin and R. D. Palmiter, *Proc. Natl. Acad. Sci. USA*, 1999, **96**, 1716–1721.

108. A. L. Friedlich, J. Y. Lee, T. van Groen, R. Cherny, I. Volitakis, T. B. Cole, R. D. Palmiter, J. Y. Koh and A. I. Bush, *J. Neurosci.*, 2004, **24**, 3453–3459.

109. J.-Y. Lee, T. B. Cole, R. D. Palmiter, S. W. Suh and J.-Y. Koh, *Proc. Natl. Acad. Sci. USA*, 2002, **99**, 7705–7710.

110. A. I. Bush and R. E. Tanzi, *Proc. Natl. Acad. Sci. USA*, 2002, **99**, 7317–7319.

111. J.-Y. Lee, J.-H. Kim, S. H. Hong, J. Y. Lee, R. A. Cherny, A. I. Bush, R. D. Palmiter and J.-Y. Koh, *J. Biol. Chem.*, 2004, **279**, 8602–8607.

112. M. A. Lovell, J. L. Smith and W. R. Markesbery, *J. Neuropathol. Exp. Neurol.*, 2006, **65**, 489–498.

113. S. Gandy, *J. Clin. Invest.*, 2005, **115**, 1121–1129.

114. Y. Ling, K. Morgan and N. Kalsheker, *Int. J. Biochem. Cell Biol.*, 2003, **35**, 1505–1535.

115. Y. H. Suh and F. Checler, *Pharmacol. Rev.*, 2002, **54**, 469–525.

116. A. I. Bush, W. H. Pettingell Jr., M. D. Paradis and R. E. Tanzi, *J. Biol. Chem.*, 1994, **269**, 12152–12158.

117. A. Caragounis, T. Du, G. F. Fi, K. M. Laughton, I. Volitakis, R. A. Sharples, R. A. Cherny, C. L. Masters, S. C. Drew, A. F. Hill, Q. X. Li, P. J. Crouch, K. J. Barnham and A. R. White, *Biochem. J.*, 2007, **407**, 435–450.

118. R. Z. Lin, X. H. Chen, W. M. Li, Y. F. Han, P. Q. Liu and R. B. Pi, *Neurosci. Lett.*, 2008, **440**, 344–347.

119. A. D. Armendariz, M. Gonzalez, A. V. Loguinov and C. D. Vulpe, *Physiol. Genomics*, 2004, **20**, 45–54.

120. S. A. Bellingham, D. K. Lahiri, B. Maloney, S. La Fontaine, G. Multhaup and J. Camakaris, *J. Biol. Chem.*, 2004, **279**, 20378–20386.

121. C. D. Davis, D. B. Milne and F. H. Nielsen, *Am. J. Clin. Nutr.*, 2000, **71**, 781–788.

122. S. A. Bellingham, G. D. Ciccotosto, B. E. Needham, L. R. Fodero, A. R. White, C. L. Masters, R. Cappai and J. Camakaris, *J. Neurochem.*, 2004, **91**, 423–428.

123. A. R. White, R. Reyes, J. F. Mercer, J. Camakaris, H. Zheng, A. I. Bush, G. Multhaup, K. Beyreuther, C. L. Masters and R. Cappai, *Brain Res.*, 1999, **842**, 439–444.

124. T. A. Bayer, S. Schafer, A. Simons, A. Kemmling, T. Kamer, R. Tepest, A. Eckert, K. Schussel, O. Eikenberg, C. Sturchler-Pierrat, D. Abramowski, M. Staufenbiel and G. Multhaup, *Proc. Natl. Acad. Sci. USA*, 2003, **100**, 14187–14192.

125. A. L. Phinney, B. Drisaldi, S. D. Schmidt, S. Lugowski, V. Coronado, Y. Liang, P. Horne, J. Yang, J. Sekoulidis, J. Coomaraswamy, M. A. Chishti, D. W. Cox, P. M. Mathews, R. A. Nixon, G. A. Carlson, P. St George-Hyslop and D. Westaway, *Proc. Natl. Acad. Sci. USA*, 2003, **100**, 14193–14198.

126. H. Kessler, F. G. Pajonk, P. Meisser, T. Schneider-Axmann, K. H. Hoffmann, T. Supprian, W. Herrmann, R. Obeid, G. Multhaup, P. Falkai and T. A. Bayer, *J. Neur. Transm.*, 2006, **113**, 1763–1769.

127. D. H. Linkous, P. A. Adlard, P. B. Wanschura, K. M. Conko and J. M. Flinn, *J. Alzheimers Dis.*, 2009, **18**, 565–579.

128. M. Stoltenberg, A. I. Bush, G. Bach, K. Smidt, A. Larsen, J. Rungby, S. Lund, P. Doering and G. Danscher, *Neuroscience*, 2005, **150**, 357–369.

129. A. Venti, T. Giordano, P. Eder, A. I. Bush, D. K. Lahiri, N. H. Greig and J. T. Rogers, *Ann. N. Y. Acad. Sci.*, 2004, **1035**, 34–48.

130. R. Roychaudhuri, M. Yang, M. M. Hoshi and D. B. Teplow, *J. Biol. Chem.*, 2009, **284**, 4749–4753.

131. P. Faller and C. Hureau, *Dalton Trans.*, 2009, 1080–1094.

132. H. Kozlowski, W. Bal, M. Dyba and T. Kowalik-Jankowska, *Coord. Chem. Rev.*, 1999, **184**, 319–346.

133. I. Sóvágó and K. Ősz, *Dalton Trans.*, 2006, 3841–3854.

134. D. R. Brown and H. Kozlowski, *Dalton Trans.*, 2004, 1907–1917.

135. E. Gaggelli, K. Kozlowski, D. Valensin and G. Valensin, *Chem. Rev.*, 2006, **106**, 1995–2044.

136. K. J. Barnham, F. Haeffner, G. D. Ciccotosto, C. C. Curtain, D. Tew, C. Mavros, K. Beyreuther, D. Carrington, C. L. Masters, R. A. Cherny, R. Cappai and A. I. Bush, *FASEB J.*, 2004, **18**, 1427–1429.

137. T. Kowalik-Jankowska, M. Ruta Dolejsz, K. Wisniewska, L. Lankiewicz and H. Kozlowski, *J. Chem. Soc. Dalton Trans.*, 2000, 4511–4519.

138. T. Kowalik-Jankowska, M. Ruta Dolejsz, K. Wisniewska and L. Lankiewicz, *J. Inorg. Biochem.*, 2001, **86**, 535–545.

139. T. Kowalik-Jankowska, M. Ruta Dolejsz, K. Wisniewska and L. Lankiewicz, *J. Inorg. Biochem.*, 2002, **92**, 1–10.

140. T. Kowalik-Jankowska, M. Ruta Dolejsz, K. Wisniewska and L. Lankiewicz, *J. Inorg. Biochem.*, 2003, **95**, 270–282.

141. D. Valensin, F. M. Mancini, M. Luczkowski, A. Janicka, K. Wisniewska, E. Gaggelli, G. Valensin, L. Lankiewicz and H. Kozlowski, *Dalton Trans.*, 2004, 16–22.

142. M. Luczkowski, K. Wisniewska, L. Lankiewicz and H. Kozlowski, *J. Chem. Soc. Dalton Trans.*, 2002, 2266–2268.

143. S. Zirah, S. Rebuffat, S. A. Kozin, P. Debey, F. Fournier, D. Lesage and J.-C. Tabet, *Int. J. Mass Spectr.*, 2003, **228**, 999–1016.

144. J. Huang, Y. Yao, Y. Ye, W. Sun and W. Tang, *J. Biol. Inorg. Chem.*, 2004, **9**, 627.

145. J. Shearer and V. A. Szalai, *J. Am. Chem. Soc.*, 2008, **130**, 17826–17835.

146. G. Drochioiu, M. Manea, M. Dragusanu, M. Murariu, E. S. Dragan, B. A. Petre, G. Mezo and M. Przybylski, *Biophys. Chem.*, 2009, **144**, 9–20.

147. C. S. Atwood, R. D. Moir, X. Huang, R. C. Scarpa, N. M. Baccara, D. M. Romano, M. A. Harthorn, R. E. Tanzi and A. I. Bush, *J. Biol. Chem.*, 1998, **273**, 12817.

148. C. S. Atwood, R. C. Scarpa, X. Huang, R. D. Moir, W. D. Jones, D. P. Fairlie, R. E. Tanzi and A. I. Bush, *J. Neurochem.*, 2000, **75**, 1219.

149. L. Guilloreau, L. Damian, Y. Coppel, H. Mazarguil, M. Winterhalter and P. Faller, *J. Biol. Inorg. Chem.*, 2006, **11**, 1024–1038.

150. A. Clements, D. Allsop, D. M. Walsh and C. H. Williams, *J. Neurochem.*, 1996, **66**, 740.

151. C. A. Damante, K. Ősz, Z. Nagy, G. Pappalardo, G. Grasso, G. Impellizzeri, E. Rizzarelli and I. Sóvágó, *Inorg. Chem.*, 2008, **47**, 9669–9683.

152. C. A. Damante, K. Ősz, Z. Nagy, G. Pappalardo, G. Grasso, G. Impellizzeri, E. Rizzarelli and I. Sóvágó, *Inorg. Chem.*, 2009, **48**, 10405–10415.

153. É. Jósza, K. Ősz, C. Kállay, P. De Bona, C. A. Damante, G. Pappalardo, E. Rizzarelli and I. Sóvágó, *Dalton Trans.*, 2010, **39**, 7046–7053.

154. A. I. Bush, W. H. Pettingel, Jr., M. D. Paradis and R. E. Tanzi, *J. Biol. Chem.*, 1994, **269**, 12152.

155. C. A. Damante, K. Ősz, Z. Nagy, G. Grasso, G. Pappalardo, E. Rizzarelli and I. Sóvágó, *Inorg. Chem.*, 2011 (in press).

156. E. D. Walter, D. J. Stevens, M. P. Visconte and G. L. Millhauser, *J. Am. Chem. Soc.*, 2007, **129**, 15440–15441.

157. L. M. Sayre, D. A. Zelasko, P. L. Harris, G. Perry, R. G. Salomon and M. A. Smith, *J. Neurochem.*, 1997, **68**, 2092–2097.

158. M. A. Smith, P. L. Harris, L. M. Sayre, J. S. Beckman and G. Perry, *J. Neurosci.*, 1997, **17**, 2653–2657.

159. M. A. Smith, P. L. Richey, S. Taneda, R. K. Kutty, L. M. Sayre, V. M. Monnier and G. Perry, *Ann. N. Y. Acad. Sci.*, 1994, **738**, 447–454.

160. C. D. Smith, J. M. Carney, P. E. Starke-Reed, C. N. Oliver, E. R. Stadtman, R. A. Floyd and W. R. Markesbery, *Proc. Natl. Acad. Sci. USA*, 1991, **88**, 10540–10543.

161. G. Perry, M. A. Taddeo, R. B. Petersen, R. J. Castellani, P. L. Harris, S. L. Siedlak, A. D. Cash, Q. Liu, A. Nunomura, C. S. Atwood and M. A. Smith, *Biometals*, 2003, **16**, 77–81.

162. C. Opazo, X. Huang, R. A. Cherny, R. D. Moir, A. E. Roher, A. R. White, R. Cappai, C. L. Masters, R. E. Tanzi, N. C. Inestrosa and A. I. Bush, *J. Biol. Chem.*, 2002, **277**, 40302–40308.

163. K. J. Barnham, F. Haeffner, G. D. Ciccotosto, C. C. Curtain, D. Tew, C. Mavros, K. Beyreuther, D. Carrington, C. L. Masters, R. A. Cherny, R. Cappai and A. I. Bush, *FASEB J.*, 2004, **18**, 1427–1429.

164. C. Behl, J. B. Davis, R. Lesley and D. Schubert, *Cell*, 1994, **77**, 817–827.

165. X. D. Huang, C. S. Atwood, M. A. Hartshorn, G. Multhaup, L. E. Goldstein, R. C. Scarpa, M. P. Cuajungco, D. N. Gray, J. Lim, R. D. Moir, R. E. Tanzi and A. I. Bush, *Biochemistry*, 1999, **38**, 7609–7616.

166. D. L. Jiang, L. J. Men, J. X. Wang, Y. Zhang, S. Chickenyen, Y. S. Wang and F. M. Zhou, *Biochemistry*, 2007, **46**, 9270–9282.

167. X. Huang, M. P. Cuajungco, C. S. Atwood, M. A. Hartshorn, J. D. Tyndall, G. R. Hanson, K. C. Stokes, M. Leopold, G. Multhaup, L. E. Goldstein, R. C. Scarpa, A. J. Saunders, J. Lim, R. D. Moir, C. Glabe, E. F. Bowden, C. L. Masters, D. P. Fairlie, R. E. Tanzi and A. I. Bush, *J. Biol. Chem.*, 1999, **274**, 37111–37116.
168. L. Galeazzi, P. Ronchi, C. Franceschi and S. Giunta, *Amyloid*, 1999, **6**, 7–13.
169. K. J. Barnham, W. J. McKinstry, G. Multhaup, D. Galatis, C. J. Morton, C. C. Curtain, N. A. Williamson, A. R. White, M. G. Hinds, R. S. Norton, K. Beyreuther, C. L. Masters, M. W. Parker and R. Cappai, *J. Biol. Chem.*, 2003, **278**, 17401–17407.
170. C. S. Atwood, G. Perry, H. Zeng, Y. Kato, W. D. Jones, K. Q. Ling, X. Huang, R. D. Moir, D. Wang, L. M. Sayre, M. A. Smith, S. G. Chen and A. I. Bush, *Biochemistry*, 2004, **43**, 560–568.
171. R. A. Himes, G. A. Park, G. S. Siluvai, N. J. Blackburn and K. D. Karlin, *Angew. Chem. Int. Ed.*, 2008, **47**, 9084–9087.
172. J. Shearer and V. A. Szalai, *J. Am. Chem. Soc.*, 2008, **130**, 17826–17835.
173. M. P. Cuajungco and K. Y. Faget, *Brain Res. Rev.*, 2003, **41**, 44–56.
174. G. Meloni, V. Sonois, T. Delaine, L. Guilloreau, A. Gillet, J. Teissie, P. Faller and M. Vasak, *Nat. Chem. Biol.*, 2008, **4**, 366–372.
175. D. Noy, I. Solomonov, O. Sinkevich, T. Arad, K. Kjaer and I. Sagi, *J. Am. Chem. Soc.*, 2008, **130**, 1376–1383.
176. C. Ha, J. Ryu and C. B. Park, *Biochemistry*, 2007, **46**, 6118–6125.
177. M. Innocenti, E. Salvietti, M. Guidotti, A. Casini, S. Bellandi, M. L. Foresti, C. Gabbiani, A. Pozzi, P. Zatta and L. Messori, *J. Alzheimers. Dis.*, 2010, **19**, 1323–1329.
178. J. Dong, J. E. Shokes, R. A. Scott and D. G. Lynn, *J. Am. Chem. Soc.*, 2006, **128**, 3540–3542.
179. P. Faller, *ChemBioChem*, 2009, **10**, 2837–2845.
180. X. Huang, C. S. Atwood, R. D. Moir, M. A. Hartshorn, R. E. Tanzi and A. I. Bush, *J. Biol. Inorg. Chem.*, 2004, **9**, 954–960.
181. J. Zou, K. Kajita and N. Sugimoto, *Angew. Chem. Int. Ed.*, 2001, **40**, 2274–2277.
182. A. I. Bush, W. H. Pettingell, Jr., G. Multhaup, M. D. Paradis, J.-P. Vonsattel, J. F. Gusella, K. Beyreuther, C. L. Masters and R. E. Tanzi, *Science*, 1994, **265**, 1464–1467.
183. A. I. Bush, R. D. Moir, K. M. Rosenkranz and R. E. Tanzi, *Science*, 1995, **268**, 1921–1923.
184. X. Huang, C. S. Atwood, R. D. Moir, M. A. Hartshorn, J.-P. Vonsattel, R. E. Tanzi and A. I. Bush, *J. Biol. Chem.*, 1997, **272**, 26464–26470.
185. S. Jun and S. Saxena, *Angew. Chem. Int. Ed. Eng.*, 2007, **46**, 3959–3961.
186. J. Dong, J. M. Canfield, A. K. Mehta, J. E. Shokes, B. Tian W. S. Childers, J. A. Simmons, Z. Mao, R. A. Scott, K. Warncke and D. G. Lynn, *Proc. Natl. Acad. Sci. USA*, 2007, **104**, 13313–13318.
187. K. Suzuki, T. Miura and H. Takeuchi, *Biochem. Biophys. Res. Commun.*, 2001, **285**, 991–996.

188. C. Ha, J. Ryu and C. B. Park, *Biochemistry*, 2007, **46**, 6118–6125.
189. J. Ryu, K. Girigoswami, C. Ha, S. Hee Ku and C. B. Park, *Biochemistry*, 2008, **47**, 5328–5335.
190. M. A. Lovell, C. Xie and W. R. Markesbery, *Brain Res.*, 1999, **823**, 88–95.
191. G. M. Bishop and S. R. Robinson, *Brain Pathol.*, 2004, **14**, 448–452.
192. Y. Yoshiike, K. Tanemura, O. Murayama, T. Akagi, M. Murayama, S. Sato, X. Sun, N. Tanaka and A. Takashima, *J. Biol. Chem.*, 2001, **276**, 32293–32299.
193. V. Tõugu, A. Karafin, K. Zovo, R. S. Chung, C. Howells, A. K. West and P. Palumaa, *J. Neurochem.*, 2009, **110**, 1784–1795.
194. M. P. Cuajungco, L. E. Goldstein, A. Nunoimura, M. A. Smith, J. T. Lim, C. S. Atwood, X. Huang, Y. W. Farrag, G. Perry and A. I. Bush, *J. Biol. Chem.*, 2000, **275**, 19439–19442.
195. P. Moreira, C. Pereira, M. S. Santos and C. Oliveira, *Antioxid. Redox Signal.*, 2000, **2**, 317–325.
196. M. P. Cuajungco and K. Y. Faget, *Brain Res. Rev.*, 2003, **41**, 44–56.
197. K. Garai, B. Sahoo, S. K. Kaushalya, R. Desai and S. Maiti, *Biochemistry*, 2007, **46**, 10655–10663.
198. Y. Miller, B. Ma and R. Nussinov, *Chem Rev.*, 2010, **110**, 4820–4838.
199. Y. Miller, B. Ma and R. Nussinov, *Proc. Natl. Acad. Sci. USA*, 2010, **107**, 9490–9495.
200. A. T. Petkova, Y. Ishii, J. J. Balbach, O. N. Antzutkin, R. D. Leapman, F. Delaglio and R. Tycko, *Proc. Natl. Acad. Sci. U.S.A.*, 2002, **99**, 16742–16747.
201. I. Kheterpal, A. Williams, C. Murphy, B. Bledsoe and R. Wetzel, *Biochemistry*, 2001, **40**, 11757–11767.
202. J. W. Karr, L. J. Kaupp and V. A. Szalai, *J. Am. Chem. Soc.*, 2004, **126**, 13534–13538.
203. J. W. Karr and V. A. Szalai, *Biochemistry*, 2008, **47**, 5006–5016.
204. C. J. Sarell, C. D. Syme, S. E. J. Rigby and J. H. Viles, *Biochemistry*, 2009, **48**, 4388–4402.
205. B. Halliwell and J. M. Gutteridge, *Free Radicals in Biology and Medicine*, 3rd ed., Oxford University Press, New York, 1999, pp. 172–183.
206. P. De Bona, G. Pappalardo and E. Rizzarelli, unpublished results.

CHAPTER 7

Zinc, Copper, Neurotrophic Factors and Neurodegeneration

G. AMADORO[a*] AND P. CALISSANO[a,b]

[a] Institute of Neurobiology and Molecular Medicine-CNR, IRCSS Fondazione Santa Lucia Via del Fosso di Fiorano 64-65, 00143, Rome, Italy; [b] European Brain Research Institute (EBRI), Via del Fosso di Fiorano 64-65, 00143, Rome, Italy

7.1 Introduction

Neurotrophic factors (NTF), such as nerve growth factor (NGF), brain-derived neurotrophic factor (BDNF), neurotrophin-3 (NT-3) and neurotrophin-4/5 (NT-4/5), are small polypeptides that maintain the survival and differentiation of specific neuronal populations and play key roles in axonal guidance, cell morphology, cognition and memory. They were first identified as survival factors for sympathetic and sensory neurons,[1,2] and their limited quantities during development control the number of surviving neurons to ensure a match between neurons and the requirement for a suitable density of target innervations. The availability of NTFs is required into adulthood, when they control formation and storage of memory, synaptic function and plasticity and sustain neuronal cell survival, morphology and differentiation.[3] In addition, secreted NTFs are involved in transmission of electrochemical signals between neurons in an activity-dependent fashion, they modulate the synaptic efficacy in the central nervous system (CNS) and peripheral nervous system (PNS) and play an important role in the glial–neuronal interactions during the myelination process. They are synthesized initially as precursor forms, or proneurotrophins

RSC Drug Discovery Series No. 7
Neurodegeneration: Metallostasis and Proteostasis
Edited by Danilo Milardi and Enrico Rizzarelli
© Royal Society of Chemistry 2011
Published by the Royal Society of Chemistry, www.rsc.org

(proNTF), and subsequently cleaved intracellularly by the proprotein convertase furin or extracellularly by the serine protease and by matrix metalloproteinases (MMPs), to release C-terminal mature forms. NTFs stimulate several intracellular signaling pathways, by activating two types of membrane-bound receptors: (i) Trks (an acronym derived from "tropomyosin-related kinase", subsequently changed tp "tyrosine receptor kinase"), and (ii) p75[NTR]. In their uncleaved inactive (proNTF) form, all NTFs bind to the p75[NTR], while each mature NT selectively activates one of three types of Trk receptor. NGF activates TrkA, NT-3 activates TrkC, while both BDNF and NT-4 activate TrkB receptors.[4] Although it has been hypothesized that NTs selectively maintain neuronal viability while proNTs trigger cell death through p75[NTR],[5,6] each is essential for maintaining the proper local architecture and function of the brain during neurodevelopment. This conclusion is suggested by the lethal phenotype characterized by severe neural defects of knockout (KO) mice for NGF, BDNF and NT-3. Upon engagement with Trk receptors, target-released NTFs stimulate G protein Ras (guanine nucleotide binding protein), PLC-γ (phospholipase C-γ) and PI3-kinase (phosphatidylinositol 3-kinase) and signaling pathways controlled through these proteins, such as those of the MAP kinases. Activation of p75[NTR] results in signaling by NF-κB (nuclear factor kappa B) and Jun kinase. Ligand–NTF receptor complexes are internalized into endosomes – termed "signaling endosomes" – and transported in a retrograde manner from neuronal cell surface to cell body. These molecular complexes, containing an NTF and its physiological receptors, travel along the axon to the nucleus, eventually influencing the selective transcription of specific target genes.[4,7] Moreover, the degeneration of selective neuronal populations, whose survival strictly depends on different NTFs,[8] is a common feature of several human neurodegenerative diseases, including Alzheimer's disease (AD), Parkinson's Disease (PD), Huntington's disease (HD) and amyotrophic lateral sclerosis (ALS).[9] For example, dysfunction of cholinergic neurons of the basal forebrain (BFCN) that provide the major cholinergic innervation to the cortex and hippocampus and play a key role in memory and attention processes is a cardinal feature of AD and correlates with cognitive decline.[10] Another NTF widely expressed in adult mammalian brain is BDNF, which has been shown to promote the survival of all major neuronal types affected in AD and PD, while its cortical production requires the proper activity of the corticostriatal synapses and the survival of the GABA-ergic medium-sized spiny striatal neurons, which are selectively lost in humans with these diseases.[11]

On the other hand, several studies suggest that dyshomeostasis of endogenous biometals such as copper (Cu) and zinc (Zn) can be involved in the etiopathogenesis of various neuropathological conditions. Despite the fact that most common human neurological disorders have a distinct etiological basis, they share striking similarities because they are all characterized by a documented impairment of brain metal homeostasis. Moreover alteration of metal metabolism, which normally occurs during physiological aging, is greatly enhanced and significantly exacerbates the extension of neurodegeneration by favoring increased oxidative stress, abnormal metal–protein interactions with aggregation, and also an impairment of NTF-mediated intracellular pathways.[12–17] In recent years, the interest in metal ion imbalance as a possible neuropathogenic factor has been strongly supported by the finding that a Cu and Zn ionophore that is able to penetrate the blood–brain barrier (BBB), Clioquinol (5-chloro-7-iodo-8-hydroxyquinoline, CQ), may be a promising disease-modifying candidate for use in therapy. Indeed, CQ treatment reduces the size and number of amyloid beta (Aβ) plaques in transgenic AD mice (Tg2576), thus delaying the related cognitive impairment and improving general health and weight parameters when compared with untreated, control mice.[18,19] In humans, a phase II clinical trial with CQ – which was initially used as an antibiotic – reveals several positive effects of the drug when administered orally in moderately severe cases of AD without any evident sign of other clinical complications.[20] The proposed mechanisms by which CQ can promote neuroprotective effects are: (i) by enhancing intracellular Cu and Zn uptake, thereby acting as an ionophore that favors the clearance of these ions from parenchymal amyloid plaques and the synaptic space, and (ii) by buffering the synaptic Zn thereby preventing Zn neurotoxicity and/or inhibiting the aggregation and oligomerization of toxic Aβ species in the synaptic cleft.[14] In addition, CQ treatment of transgenic HD mice (R6/2) also improved the behavioral and pathological phenotypes, by decreasing the accumulation of polyglutamine-expanded huntingtin and the striatal atrophy and by enhancing the cognitive performance with a significant lifespan extension.[21] In a similar way, the Cu chelator D-penicillamine seems to delay the onset of prion protein disease (PrPD).[22]

Here we discuss recent findings that strongly link metal ion imbalance, NTFs signaling and neuronal degeneration. Moreover, although further experimentation will be necessary to elucidate the underlying molecular mechanisms, future pharmacological strategies that are aimed at restoring metal homeostasis and the NTF pathway in neurological disorders are already assumed to have clinical potential.

7.2 Zinc and Neurotrophins in the Physiology and Pathology of the CNS

Zn is a redox-active metal that is predominantly used by organisms living in oxygen-rich environments. In its ionic form (Zn^{2+}) it is enriched at many glutamatergic nerve terminals in specific brain areas, such as the neocortex, amygdala and hippocampus, and is involved in the control of several physiological and pathological cerebral functions. Although cytosolic $[Zn^{2+}]$ is in picomolar concentrations, its intracellular level rises to micromolar values in the proximity of axon terminals and – following release from synaptic vesicles – it interacts with various neuronal ion channels, receptors and transporters, thus modulating synaptic transmission and plasticity.[16,23] An altered Zn^{2+} balance – due to a breakdown in the homeostatic mechanisms that intracellularly compartmentalize and regulate this biometal – has been described during normal aging and in major neurological disorders, such as AD, PD, ALS, and HD.[12,13,16,23,24]

Moreover, a possible modulatory role of vesicular Zn^{2+} in brain involves NTF-mediated signaling. Although 20 years of research have so far shown controversial results, a very recent study provides an intriguing picture of how Zn ions may contribute to synaptic plasticity by activating Trk signaling in an NT-independent manner. Transactivation refers to the process whereby a given receptor and its downstream signaling is activated by a stimulus that does not act directly on the receptor,[25] but participates in the crosstalk among diverse intracellular inputs and allows combination and diversification of the intracellular transduction pathways. In fact, Zn ions can transactivate TrkB – in the absence of BDNF – and potentiate the hippocampal long-term potentiation (LTP) at mossy fiber–CA3 pyramidal synapses, the most Zn^{2+}-enriched synapses in the brain. Zn^{2+} – but not other divalent or monovalent cations (such as Mg^{2+}, Ca^{2+}, Na^+, K^+) – selectively upregulates TrkB (and not TrkA or TrkC)-mediated signaling, through increasing the activity of Src tyrosine family kinase at the postsynaptic density (PSD) of excitatory synapses. LTP at these synapses, a process implicated in the storage and recall of amnestic information, is actually impaired by genetic deletion of TrkB in conditional null mutant mice, or by pharmacological inhibition of its kinase activity with K252a and by CaEDTA, a selective Zn chelator. Regarding the possible molecular mechanism of action, relevant studies have proved that, upon high-frequency neuronal activity, the Zn ion acts as a trans-synaptic messenger because it enters postsynaptic terminals through voltage-gated calcium channels (VGCCs) and *N*-methyl D-aspartate (NMDA) receptors. Here it derepresses the Src kinase which, in turn, transactivates TrkB by phosphorylation at specific Y515 residue. The discovery that the Zn ion-induced TrkB activation may also occur in BDNF-independent mechanisms has important pathological implications, since enhanced TrkB stimulation promotes limbic epileptogenesis, an event known to be characterized by a massive sprouting of Zn^{2+}-rich mossy fibers.[26–29] On the other hand, Zn (and Cu) ions can activate TrkB through alternative, indirect, mechanism involving the maturation of its physiological ligand. Both endogenous metals modulate the extracellular activity of MMP-2 and MMP-9, two metalloproteinases that are localized at the synaptic cleft and act in the extracellular space by converting immature pro-BDNF to functional BDNF.[30,31] Indeed, it has been proposed that extracellular Zn ions can activate such metalloproteinases either directly, by binding to the inhibitory cysteine residues, or indirectly by activating oxidative pathways[32,33] or upstream proteases. Definitive evidence about the physiological function of Zn ions in sustaining synaptic function in adulthood was reported from a recent study in ZnT3 knockout (KO) transgenic mice. The genetic ablation of Zn transported-3 (ZnT3) – a protein essential for its loading into synaptic vesicles – evokes an age-dependent cognitive loss, since ZnT3 KO mice phenocopy the synaptic and memory deficits typically observed in AD patients, in correlation with a significant decrease of the proBDNF/BDNF conversion.[34] Moreover, as reported by Deshpande *et al.*,[35] the sequestration of

Zn ions in oligomeric Aβ–Zn^{2+} complexes in AD cases may also induce a drastic reduction in biometal availability at the level of the synaptic cleft, thus resulting in a loss of its local neuromodulatory activity that, in turn, eventually induces the disease-related cognitive deficits.[35] A role of Zn in insulin and insulin-like growth factor-1 (IGF-1) signaling underlies its essential role in the induction of cell proliferation by IGF-1.[36] An increased cellular influx of Zn augments the phosphorylation of the activating tyrosine residues of the IGF-1 receptor and insulin receptor substrate 1 (IRS-1). In contrast, the chelation of cellular Zn attenuates the insulin/IGF-1-induced phosphorylation signals, indicating that this biometal probably exerts its action by inhibiting protein tyrosine phosphatases (PTPs). In fact, an overload of cellular Zn after incubation with both Zn and the ionophore pyrithione augments the phosphorylation: (i) of three tyrosine residues, 1158, 1162, and 1163, in the autocatalytic region of the β-domain of the insulin and IGF-1 receptors (pY^3IR/IGF-1R), and (ii) of tyrosine 856 of insulin receptor substrate 1 (pY IRS-1). However, the specific chelation of cellular Zn with membrane-permeable N,N,N',N'-tetrakis (2-pyridylmethyl)ethylenediamine suppresses the insulin- and IGF-1-stimulated phosphorylation.[37]

Besides its physiological functions, Zn^{2+} is also a potent neurotoxin that is implicated in neuronal and glial death in epilepsy, ischemia, brain trauma, ALS and – as a prominent component of senile plaques – in AD and cerebral amyloid angiopathy (CAA). The Zn ion promotes neuronal apoptosis: (i) by increasing the expression and secretion into the medium of NGF, the agonist of the low-affinity NTF receptor p75 ($p75^{NTR}$), and (ii) by modulating $p75^{NTR}$ and its associated death executor, NADE.[38] The 22 kDa protein termed $p75^{NTR}$-associated death executor (NADE) was discovered to be a necessary factor for $p75^{NTR}$-mediated apoptosis in several cells. In fact, the co-induction of $p75^{NTR}$ and NADE plays a role in Zn-triggered neuronal death in *in vitro* and *in vivo* models. In this regard, it has been proven that a function-blocking antibody, $p75^{NTR}$(REX) – which inhibits the association between $p75^{NTR}$ and NADE – blocks the neuronal death induced by Zn in cultured cortical neurons. Conversely, NGF augments Zn-evoked neurodegeneration. The reduction of NADE expression with cycloheximide or NADE antisense oligonucleotides also attenuates Zn-dependent neuronal death, which is efficiently prevented by caspase(s) inhibitor, suggesting that these apoptotic proteases are also involved. Finally, Zn-sustained neurotoxicity is also involved in neurodegeneration of rat hippocampal CA1 neurons after transient forebrain ischemia, proving that there is a close *in vivo* correlation between Zn accumulation and $p75^{NTR}$/NADE induction. In addition, Zn^{2+} stimulates the expression of early growth response factor 1 (EGR1), which is induced after various cerebral insults such as ischemia and seizures. EGR1 was first discovered as one of the immediate-early gene transcription factors induced by NGF in PC12 cells,[39] a pheochromocytoma cell line in which NGF treatment induces the acquisition of a neuronal-like phenotype. Further studies revealed that Egr-1 can also activate genes for platelet-derived growth factor (PDGF), the p75 NTF receptor, urokinase-type plasminogen activator, transforming growth factor-β (TGF-β), insulin-like growth factor (IGF)-II, and tumor necrosis factor (TNF)-alpha in several cells.[40] An interesting study demonstrated that the Zn ion, rather than calcium, induces lasting expression of Egr-1 in cortical cultures, by activating the extracellular signal-regulated kinase (Erk) Erk1/2.

PD098059, a pharmacological inhibitor of the Erk 1/2 upstream kinase mitogen-activated protein kinase 1 (MEK1) significantly blocks the Erk 1/2 activation, the Egr-1 induction and the neuronal death evoked by Zn treatment. The neurotoxic potential of Zn^{2+} *in vitro* and *in vivo* has also been reported in ALS models, by assessing the selectivity of the divalent cation in altering the BDNF-dependent survival of mouse motor neurons. Indeed, BDNF promoter cross-linking efficiency, and TrkB receptor cross-linking to BDNF, are significantly inhibited by Zn^{2+}, suggesting that the cation-induced change in BDNF conformation inhibits the receptor-binding activity.[41] Finally, in the presence of extracellular Zn, the NGF-differentiated PC12 are positive for propidium iodide (PI) and show nuclear fragmentation, caspase-3 activation and reactive oxygen species (ROS) production, suggesting that in this neuronal model this biometal induces mainly a necrotic process, instead of a classical apoptosis.[42] On the other hand, CQ may cause perturbation of the intracellular NGF-dependent survival pathway, by inhibiting the Trk-initiated signaling pathway in a dose-dependent manner. With regard to the mechanism, recently it has been reported that CQ inhibits NGF-induced Trk autophosphorylation, probably by acting at the level of mitogen-activated protein kinase (MAPK) phosphorylation, which is located downstream in the NGF–Trk intracellular signaling pathway. NGF differentiated cells are more vulnerable than naïve cells to CQ treatment, which also induces neuritis retraction and cell death.[43]

However, Zn^{2+} can also exert a neuroprotective action by modulating the specific intracellular signaling pathways, as shown during ischemic preconditioning. This term refers to a physiological process in which a brief sublethal ischemic hit is able to protect neurons from a subsequent stronger insult. In *in vitro* and *in vivo* models of ischemic preconditioning, a recent study demonstrated that a sublethal insult triggers a mild postsynaptic Zn^{2+} rise in degenerating neurons, while an intraventricular administration of CaEDTA abrogates both Zn accumulation and the protective effect against subsequent full ischemia. This event shows the classical hallmarks of ischemic preconditioning, such as caspase-3 activation, poly(ADP-ribose) polymerase-1 (PARP-1) cleavage and heat shock protein 70 (HSP70) induction, but it is also modulated by p75^{NTR} activation. It is tempting to hypothesize that a mild accumulation of Zn ions may induce p75^{NTR}-dependent caspase-3 activation at levels that do not induce full, severe apoptosis but are sufficient to promote the beneficial cleavage of PARP1, thereby blocking the downstream damaging effect of this enzyme. The resultant caspase-3-mediated reduction in the intracellular level of PARP-1 and the Zn^{2+}-dependent induction of HSP70 – before the neuronal apoptosis develops – provides broad-spectrum protection against the induction of neuronal death.[44] Moreover, the finding that chronic treatment with Zn^{2+} induces an increase in cortical levels of BDNF mRNA suggests that the positive effect of this divalent cation on NTF transcription might also represent an alternative and/or synergistic trophic pathway which is tonically operated by the metal localized at synapses.[45]

7.3 Copper and Neurotrophins in the Physiology and Pathology of the CNS

Copper (Cu) is another redox-active metal that is abundantly present in all animal tissues and exists either in the oxidated (Cu^{2+}) or reduced (Cu^{+}) valence states.[23] In contrast with Zn^{2+}, the dichotomous nature of Cu demands that its intricate homeostatic mechanisms *aim* to maintain the appropriate neuronal levels essential to the integrity of normal brain functions.[46] Under normal physiological conditions, its brain concentration is an order of magnitude higher than that in blood, indicating the importance of this ion in cerebral functions.[47,48] In brain, Cu^{2+} may be released into the synaptic cleft following neuronal activity.[49–51] Secreted Cu^{2+} can not only modulate the activity of NMDA, gamma-aminobutyric acid (GABA) and glycine receptors, thus affecting neuronal excitability,[50–54] but it also triggers exocytosis.[55] As for Zn^{2+}, since the free forms of this biometal are potentially damaging, its adsorption, distribution and excretion are tightly controlled and orchestrated by several proteins. In fact, although the brain possesses efficient buffering mechanisms to prevent abnormal biometal homeostasis, changes in the physiological concentration/availability of Cu have been implicated in the pathogenesis of AD, PD, ALS and PrPD.[15,56–58]

In parallel with Zn^{2+}, Cu^{2+} also may contribute to synaptic plasticity by modulating NTF-mediated signaling. An interesting study reported that Cu^{2+} stimulates Erk 1/2 and Src tyrosine kinase signaling; these are two enzymes activated downstream of TrkB in cortical neurons. To this regard it has been clearly demonstrated that Cu^{2+} increases the levels of pro- and mature BDNF in culture media, by affecting the activity of matrix metalloproteinases 2 and 9 (MMP2 and MMP9). This finding suggests that, like Zn^{2+}, Cu^{2+} induces metalloproteinase activity, releases pro-BDNF from cells and phosphorylates TrkB.[31] Moreover, an additional role for Cu in NTF-dependent neuronal survival and differentiation also involves the NGF signaling pathway. Interesting experiments have demonstrated that NGF promotes the Cu accumulation required for optimum neurite outgrowth in PC12 cells, since a specific Cu-chelator, TEPA (tetra-ethylene pentamine), significantly reduces NTF-mediated neuritogenesis.[59] Moreover, Cu can activate epidermal growth factor receptor (EGFR), a membrane-spanning protein activated upon binding of its physiological ligand epidermal growth factor (EGF), which is a growth factor whose possible role in neurodegenerative disease is emerging. Indeed, EGFR activation stimulates the survival, development and growth of cortical astrocytes, and neuronal migration and synaptic plasticity in hippocampal neurons. In addition, aged brains can respond to exogenous EGF with an increased neurogenesis.[60] As reported above for the Zn ion, a recent paper reported Cu-mediated EGFR activation through a ligand-independent mechanism, resulting in intracellular stimulation of downstream pathways, including PI3K and MAPK.[61] Although physiological EGFR activation involves EGF binding to its extracellular domain, Zn and Cu can also stimulate this process.[62] It has been found that Zn-induced EGFR phosphorylation at tyrosines 1068, 845 and 1173 results from the transactivation of the receptor via the non-receptor tyrosine kinase c-Src. This finding has been demonstrated by the fact that pretreatment with the specific c-Src inhibitor PP2 abolishes the metal-dependent Tyr phosphorylation of the receptor.[63] Additional studies suggest that EGFR activation by the metal may also occur through an autocrine mechanism in cells. In fact, exposure of Zn-treated human bronchial epithelial cells to an antibody against the ligand of EGF, HB-EGF, prevents the metal-induced tyrosine phosphorylation by blocking the release of HB-EGF in cell culture media. These findings suggest that metal-driven activation of EGFR may occur directly, by interaction with the receptor, or indirectly, by activation of upstream signaling molecules (c-Src) or, finally, by the cellular release of its physiological ligand which, in turn, acts on the extracellular binding domain of EGFR.

7.4   Conclusions

A growing number of studies suggest that normal aging is characterized by a significant brain dysmetabolism of metal ions, such as Zn and Cu, probably due to a progressive deterioration of the cellular regulatory systems (transport, uptake, distribution, export). Although imbalance of metal homeostasis occurs to some extent in normal aging, it appears to be greatly enhanced under various neuropathological conditions, thus causing increased oxidative stress, abnormal metal–protein interactions and altered neurotrophic-mediated signaling. On the other hand, NTF dysfunction is one of the etiopathogenetic factors of AD, PD, ALS and HD, a heterogeneous group of neurodegenerative disorders characterized by the abnormal deposition of misfolded proteins within neurons or brain parenchyma and/or by increased ROS production. Metal-targeted pharmacological strategies, aimed at delaying and modifying the progression of these human diseases, are currently and successfully exploited as an effective treatment alternative. This suggests that a tight link between NTF-dependent signaling and metal ion dyshomeostasis plays a key role in CNS physiopathology.

References

1. R. Levi Montalcini, *Science*, 1987, **237**, 1154–1162.
2. R. Levi Montalcini and G. Levi, *Arch. Biol.*, 1942, **LIII**, 537–545.
3. M. V. Chao, *Nat. Rev. Neurosci.*, 2003, **4**(4), 299–309.
4. A. Patapoutian and L. F. Reichardt, *Curr. Opin. Neurobiol.*, 2001, **11**(3), 272–280.

5. M. V. Chao, R. Rajagopal and F. S. Lee, *Clin. Sci. (Lond).*, 2006, **110**(2), 167–173.
6. W. J. Friedman, *J. Neurosci.*, 2000, **20**(17), 6340–6346.
7. E. J. Huang and L. F. Reichardt, *Annu. Rev. Biochem.*, 2003, **72**, 609–642.
8. E. Huang and L. Reichardt, *Annu. Rev. Neurosci.*, 2001, **24**, 677–736.
9. M. P. Mattson, *Nat. Rev. Mol. Cell Biol.*, 2000, **1**(2), 120–129.
10. G. Niewiadomska, A. Mietelska-Porowska and M. Mazurkiewicz, *Behav. Brain Res.*, 2010, [Epub ahead of print].
11. C. Zuccato and E. Cattaneo, *Prog. Neurobiol.*, 2007, **81**(5–6), 294–330.
12. K. J. Barnham and A. I. Bush, *Curr. Opin. Chem. Biol.*, 2008, **12**(2), 222–228.
13. A. I. Bush and R. E. Tanzi, *Neurotherapeutics*, 2008, **5**(3), 421–432.
14. P. Zatta, D. Drago, S. Bolognin and S. L. Sensi, *Trends Pharmacol. Sci.*, 2009, **30**(7), 346–355.
15. S. Bolognin, L. Messori and P. Zatta, *Neuromol. Med.*, 2009, **11**(4), 223–238.
16. S. L. Sensi, P. Paoletti, A. I. Bush and I. Sekler, *Nat. Rev. Neurosci.*, 2009, **10**(11), 780–791.
17. M. S. Clegg, L. A. Hanna, B. J. Niles, T. Y. Momma and C. L. Keen, *IUBMB Life*, 2005, **57**(10), 661–669.
18. P. A. Adlard, R. A. Cherny, D. I. Finkelstein, E. Gautier, E. Robb, M. Cortes, I. Volitakis, X. Liu, J. P. Smith, K. Perez, K. Laughton, Q. X. Li, S. A. Charman, J. A. Nicolazzo, S. Wilkins, K. Deleva, T. Lynch, G. Kok, C. W. Ritchie, R. E. Tanzi, R. Cappai, C. L. Masters, K. J. Barnham and A. I. Bush, *Neuron*, 2008, **59**(1), 43–55.
19. R. A. Cherny, C. S. Atwood, M. E. Xilinas, D. N. Gray, W. D. Jones, C. A. McLean, K. J. Barnham, I. Volitakis, F. W. Fraser, Y. Kim, X. Huang, L. E. Goldstein, R. D. Moir, J. T. Lim, K. Beyreuther, H. Zheng, R. E. Tanzi, C. L. Masters and A.I. Bush, *Neuron*, 2001, **30**(3), 665–676.
20. C. W. Ritchie, A. I. Bush, A. Mackinnon, S. Macfarlane, M. Mastwyk, L. MacGregor, L. Kiers, R. Cherny, Q. X. Li, A. Tammer, D. Carrington, C. Mavros, I. Volitakis, M. Xilinas, D. Ames, S. Davis, K. Beyreuther, R. E. Tanzi and C. L. Masters, *Arch. Neurol.*, 2003, **60**, 1685–1691.
21. T. Nguyen, A. Hamby and S. M. Massa, *Proc. Natl. Acad. Sci. USA*, 2005, **102**(33), 11840–11845.
22. E. M. Sigurdsson, D. R. Brown, M. A. Alim, H. Scholtzova, R. Carp and H. C. Meeker, *J. Biol. Chem.*, 2003, **278**, 46199–46202.
23. J. A. Duce and A. I. Bush, *Prog. Neurobiol.*, 2010 [Epub ahead of print].
24. W. I. Vonk and L. W. Klomp, *Biochem. Soc. Trans.*, 2008, **36**(Pt 6), 1322–1328.
25. G. Carpenter, *J. Cell Biol.*, 1999, **146**(4), 697–702.
26. X. P. He, L. Minichiello, R. Klein and J. O. McNamara, *J. Neurosci.*, 2002, **22**, 7502–7508.
27. X. P. He, R. Kotloski, S. Nef, B. W. Luikart, L. F. Parada and J. O. McNamara, *Neuron*, 2004, **43**, 31–42.
28. D. K. Binder, M. J. Routbort and J. O. McNamara, *J. Neurosci.*, 1999, **19**, 4616–4626.

29. O. Timofeeva and J. V. Nadler, *Brain Res.*, 2006, **1078**, 227–234.
30. J. J. Hwang, M. H. Park, S. Y. Choi and J. Y. Koh, *J. Biol. Chem.*, 2005, **280**, 11995–12001.
31. J. J. Hwang, M. H. Park and J. Y. Koh, *J. Neurosci. Res.*, 2007, **85**(10), 2160–2166.
32. Z. Gu, M. Kaul, B. Yan, S. J. Kridel, J. Cui, A. Strongin, J. W. Smith, R. C. Liddington and S. A. Lipton, *Science*, 2002, **297**, 1186–1190.
33. S. Rajagopalan, X. P. Meng, S. Ramasamy, D. G. Harrison and Z. S. Galis, *J. Clin. Invest.*, 1996, **98**, 2572–2579.
34. P. A. Adlard, J. M. Parncutt, D. I. Finkelstein and A. I. Bush, *J. Neurosci.*, 2010, **30**(5), 1631–1636.
35. A. Deshpande, H. Kawai, R. Metherate, C. G. Glabe and J. Busciglio, *J. Neurosci.*, 2009, **29**(13), 4004–4015.
36. D. Lefebvre, C. M. Boney, J. M. Ketelslegers and J. P. Thissen, *FEBS Lett.*, 1999, **449**, 284–288.
37. H. Haase and W. Maret, *Exp. Cell Res.*, 2003, **291**(2), 289–298.
38. J. A. Park, J. Y. Lee, T. A. Sato and J. Y. Koh, *J. Neurosci.*, 2000, **20**, 9096–9103.
39. J. Milbrandt, *Science*, 1987, **238**(4828), 797–799.
40. A. M. Beckmann and P. A. Wilce, *Neurochem. Int.*, 1997, **31**(4), 477–510; discussion 517–516.
41. J. I. Post, J. K. Eibl and G. M. Ross, *Amyotroph. Lateral Scler.*, 2008, **9**(3), 149–155.
42. F. J. Sánchez-Martín, E. Valera, I. Casimiro and J. M. Merino, *Brain Res. Bull.*, 2010, **81**(4–5), 458–466.
43. K. Asakura, A. Ueda, N. Kawamura, M. Ueda, T. Mihara and T. Mutoh, *Brain Res.*, 2009, **8**(1301), 110–115.
44. J. Y. Lee, Y. J. Kim, T. Y. Kim, J. Y. Koh and Y. H. Kim, *J. Neurosci.*, 2008, **28**(43), 10919–10927.
45. G. Nowak, B. Legutko, B. Szewczyk, M. Papp, M. Sanak and A. Pilc, *Eur. J. Pharmacol.*, 2004, **492**(1), 57–59.
46. Y. H. Hung, A. I. Bush and R. A. Cherny, *J. Biol. Inorg. Chem.*, 2009, **15**(1), 61–76.
47. J. Donaldson, T. S. Pierre, J. L. Minnich and A. Barbeau, *Can. J. Biochem.*, 1973, **51**, 87–92.
48. G. R. Merriam, L. L. Nunnelley, J. W. Trish and F. Naftolin, *Brain Res.*, 1979, **171**, 503–510.
49. D. E. Hartter and A. Barnea, *Synapse*, 1988, **2**(4), 412–415.
50. J. Kardos, I. Kovacs, F. Hajos, M. Kalman and M. Simonyi, *Neurosci. Lett.*, 1989, **103**, 139–144.
51. A. Hopt, S. Korte, H. Fink, U. Panne, R. Niessner, R. Jahn, H. Kretzschmar and J. Herms, *J. Neurosci. Meth.*, 2003, **128**, 159–172.
52. P. Q. Trombley and G. M. Shepherd, *J. Neurophysiol.*, 1996, **76**, 2536–2546.
53. M. L. Schlief, A. M. Craig and J. D. Gitlin, *J. Neurosci.*, 2005, **25**(1), 239–246.

54. M. L. Schlief, T. West, A. M. Craig, D. M. Holtzman and J. D. Gitlin, *Proc. Natl. Acad. Sci. USA*, 2006, **103**(40), 14919–14924.

55. D. E. Hartter and A. Barnea, *Synapse*, 1988, **2**(4), 412–415.

56. D. R. Brown and H. Kozlowski, *Dalton Trans.*, 2004, **7**(13), 1907–1917.

57. J. F. Mercer, *Trends Mol. Med.*, 2001, **7**, 64–69.

58. G. Torsdottir, J. Kristinsson, S. Sveinbjornsdottir, J. Snaedal and T. Johannesson, *Pharmacol. Toxicol.*, 1999, **85**, 239–243.

59. B. Birkaya and J. M. Aletta, *J. Neurobiol.*, 2005, **63**(1), 49–61.

60. K. Abe and H. Saito, *Brain Res.*, 1992, **587**, 102–108.

61. G. Filiz, K. A. Price, A. Caragounis, T. Du and P. J. Crouch, *Eur. Biophys. J.*, 2008, **37**(3), 315–321.

62. W. Wu, J. M. Samet, R. Silbajoris, L. A. Dailey, D. Sheppard, P. A. Bromberg and L. M. Graves, *Am. J. Respir. Cell Mol. Biol.*, 2004, **30**, 540–547.

63. J. M. Samet, B. J. Dewar, W. Wu and L. M. Graves, *Toxicol. Appl. Pharmacol.*, 2003, **191**, 86–93.

CHAPTER 8

Biological Metals: Metallostasis and Alzheimer's Disease

A. REMBACH,[a] J. A. DUCE,[a,b] L. A. O'SULLIVAN,[a]
R. E. TANZI[c,d] AND A. I. BUSH[a,e*]

[a] The Mental Health Research Institute, Parkville, Victoria 3052, Australia;
[b] Centre for Neuroscience, The University of Melbourne, Victoria 3010,
Australia; [c] Genetics and Ageing Research Unit, Department of Neurology,
Massachusetts General Hospital, Charlestown, MA, 02129-2060, USA;
[d] Harvard Medical School, Charlestown, MA, 02129-2060, USA;
[e] The Department of Pathology, The University of Melbourne, Victoria
3010, Australia

8.1 Introduction

Alzheimer's disease (AD) is a devastating neurological condition resulting in progressive, irreversible and debilitating dementia that accounts for the third leading cause of death in the industrialized world's population.[1] The burden of AD to the community socially and economically is immense, and with the incidence rising with the aging population, represents a major healthcare crisis. Although AD was first described over 100 years ago, there is still a lack of practical diagnostic methods, disease management programs or effective therapeutic strategies available to clinicians.

There are several pathological hallmarks of the AD brain including the deposition of extracellular plaques and intracellular neurofibrillary tangles (NFTs). The extracellular plaques are principally composed of aggregated β-amyloid (Aβ), a cleavage product of the β-amyloid precursor protein (APP).[2]

RSC Drug Discovery Series No. 7
Neurodegeneration: Metallostasis and Proteostasis
Edited by Danilo Milardi and Enrico Rizzarelli
© Royal Society of Chemistry 2011
Published by the Royal Society of Chemistry, www.rsc.org

In contrast, intracellular NFTs are composed of insoluble inclusions of the hyperphosphorylated tau protein.[3] Despite enormous advances in the study of AD, the etiology of this disease is still not well understood. Although several genetic risk factors have been identified, over 90% of all cases are sporadic, suggesting there may be environmental and genetic interactions underlying the pathogenesis. Another hallmark of the disease, first proposed decades ago,[4] is perturbed metal homeostasis, which is evident in both the development and progression of AD.

8.2   Biometals

Biological systems integrate metals for a range of chemical reactions. However, a disturbance in metabolism of any metal ion will result in deleterious consequences leading to cellular damage.[5] Due to the toxic propensity of metals, biological systems, in particular the central nervous system (CNS), maintain and restrict the concentration and movement of metal ions from the circulation and across the blood–brain barrier (BBB). Despite increasing evidence that iron, copper and zinc are implicated in neurodegenerative diseases, it appears that this toxicity is not merely due to increased exposure to metal concentration but rather a failure of homeostatic mechanisms that process and control metals.

The requirement for metals for catalytic biochemical reactions is so critical that it has been estimated that >30% of all enzymes require metals to carry out their reactions. This percentage may have been underestimated, as the completed genomes are still being analyzed.[6] The crucial functions of metal ions are also governed by the binding partners or ligands that facilitate transport or cellular distribution.[7] It is now accepted that a labile pool of loosely bound ions develops in pathological situations, and that these are exchangeable and prone to disturbances in homeostatic regulation.

Within the brain, the three biological transition metals, copper, zinc and iron, comprise a major endogenous pool of transition metals. While copper and zinc are used as secondary messengers in synaptic signaling, labile iron is thought to govern the expression of proteins and to regulate various pathways.

This chapter focuses on a new concept termed "metallostasis" which can be described as an imbalance of metal homeostasis. This may occur at any point in the metal ion pathway including uptake, release, storage, transport and metabolic regulation. The origin of such disturbances can arise from one or more of the following factors, including the consequence of normal aging processes, genetic mutations, environmental endotoxicity, deficiencies in diet and the consequences of drug interactions.

It is crucial to understand the biological implications of metallostasis in the context of an ever increasing catalogue of human diseases related to metal ion physiology. In AD, a "metal hypothesis" has been proposed to account for the neurotoxicitiy of Aβ. This has been reinforced by the emergence of therapeutic ionophores, which have shown promising results in ablating or reversing metal-lostasis. These compounds, which are designed to sequester and/or redistribute metal ions away from critical cellular compartments, as a potential therapeutic intervention against AD, are shedding light on the mechanism of how metal–Aβ interactions may lead to inevitable neurotoxicity in disease states.

8.3   Copper

Copper is an essential metal ion critical for normal cellular biochemistry. Copper serves as a catalytic cofactor in redox chemistry for normal growth and development, however this redox propensity can also facilitate the production of highly reactive oxygen species (ROS). Copper exists in the oxidized (Cu^{2+}) and reduced (Cu^+) valence states,[8] can coordinate with a range of ligands that include carboxylate oxygen, imidazole nitrogen, cysteine thiolate, methionine thioether groups, and engage in cation-interactions.[9,10]

Due to the damaging propensity of free metal ions, cellular systems employ control mechanisms to facilitate their absorption, distribution and excretion by a number of key proteins. Ceruloplasmin, albumin and transcuprein are some of the key binding proteins involved in the copper pathway in peripheral tissues, including plasma. Moving across membranes, copper import is driven by the copper transporter Ctr1 (which facilitates reuptake in neurons)[11] and it is exported by ATP7a or ATP7b. Copper is imported across the BBB through endothelial cells by ATP7a.[12–14] Intracellular copper is then delivered to various targets by copper chaperone proteins.[15] Various disease states have been linked to disturbances in these transport and delivery pathways.

The only established cellular environment where free copper has been observed is in neuronal synapses. Hippocampal neurites expressing post-synaptic N-methyl-D-aspartic acid (NMDA) receptors, when activated, trigger the mobilization of ATP7a and the release of free copper into the synaptic cleft to facilitate signal transduction.[16] Copper also modulates NMDA receptor function by promoting the reaction of nitric oxide with thiols to control extracellular S-nitrosylaton of the NMDA receptor.[16] It has also been reported that appropriate copper delivery can protect primary hippocampal neurons from NMDA-mediated excitotoxicity, a pathway dependent upon the local production of nitric oxide.[17]

Perturbed copper homeostasis in the CNS is implicated in several neurode-generative diseases.[18–20] In AD, CSF copper and peripheral copper (serum) levels are elevated,[21–23] and may correlate with changes in copper-dependent proteins such as ceruloplasmin.[24] Copper also appears to correlate with other AD related cellular elements. For instance, in CSF, there is a strong inverse correlation of Aβ42 with copper levels,[23] and copper mediates low-density lipoprotein (LDL) oxidation by homocysteine;[25,26] these are well established risk factors for developing AD.

In AD, copper appears to be improperly distributed and accumulates extracellularly, which may account for observations that AD-affected CNS tissue is copper deficient.[27,28] Indeed there is a five-fold increase in the copper content of extracellular plaques when compared to normal age-matched parenchyma.[29] Copper associated with the lipid component of the plasma membrane may accelerate plaque deposition through aggregation of Aβ and oxidative stress in the brain.[30] Confirmation of this stems from studies where rabbits fed excess copper and cholesterol diets demonstrate accelerated plaque formation.[30] This observation is also supported by high cholesterol diets being a major risk factor for developing AD.[31]

8.3.1 β-Amyloid Precursor Protein

The APP sequence expresses several putative metal binding sites.[32–36] The binding sites for copper are located at the amino-terminal ectodomain of APP as well as within the Aβ sequence of APP. The copper binding domain within the ectodomain shows structural homology to other copper chaperones and consists of four ligands (His-147, His-151, Tyr-168 and Met-170).[35]

APP plays a critical role in modulating metal ion homeostasis. *In vitro* studies have shown that APP mRNA expression can be regulated by interfering with the copper efflux and cellular transport machinery.[37,38] In contrast, *in vivo* and *in vitro* studies have shown that mice lacking APP demonstrated an increase in systemic copper levels,[39,40] while an overexpression of APP led to a decrease in copper levels.[41–43]

The processing of APP is also governed by the activities of alpha-, beta- and gamma-secretase which are, in turn, modulated by different metal ions (for review, see references 44–46). The proposed alpha-secretases, ADAM-17 and ADAM-10,[47] are members of the cell surface metalloproteinase family of disintegrin and metalloprotease (ADAM) proteins, and ADAM-10 can be inhibited by a dominant negative form of the enzyme that has a point mutation in its zinc-binding site.[48] Conversely, beta-secretase has been identified as a membrane anchored aspartyl protease termed β-APP-site cleaving enzyme or BACE,[49–53] and the high molecular weight active gamma-secretase complex is comprised of several components including presenilin, APH-1, PEN-2 and nicastrin.[54] Overall it is beta- and gamma-secretases that regulate the specific cleavage of APP to yield various lengths of the Aβ peptide.

BACE1 is thought to contribute to the pathogenesis of AD by regulating β-cleavage of APP, and its activity is influenced by the binding of copper. BACE1 binds copper in its C-terminal domain and co-immunoprecipitates in brain homogenates with domain I of the copper chaperone CCS,[55] where it may also regulate the activity of superoxide dismutase 1 (SOD1) through its competition for copper.[55] Interestingly, it has been recently shown that CCS levels also directly influence the production of Aβ.[56] Endocytosis of APP to BACE1-rich endosomes is a prerequisite for β-cleavage of APP,[50] and this process is dependent upon the expression of flotillin-2 and cholesterol levels, both concentrated within lipid rafts along with cellular APP.[57,58] It is therefore no surprise that elevated copper reduces flotillin-2 levels in lipid rafts, reducing the endocytosis of APP.[59] Subsequent *in vitro* studies have shown that a copper deficiency reduces the level of APP mRNA while the APP protein level remains constant; this indicates that copper deficiency may accelerate APP translation.[60] In summary, these studies show that the expression and cleavage of APP are directly influenced by upstream and local metal ion processes.

8.3.2   β-Amyloid

Aβ copurifies with copper from senile plaques extracted from the human brain.[61,62] Aβ has an affinity for copper below physiological levels in the CNS and the affinity is considered to be as high as attomolar for aggregated $Aβ_{1-42}$;[36] however, some studies have shown lower affinities towards the submicromolar range for soluble Aβ[63,64] (reviewed in references 65,66). In addition, copper can induce the aggregation of Aβ in mildly acidotic conditions (pH 6.8–7).[36,67,68] This is highlighted by a decrease in CSF $Aβ_{42}$ concentration in postmortem CSF, together with an increase in copper concentration.[23]

There could be more than one binding site for metals in Aβ, especially as an oligomer.[36] The binding of copper to Aβ is mediated by nitrogen ligands from the three histidine residues at positions 6, 13 and 14, along with an oxygen ligand,[69,70] coordinating either the oxidized[36,63,71–74] or reduced[75–77] form of copper.

Significant insight about metal binding sites on Aβ has been achieved through the study of animal models. For instance, mouse/rat Aβ, comprising amino acid substitutions for histidine at position 13, arginine at position 5 and tyrosine at position 10, imparts structural variations and changes in metal binding propensity when compared with human Aβ.[78] Interestingly, these animals show no evidence for the formation of cerebral Aβ deposits with advanced aging, supporting the idea that histidine at position 13 plays a major role in metal coordination and the predilection for Aβ to form aggregates in the presence of copper.[79]

Several oxygen ligands for metal coordination have been proposed for Aβ, including the aspartate at position 1,[80–82] and the alanine at position 2.[82,83] Other oxygen ligands have been suggested including tyrosine at position 10, glutamate at position 11 and H_2O, however these still need to be confirmed.[63,84,85] *In vitro* studies have demonstrated that the copper-binding domain in the 3-42 truncated Aβ ($Aβ_{3-42}$) self-aggregates more rapidly than $Aβ_{1-42}$ and in the presence of copper will elicit $Aβ_{1-40}$ aggregation.[86] Although the metal in Aβ is likely to coordinate in a square-planar configuration, involving three histidine imidazoles, the fourth position is yet to be identified. For extensive reviews of the metal binding chemistry of Aβ see references 65,66,87.

At pH 7.4, Aβ binds copper and zinc with equal affinity, however in the AD brain, which is considered more acidic (pH 6.6), copper will displace zinc from Aβ.[36] Even though copper promotes the aggregation of Aβ *in vitro* and binds copper in the AD brain,[29,61,62] it is interesting to note that increasing brain copper levels in transgenic models of AD inhibit Aβ aggregation.

Transgenic APP23 mice, overexpressing human APP with the AD-related Swedish mutation, had copper added via dietary supplementation. After 3 months, significant increases in brain copper levels were observed when compared to non-supplemented APP23 littermates.[43] This also restored the activity of SOD1, a copper-dependent enzyme. Overall the increase in brain copper also facilitated a decrease of soluble and insoluble $A\beta$.[43] This supports the proposition that copper is inappropriately distributed in extracellular compartments in AD. This metallostasis in the copper pathway may account for the deficiency of copper-dependent enzymatic activities in AD such as cytochrome c oxidase[88–91] and SOD1,[92,93] in addition to reported elevations in serum copper.[21,23,24]

A similar observation was made when the TgCRND8 transgenic mouse (overexpressing human APP with the AD-related Swedish and Indiana mutations) was crossed with the "toxic-milk" transgenic model that maintains a mutation in the gene encoding the copper transport protein, ATP7b.[41] This mutation interferes with copper transport by preventing copper loading into secretory vesicles and results in an increase in intracellular copper. In addition, TgCRND8 mice with the ATP7b mutation demonstrated elevated copper within the brain, and a decrease in amyloid plaque content, in addition to lowered soluble and insoluble $A\beta$ levels. Peripheral levels of $A\beta$ were also reduced in the plasma, including endogenous murine $A\beta$ levels. *In vitro* studies have confirmed that a deficiency in copper can elevate $A\beta$ secretion by either influencing APP cleavage or by inhibiting its degradation.[94] These data add support to the idea that an intracellular shift in copper can significantly influence $A\beta$ aggregation.

Oxidation injury is a prominent hallmark of AD. Under certain conditions *in vitro*, $A\beta$ has demonstrated antioxidant activity,[75,95,96] with both neurotrophic and neuroprotective effects.[97] However, significant evidence has shown that the redox activity for variants of $A\beta$ is greatest for $A\beta_{1-42 \text{ human}} > A\beta_{1-40 \text{ human}} \gg A\beta_{1-40 \text{ mouse}} \approx 0$,[98] indicating that $A\beta$ toxicity is catalyzed in the presence of copper.[61,99]

Hureau and Faller recently reviewed studies that show that the coordination of oxidized copper (Cu^{2+}) to $A\beta$ results in the ion's reduction.[100] The redox activity of $A\beta$ can elicit the generation of ROS when not in the presence of an additional moiety,[101] and it can also facilitate the production of hydrogen peroxide (H_2O_2) when exposed to biological reductants.[61,101–107] Electron donation can come from a range of cellular components, including: cholesterol, long-chain fatty acids, 3,4-dihydroxyphenylalanine (DOPA), dopamine, dopamine quinone, dihydroxyindol, isodityrosine and ascorbate.[61,62,98,101,105–110] Exposure to these reductants can trigger oxidation of $A\beta$ side-chains leading to covalent oligomerization. A potential candidate for the copper-mediated redox activity is the sulfur atom of methionine-35 becoming methionine sulfoxide;[111,112] yet this is not a prerequisite for toxicity.[113,114] In addition, aldehyde adducts can be generated to lysine residues[115] and the tyrosine at position 10.[107,116]

The conjugated aromatic ring of tyrosine at position 10 can form dityrosine cross-linked oligomers in the presence of Cu^{2+} and H_2O_2.[117] Dityrosine and 3-nitrotyrosine are increased in AD brain lesions,[118] and peptide aggregation can be driven by the formation of dityrosine linkage of $A\beta$ through non-proteolytic degradation,[68] leading to the formation of higher order oligomers.[116] After the reduction of copper, Aβ generated radicals can form covalent adducts onto other proteins. For example, cyclooxygenase 2 (COX2) is susceptible to the development of dityrosine bridges, and COX2 : Aβ covalent complexes have been shown to be elevated in the AD brain.[119] 2-Oxo-histidine adducts[120] and N3-pyroglutamate modified forms of Aβ have been extracted from AD plaques, which represent the target ligands for visualization of amyloid in the living brain by the positron emission tomography (PET) reagent Pittsburgh Compound-B (PIB).[121]

Hydrogen peroxide is also bioavailable across all cellular compartments to react with reduced metal ions (Cu^+) to generate hydroxyl radicals via Fenton chemistry. This results in the generation of lipid peroxidation products including the emergence of: 4-hydroxy-2-nonenal (HNE), protein carbonyl modifications, and nucleic acid adducts, all of which are hallmarks of AD neuropathology.[61,101,104,105,107,108,122–125] Antioxidant defenses may also be compromised in AD, for instance H_2O_2-scavenging catalase and glutathione peroxidase are overwhelmed by the catalytic generation of H_2O_2 from the Aβ : metal complexes.

There appear to be pleiotropic consequences of copper-specific metallostasis in AD. Perturbations in copper regulation appear to contribute not only to the amyloidogenic pathway but also to a number of other pathways where metal biology governs downstream cellular events. One such example is with the ability of copper to alter tau phosphorylation by rapidly activating phosphoinositol-3-kinase (PI3K), which then activates serine/threonine protein kinase (Akt) and phosphorylation of GSK3.[126] In addition, copper may activate membrane receptors, including the epidermal growth factor receptor (EGFR), and subsequent activation of mitogen-activated protein kinases (MAPK), including ERK1/2.[127]

8.4   Zinc

Zinc is also an essential metal ion. A range of proteins require zinc for normal biochemical functions, including: incorporation into zinc metalloenzymes (*e.g.* SOD1), transcription factors containing zinc-binding motifs such as zinc fingers (*e.g.* p53 and GAL4),[128] signaling (*e.g.* protein kinase C)[129] and storage to buffer cytosolic zinc pools (*e.g.* metallothioneins).[130] Other than pancreatic β islets,[131] zinc is most abundant in the brain. Glutamatergic nerve terminals are enriched with zinc ions that are released during synaptic transmission. Exchangeable cytosolic zinc is in the picomolar range,[132] but it can reach high micromolar concentrations within the synapse and synaptic vesicles.

In the event of an action potential and membrane depolarization, zinc release increases the synaptic cleft concentration to 300 μM. Synaptic zinc interacts with NMDA receptors in a similar manner to synaptic copper,[133] in addition to a range of neuronal ion channels and transporters. Factors influencing zinc metallostasis have been studied.[134] A range of cellular mechanisms are employed to ensure free zinc does not accumulate and thus become toxic. This includes several zinc transporters (ZnTs), zinc-importing proteins (ZIPs) and buffering proteins such as the metallothioneins.[134] The zinc transporter ZnT3 mediates synaptic packaging of zinc into vesicles.[135,136] Neocortex and hippocampal glutamatergic synapses are densely concentrated with ZnT3.[137] Therefore, ZnT3 is a critical upstream effector for the flux of synaptic zinc in signaling pathways important in learning, memory and high order cognitive functions.

Brain levels of zinc have been reported to gradually decrease with age in adulthood.[138–142] Zinc is one of the more abundant metals in the CSF,[23] and both CSF and plasma zinc are decreased in AD when compared with age-matched healthy controls.[22,143,144] Zinc sequestration by amyloid plaques may enhance amyloid formation. As discussed with copper, zinc is also highly enriched within AD plaques (1055 µM) compared to normal age-matched neuropil (350 µM).[29] Histological examinations demonstrate reactive zinc deposits co-localized with cerebral amyloid angiopathy deposits and NFT-bearing neurons.[145,146]

Inhibition of zinc export by 4-hydroxynonenal,[147] a peroxidation product of Aβ : copper redox activity, which is reported as elevated in AD tissue,[147] may contribute to zinc homeostasis in the AD brain. Other studies have reported that levels of ZnT1, ZnT3, ZnT4 and ZnT6 are all altered in the AD brain.[148,149]

8.4.1 β-Amyloid Precursor Protein

As with copper the APP sequence contains several putative zinc binding sites.[71,150–152] One conserved domain of APP is between positions 170 and 188,[151,152] and is critical for zinc binding through two key cysteine ligands at positions 186 and 187, as well as other possible ligands (*e.g.* Cys-174, Met-170, Asp-177 and Glu-184).[153] It has been suggested that the binding of zinc plays a role in the dimerization of APP *in vivo*.[153,154]

In addition, zinc may also govern gamma-secretase activity through its subunit, presenilin. For instance, exogenous zinc administration (but not copper or iron) increases the C-terminal fragments of presenilin 1 in neonatal primary mouse cortical cultures.[155] Recently it has been shown that zinc can trigger the oligomerization of an APP gamma-secretase substrate and inhibit its cleavage *in vitro*, supporting the notion that a disruption of zinc levels may influence abnormal Aβ processing.[156]

Similar to copper, evidence shows that zinc can influence APP processing by directly modulating secretase activity, however this seems to be through α- and γ- rather than β-secretase. In addition, zinc can influence the cleavage of Aβ from APP because the zinc binding domain in the Aβ region of the APP sequence comprises the alpha-secretase cleavage site, thereby protecting Aβ from proteolytic degradation.[150]

8.4.2 β-Amyloid

Zinc will rapidly precipitate Aβ at a physiological pH (pH 7.4).[157] The affinity of Aβ for zinc, through a low and high affinity binding site, is in the range 0.1–10 µM.[71,158] In the context of neurotransmission, this concentration of free ionic zinc in addition to copper[16,159] could render Aβ vulnerable to precipitation within synaptic cleft.[160] CSF Aβ and zinc are inversely correlated, with the association stronger in a group of subjects expressing high levels of both zinc and copper.[23] Like copper, zinc is co-purifed with Aβ extracted from the human brain suggesting that it is also associated with senile plaques.[61,62] *In vitro* studies have shown that the oxidized forms of zinc bind to Aβ in a similar fashion to copper.[36,63,71–74]

A number of APP transgenic models have been used to investigate to role of zinc in APP processing and Aβ generation. Again, the effect of dietary zinc supplementation was studied in Tg2576 (overexpressing human APP with the AD-related Swedish mutation) and TgCRND8 mice. The effect of a zinc-enriched diet resulted in an AD-like spatial memory impairment and a reduction in the presence of Aβ plaque deposits.[161] Conversely, a diet deficient in zinc triggered a significant increase in plaque volume in another transgenic model of AD.[162]

As previously mentioned, in the brain, the zinc concentration is highest within glutamatergic synaptic vesicles, from where it is subsequently released into the synaptic cleft upon neuronal depolarization.[163] As ZnT3 is responsible for the transfer and packaging of zinc in these pre-synaptic vesicles, it is not surprising that a Tg2576 mouse model that also maintains a genetic deletion of ZnT3 results in a significant decrease in cerebral plaque load.[159] In addition, there is a significant reduction in cerebral amyloid angiopathy,[146] as a consequence of the depletion of an exchangeable zinc pool in the cerebrovascular wall. It is proposed that the high level of zinc released into the synaptic cleft is a critical step toward the formation of amyloid in AD transgenic mice and rats.

Aβ catabolism is also influenced by metal–ion interactions,[45,164] and so the aggregation of Aβ by metals within the brain represents only one element of the disease pathway. Insulin-degrading enzyme (IDE) and neprilysin (NEP), two of the principal Aβ-degrading enzymes, are zinc dependent. When these enzymes are genetically ablated, Aβ catabolism is significantly diminished leading to an increase of endogenous Aβ.[165,166] Conversely, NEP can degrade both monomeric and oligomeric forms of Aβ$_{1-40}$ and Aβ$_{1-42}$,[167] and the overexpression of either IDE or NEP in transgenic mice can reduce Aβ levels, inhibit plaque formation and extend lifespan.[168]

Both IDE and NEP express a consensus sequence (HEXXH) in which the two histidines act as the ligand for the catalytic zinc. Some family members also express an additional zinc-binding histidine (or aspartate) in elongated C-terminal motifs (HEXXHXXGXXH/D).[169] This suggests that IDE and NEP will be sensitive to perturbations in zinc metabolism.[170] Both IDE and NEP are sensitive to treatment with H_2O_2, and in the AD brain, these two enzymes are also found with oxidative modifications.[171–173] Other sources of ROS can differentially alter the activity of NEP and IDE. For instance treatment with HNE, an oxidative hallmark of AD brain pathology, also differentially alters the activity of these metallopeptidases. These studies conclude that ROS generation, perhaps as a function of metal interactions with Aβ, may be inhibiting the clearance pathways for Aβ regulation.

Plasmin is another enzyme identified in the Aβ catabolism pathway. Aβ induces plasmin activation by upregulation of tissue-plasminogen activator (tPA) (for review see reference 45). Plasmin is regulated by oxidation of the histidine residue in the active site, and inhibition of substrate cleavage by tPA can occur when exposed to copper in the presence of ascorbate.[174] In addition, the tPA/plasmin system is also involved in the activation of matrix metallo-proteinases (MMPs), which are zinc-dependent enzymes also involved in the clearance of Aβ. *In vitro* studies have shown that in the presence of zinc, degradation of Aβ is inhibited.[175]

These studies illustrate the critical influence that metal ions have on the processing of APP, generation of Aβ and its subsequent distribution and clearance. The transgenic rodent models are supported by extensive *in vitro* data that highlight the need for neuronal systems to tightly control metal ion homeostasis to prevent pathological events that are susceptible to changes in metal ion physiology.

8.4.3 Tau

NFTs are comprised predominately of Tau but also consist of other cytoskeletal components, which are also dependent on metal–ion interactions. For instance, neurofilament (NF) assembly is mediated by metal ions,[176] and phosphorylated NFs are closely associated with NFT in association with the dystrophic neuritis and in the vicinity of Aβ plaques in AD.[177] Tangle bearing neurons from the AD brain are also found with high concentrations of zinc.[178,179]

The phosphorylation and subsequent hyperphosphorylation of tau are mediated by kinases that are modulated by metal ions. The kinases: rac-beta serine/threonine protein kinase (PKB), glycogen synthase kinase 3β (GSK-3β), extracellular signal-regulated kinase (ERK1/2), c-Jun NH2-terminal kinase (JNK), p38 and p70 S6 kinase, are all activated by zinc and can elicit tau hyperphosphorylation in SH-SY5Y and N2a cells.[179,180]

8.5 Iron

Iron is an essential element used across most biological systems in a range of biochemical reactions. It is a fundamental component of many enzyme systems and is critical for life. However, iron levels need to be tightly controlled because, in excess, it can impart toxicity by the production of superoxide anions and hydroxyl radicals. These molecules readily react with and cause damage to biological molecules, including proteins, lipids, and DNA. Therefore, iron homeostasis is very carefully controlled by mechanisms that limit its potential toxicity. However, iron metallostasis can result in either iron deficiencies or disorders of iron excess.[181]

The brain relies on iron not only for oxygen delivery and electron transfer but also for synthesis and myelin production.[182] Certain brain regions have particularly high levels of iron, including the substantia nigra, globus pallidus, caudate nucleus and putamen.[183,184] In addition, iron is the most abundant metal in the CSF.[23] The pathway used by iron to get into the brain is not completely understood but is suggested to be through the transferrin receptor. Iron is oxidized to a ferric state by ferroxidases such as ceruloplasmin, and then associated with transferrin which binds the transferrin receptor to facilitate its transport across the BBB.[185]

Goodman *et al.* (1953) first reported that AD senile plaques were associated with an increase in iron levels.[186] Iron levels within plaques are three-fold higher when compared to levels in healthy control neuropil.[29] This finding, that excess iron is predominant in AD, has been subsequently confirmed in several studies using a range of specific and sensitive techniques.[187–191]

Metallostasis of brain iron metabolism is a consequence of genetic and sporadic factors and can alter import, export, uptake, release or intracellular compartmentalization within the iron pathway.[192,193] In AD, excess iron is associated with plaque deposits and neurofibrillary tangles.[194–197] In addition, tissue ferritin levels are reported as increased in AD, and transferrin levels are decreased.[188]

Oxidative stress is one of the earliest hallmarks of AD pathology and is clearly involved in iron-related pathogenesis, causing progression of the disease.[192,198] It is proposed that subtle changes in iron homeostasis can render neurons vulnerable to oxidative stress.[188,199] For instance, alterations in SOD1 activity lead to changes in superoxide detoxification and iron metabolism in glial and neuronal cells.[200,201] Lactoferrin, a transferrin homologue that inhibits the formation of hydroxyl radicals, is also upregulated in AD, which is thought act in cell protection against DNA damage.[202]

Other studies have shown that there are genetic risk factors for AD that are critical for iron metabolism. These include alterations of the hereditary hemochromatosis gene (HFE), which is responsible for excess iron in hemochromatosis, and the transferrin C2 allele.[203,204] In a similar way to copper, cholesterol associated with iron can also elicit an AD-like phenotype in transgenic animals,[205] and may be another risk factor for humans.[206]

Other elements critical for maintaining iron metabolism have been implicated in AD, including a significant correlation between gender and age related risk of developing AD and brain ferritin iron levels.[207] In addition, melanotransferrin (p97), from the same family of proteins as transferrin and lactoferrin, has been proposed as a biomarker for AD.[208,209]

8.5.1 β-Amyloid Precursor Protein

APP translation is regulated by an iron response element (IRE) type II motif present in its 5′ untranslated region.[210] This system detects increases in iron levels, and elicits the production of APP to mobilize iron for export. Studies have shown that APP 5′-UTR translation can be downregulated by intracellular metal chelation and then reversed by the addition of iron.[211]

Recently we observed that APP contains a REXXE ferroxidase consensus motif as found in the ferroxidase active site of H-ferritin.[212] Like most ferroxidases, APP's ability to oxidize Fe^{2+} to Fe^{3+} facilitates the efflux of iron out of the cell and suppression of APP in neurons, as well as other specific cell-types, induces marked iron retention. Interestingly, APP-specific ferroxidase activity is inhibited by zinc, and APP ferroxidase activity in AD post-mortem specimens is inhibited by endogenous Zn^{2+}, transferred from extracellular amyloid aggregates.[212]

8.5.2 Tau

Iron also appears to influence the pathway leading to the formation of abnormally phosphorylated tau in NFTs. For example, Fe^{3+} can mediate the aggregation of hyperphosphorylated tau, a process that can be reversed by reduction back to Fe^{2+}.[213] In fact, tau aggregates extracted from the AD brain can be resolubilized by incubation with reductants, clearly highlighting a role of iron in NFT pathology.[213] However, contradictory reports have indicated that cultures of primary hippocampal neurons exposed to iron citrate ablate tau phosphorylation, possibly from a downregulation in the activity of the Cdk5/p25 complex, where p25 may be a regulator of the activity of protein kinase Cdk5.[214]

8.6 Conclusions

In conclusion, this chapter has outlined the pleiotropic roles of biologically relevant metals in AD and the functional implications of uncoupled homeo-static mechanisms. Metallostasis appears to be an obligatory step in the pathophysiological events leading to AD; whether up- or downstream of disease initiation, metal pathways represent a logical target for the application of therapeutic strategies. When developing novel drugs it is important to consider issues of specificity, selectivity and the downstream consequence of disease pathway intervention. Considerable research is necessary to enhance our understanding of these pathways, leading to optimized metal targeted pharmacological strategies to delay the progression or ultimately prevent the onset of AD.

References

1. M. P. Mattson, *Nature*, 2004, **430**, 631–639.
2. R. E. Tanzi, E. D. Bird, S. A. Latt and R. L. Neve, *Science*, 1987, **238**, 666–669.
3. C. M. Clark, D. Ewbank, V. M. Lee and J. Q. Trojanowski, *Molecular Pathology of Alzheimer's Disease: Neuronal Cytoskeletal Abnormalities*, Butterworth–Heinemann, Boston, 1998.
4. D. R. Crapper, S. S. Krishnan and A. J. Dalton, *Science*, 1973, **180**, 511–513.
5. J. J. R. Frausto da Silva and R. J. P. Williams, *The Biological Chemistry of the Elements*, Oxford University Press, Oxford, 2001.

6. F. P. Guengerich, *J. Biol. Chem.*, 2009, **284**, 18557.

7. T. D. Rae, P. J. Schmidt, R. A. Pufahl, V. C. Culotta and T. V. O'Halloran, *Science*, 1999, **284**, 805–808.

8. P. G. Ridge, Y. Zhang and V. N. *PLoS One*, 2008, **3**, e1378.

9. I. Bertini, H. B. Gray, E. I. Stiefel and J. S. Valentine, *Biological Inorganic Chemistry: Structure and Reactivity*, University Science Books, Sausalito, CA, 2007.

10. K. J. Franz, *Nat. Chem. Biol.*, 2008, **4**, 85–86.

11. A. Giese, M. Buchholz, J. Herms and H. A. Kretzschmar, *J. Mol. Neurosci.*, 2005, **27**, 347–354.

12. Y. Qian, E. Tiffany-Castiglioni, J. Welsh and E. D. Harris, *J. Nutr.*, 1998, **128**, 1276–1282.

13. R. El Meskini, K. L. Crabtree, L. B. Cline, R. E. Mains, B. A. Eipper and G. V. Ronnett, *Mol. Cell. Neurosci.*, 2007, **34**, 409–421.

14. M. J. Niciu, X. M. Ma, R. El Meskini, J. S. Pachter, R. E. Mains and B. A. Eipper, *Neurobiol. Dis.*, 2007, **27**, 278–291.

15. H. Tapiero, D. M. Townsend and K. D. Tew, *Biomed. Pharmacother.*, 2003, **57**, 386–398.

16. M. L. Schlief, A. M. Craig and J. D. Gitlin, *J. Neurosci.*, 2005, **25**, 239–246.

17. M. L. Schlief, T. West, A. M. Craig, D. M. Holtzman and J. D. Gitlin, *Proc. Natl. Acad. Sci. USA*, 2006, **103**, 14919–14924.

18. G. Torsdottir, J. Kristinsson, S. Sveinbjornsdottir, J. Snaedal and T. Johannesson, *Pharmacol. Toxicol.*, 1999, **85**, 239–243.

19. J. F. Mercer, *Trends Mol. Med.*, 2001, **7**, 64–69.

20. D. R. Brown and H. Kozlowski, *Dalton Trans*, 2004, 1907–1917.

21. R. Squitti, D. Lupoi, P. Pasqualetti, G. Dal Forno, F. Vernieri, P. Chiovenda, L. Rossi, M. Cortesi, E. Cassetta and P. M. Rossini, *Neurology*, 2002, **59**, 1153–1161.

22. H. Basun, L. G. Forssell, L. Wetterberg and B. Winblad, *J. Neural. Transm. Park. Dis. Dement. Sect.*, 1991, **3**, 231–258.

23. D. Strozyk, L. J. Launer, P. A. Adlard, R. A. Cherny, A. Tsatsanis, I. Volitakis, K. Blennow, H. Petrovitch, L. R. White and A. I. Bush, *Neurobiol. Aging*, 2009, **30**, 1069–1077.

24. R. Squitti, P. Pasqualetti, G. Dal Forno, F. Moffa, E. Cassetta, D. Lupoi, F. Vernieri, L. Rossi, M. Baldassini and P. M. Rossini, *Neurology*, 2005, **64**, 1040–1046.

25. E. Nakano, M. P. Williamson, N. H. Williams and H. J. Powers, *Biochim. Biophys. Acta*, 2004, **1688**, 33–42.

26. S. Seshadri, A. Beiser, J. Selhub, P. F. Jacques, I. H. Rosenberg, R. B. D'Agostino, P. W. Wilson and P. A. Wolf, *N. Engl. J. Med.*, 2002, **346**, 476–483.

27. D. A. Loeffler, P. A. LeWitt, P. L. Juneau, A. A. Sima, H. U. Nguyen, A. J. DeMaggio, C. M. Brickman, G. J. Brewer, R. D. Dick, M. D. Troyer and L. Kanaley, *Brain Res.*, 1996, **738**, 265–274.

28. M. A. Deibel, W. D. Ehmann and W. R. Markesbery, *J. Neurol. Sci.*, 1996, **143**, 137–142.
29. M. A. Lovell, J. D. Robertson, W. J. Teesdale, J. L. Campbell and W. R. Markesbery, *J. Neurol. Sci.*, 1998, **158**, 47–52.
30. D. L. Sparks and B. G. Schreurs, *Proc. Natl. Acad. Sci. USA*, 2003, **100**, 11065–11069.
31. M. C. Morris, D. A. Evans, C. C. Tangney, J. L. Bienias, J. A. Schneider, R. S. Wilson and P. A. Scherr, *Arch. Neurol.*, 2006, **63**, 1085–1088.
32. D. Valensin, F. M. Mancini, M. Luczkowski, A. Janicka, K. Wisniewska, E. Gaggelli, G. Valensin, L. Lankiewicz and H. Kozlowski, *Dalton Trans.*, 2004, 16–22.
33. A. Simons, T. Ruppert, C. Schmidt, A. Schlicksupp, R. Pipkorn, J. Reed, C. L. Masters, A. R. White, R. Cappai, K. Beyreuther, T. A. Bayer and G. Multhaup, *Biochemistry*, 2002, **41**, 9310–9320.
34. L. Hesse, D. Beher, C. L. Masters and G. Multhaup, *FEBS Lett.*, 1994, **349**, 109–116.
35. K. J. Barnham, W. J. McKinstry, G. Multhaup, D. Galatis, C. J. Morton, C. C. Curtain, N. A. Williamson, A. R. White, M. G. Hinds, R. S. Norton, K. Beyreuther, C. L. Masters, M. W. Parker and R. Cappai, *J. Biol. Chem.*, 2003, **278**, 17401–17407.
36. C. S. Atwood, R. C. Scarpa, X. Huang, R. D. Moir, W. D. Jones, D. P. Fairlie, R. E. Tanzi and A. I. Bush, *J. Neurochem.*, 2000, **75**, 1219–1233.
37. S. A. Bellingham, D. K. Lahiri, B. Maloney, S. La Fontaine, G. Multhaup and J. Camakaris, *J. Biol. Chem.*, 2004, **279**, 20378–20386.
38. A. D. Armendariz, M. Gonzalez, A. V. Loguinov and C. D. Vulpe, *Physiol. Genomics*, 2004, **20**, 45–54.
39. A. R. White, R. Reyes, J. F. Mercer, J. Camakaris, H. Zheng, A. I. Bush, G. Multhaup, K. Beyreuther, C. L. Masters and R. Cappai, *Brain Res.*, 1999, **842**, 439–444.
40. S. A. Bellingham, G. D. Ciccotosto, B. E. Needham, L. R. Fodero, A. R. White, C. L. Masters, R. Cappai and J. Camakaris, *J. Neurochem.*, 2004, **91**, 423–428.
41. A. L. Phinney, B. Drisaldi, S. D. Schmidt, S. Lugowski, V. Coronado, Y. Liang, P. Horne, J. Yang, J. Sekoulidis, J. Coomaraswamy, M. A. Chishti, D. W. Cox, P. M. Mathews, R. A. Nixon, G. A. Carlson, P. St George-Hyslop and D. Westaway, *Proc. Natl. Acad. Sci. USA,* 2003, **100**, 14193–14198.
42. C. J. Maynard, R. Cappai, I. Volitakis, R. A. Cherny, A. R. White, K. Beyreuther, C. L. Masters, A. I. Bush and Q. X. Li, *J. Biol. Chem.*, 2002, **277**, 44670–44676.
43. T. A. Bayer, S. Schafer, A. Simons, A. Kemmling, T. Kamer, R. Tepest, A. Eckert, K. Schussel, O. Eikenberg, C. Sturchler-Pierrat, D. Abramowski, M. Staufenbiel and G. Multhaup, *Proc. Natl. Acad. Sci. USA*, 2003, **100**, 14187–14192.
44. Y. H. Suh and F. Checler, *Pharmacol. Rev.*, 2002, **54**, 469–525.

45. Y. Ling, K. Morgan and N. Kalsheker, *Int. J. Biochem. Cell Biol.*, 2003, **35**, 1505–1535.

46. S. Gandy, *J. Clin. Invest.*, 2005, **115**, 1121–1129.

47. J. D. Buxbaum, K. N. Liu, Y. Luo, J. L. Slack, K. L. Stocking, J. J. Peschon, R. S. Johnson, B. J. Castner, D. P. Cerretti and R. A. Black, *J. Biol. Chem.*, 1998, **273**, 27765–27767.

48. S. Lammich, E. Kojro, R. Postina, S. Gilbert, R. Pfeiffer, M. Jasionowski, C. Haass and F. Fahrenholz, *Proc. Natl. Acad. Sci. USA*, 1999, **96**, 3922–3927.

49. R. Yan, M. J. Bienkowski, M. E. Shuck, H. Miao, M. C. Tory, A. M. Pauley, J. R. Brashier, N. C. Stratman, W. R. Mathews, A. E. Buhl, D. B. Carter, A. G. Tomaselli, L. A. Parodi, R. L. Heinrikson and M. E. Gurney, *Nature*, 1999, **402**, 533–537.

50. R. Vassar, B. D. Bennett, S. Babu-Khan, S. Kahn, E. A. Mendiaz, P. Denis, D. B. Teplow, S. Ross, P. Amarante, R. Loeloff, Y. Luo, S. Fisher, J. Fuller, S. Edenson, J. Lile, M. A. Jarosinski, A. L., Biere, E. Curran, T. Burgess, J. C. Louis, F. Collins, J. Treanor, G. Rogers and M. Citron, *Science*, 1999, **286**, 735–741.

51. S. Sinha, J. P. Anderson, R. Barbour, G. S. Basi, R. Caccavello, D. Davis, M. Doan, H. F. Dovey, N. Frigon, J. Hong, K. Jacobson-Croak, N. Jewett, P. Keim, J. Knops, I. Lieberburg, M. Power, H. Tan, G. Tatsuno, J. Tung, D. Schenk, P. Seubert, S. M. Suomensaari, S. Wang, D. Walker, J. Zhao, L. McConlogue and V. John, *Nature*, 1999, **402**, 537–540.

52. Y. Luo, B. Bolon, S. Kahn, B. D. Bennett, S. Babu-Khan, P. Denis, W. Fan, H. Kha, J. Zhang, Y. Gong, L. Martin, J. C. Louis, Q. Yan, W. G. Richards, M. Citron and R. Vassar, *Nat. Neurosci.*, 2001, **4**, 231–232.

53. H. Cai, Y. Wang, D. McCarthy, H. Wen, D. R. Borchelt, D. L. Price and P. C. Wong, *Nat. Neurosci.*, 2001, **4**, 233–234.

54. B. De Strooper, *Neuron*, 2003, **38**, 9–12.

55. B. Angeletti, K. J. Waldron, K. B. Freeman, H. Bawagan, I. Hussain, C. C. Miller, K. F. Lau, M. E. Tennant, C. Dennison, N. J. Robinson and C. Dingwall, *J. Biol. Chem.*, 2005, **280**, 17930–17937.

56. E. H. Gray, K. J. De Vos, C. Dingwall, M. S. Perkinton and C. C. Miller, *J. Alzheimers Dis.*, 2012, **21**, 1101–1105.

57. R. Ehehalt, P. Keller, C. Haass, C. Thiele and K. Simons, *J. Cell Biol.*, 2003, **160**, 113–123.

58. A. Schneider, L. Rajendran, M. Honsho, M. Gralle, G. Donnert, F. Wouters, S. W. Hell and M. Simons, *J. Neurosci.*, 2008, **28**, 2874–2882.

59. Y. H. Hung, E. L. Robb, I. Volitakis, M. Ho, G. Evin, Q. X. Li, J. G. Culvenor, C. L. Masters, R. A. Cherny and A. I. Bush, *J. Biol. Chem.*, 2009, **284**, 21899–21907.

60. P. E. Vardas, A. Auricchio, J. J. Blanc, J. C. Daubert, H. Drexler, H. Ector, M. Gasparini, C. Linde, F. B. Morgado, A. Oto, R. Sutton, M. Trusz-Gluza, A. Vahanian, J. Camm, R. De Caterina, V. Dean, K. Dickstein, C. Funck-Brentano, G. Filippatos, I. Hellemans, S. D.

Kristensen, K. McGregor, U. Sechtem, S. Silber, M. Tendera, P. Widimsky, J. L. Zamorano, S. G. Priori, C. Blomstrom-Lundqvist, M. Brignole, J. B. Terradellas, P. Castellano, J. Cleland, J. Farre, M. Fromer, J. Y. Le Heuzey, G. Y. Lip, J. L. Merino, A. S. Montenero, P. Ritter, M. J. Schalij and C. Stellbrink, *Rev. Port. Cardiol.*, 2008, **27**, 639–687.

61. C. Opazo, X. Huang, R. A. Cherny, R. D. Moir, A. E. Roher, A. R. White, R. Cappai, C. L. Masters, R. E. Tanzi, N. C. Inestrosa and A. I. Bush, *J. Biol. Chem.*, 2002, **277**, 40302–40308.

62. J. Dong, C. S. Atwood, V. E. Anderson, S. L. Siedlak, M. A. Smith, G. Perry and P. R. Carey, *Biochemistry*, 2003, **42**, 2768–2773.

63. C. D. Syme, R. C. Nadal, S. E. Rigby and J. H. Viles, *J. Biol. Chem.*, 2004, **279**, 18169–18177.

64. W. Garzon-Rodriguez, A. K. Yatsimirsky and C. G. Glabe, *Bioorg. Med. Chem. Lett.*, 1999, **9**, 2243–2248.

65. P. Faller and C. Hureau, *Dalton Trans.*, 2009, 1080–1094.

66. M. Rozga and W. Bal, *Chem. Res. Toxicol.*, 2009, **23**, 298–308.

67. C. Ha, J. Ryu and C. B. Park, *Biochemistry*, 2007, **46**, 6118–6125.

68. C. S. Atwood, R. D. Moir, X. Huang, R. C. Scarpa, N. M. Bacarra, D. M. Romano, M. A. Hartshorn, R. E. Tanzi and A. I. Bush, *J. Biol. Chem.*, 1998, **273**, 12817–12826.

69. C. C. Curtain, F. Ali, I. Volitakis, R. A. Cherny, R. S. Norton, K. Beyreuther, C. J. Barrow, C. L. Masters, A. I. Bush and K. J. Barnham, *J. Biol. Chem.*, 2001, **276**, 20466–20473.

70. C. C. Curtain, F. E. Ali, D. G. Smith, A. I. Bush, C. L. Masters and K. J. Barnham, *J. Biol. Chem.*, 2003, **278**, 2977–2982.

71. A. I. Bush, W. H. Pettingell, G. Multhaup, M. deParadis, J. P. Vonsattel, J. F. Gusella, K. Beyreuther, C. L. Masters and R. E. Tanzi, *Science*, 1994, **265**, 1464–1467.

72. J. Danielsson, R. Pierattelli, L. Banci and A. Graslund, *FEBS J.*, 2007, **274**, 46–59.

73. C. D. Syme and J. H. Viles, *Biochim. Biophys. Acta*, 2006, **1764**, 246–256.

74. J. W. Karr and V. A. Szalai, *Biochemistry*, 2008, **47**, 5006–5016.

75. J. Shearer and V. A. Szalai, *J. Am. Chem. Soc.*, 2008, **130**, 17826–17835.

76. R. A. Himes, G. Y. Park, G. S. Siluvai, N. J. Blackburn and K. D. Karlin, *Angew. Chem. Int. Ed. Engl.*, 2008, **47**, 9084–9087.

77. C. Hureau, V. Balland, Y. Coppel, P. L. Solari, E. Fonda and P. Faller, *J. Bio. Inorg. Chem.*, 2009, **14**, 995–1000.

78. E. Gaggelli, Z. Grzonka, H. Kozlowski, C. Migliorini, E. Molteni, D. Valensin and G. Valensin, *Chem. Commun. (Camb.)*, 2008, 341–343.

79. D. W. Vaughan and A. Peters, *J. Neuropathol. Exp. Neurol.*, 1981, **40**, 472–487.

80. Y. Mekmouche, Y. Coppel, K. Hochgrafe, L. Guilloreau, C. Talmard, H. Mazarguil and P. Faller, *Chembiochem.*, 2005, **6**, 1663–1671.

81. S. C. Drew, C. J. Noble, C. L. Masters, G. R. Hanson and K. J. Barnham, *J. Am. Chem. Soc.*, 2009, **131**, 1195–1207.

82. P. Dorlet, S. Gambarelli, P. Faller and C. Hureau, *Angew. Chem. Int. Ed. Engl.*, 2009, **48**, 9273–9276.

83. S. C. Drew, C. L. Masters and K. J. Barnham, *J. Am. Chem. Soc.*, 2009, **131**, 8760–8761.

84. J. W. Karr and V. A. Szalai, *J. Am. Chem. Soc.*, 2007, **129**, 3796–3797.

85. J. W. Karr, H. Akintoye, L. J. Kaupp and V. A. Szalai, *Biochemistry*, 2005, **44**, 5478–5487.

86. G. McColl, B. R. Roberts, A. P. Gunn, K. A. Perez, D. J. Tew, C. L. Masters, K. J. Barnham, R. A. Cherny and A. I. Bush, *J. Biol. Chem.*, 2009, **284**, 22697–22702.

87. A. Rauk, *Chem. Soc. Rev.*, 2009, **38**, 2698–2715.

88. P. G. Sullivan and M. R. Brown, *Prog. Neuropsychopharmacol. Biol. Psychiatry*, 2005, **29**, 407–410.

89. I. Maurer, S. Zierz and H. J. Moller, *Neurobiol. Aging*, 2000, **21**, 455–462.

90. D. A. Cottrell, E. L. Blakely, M. A. Johnson, P. G. Ince and D. M. Turnbull, *Neurology*, 2001, **57**, 260–264.

91. S. M. Cardoso, M. T. Proenca, S. Santos, I. Santana and C. R. Oliveira, *Neurobiol. Aging*, 2004, **25**, 105–110.

92. P. P. De Deyn, M. Hiramatsu, F. Borggreve, J. Goeman, R. D'Hooge, J. Saerens and A. Mori, *Alzheimer Dis. Assoc. Disord.*, 1998, **12**, 26–32.

93. R. A. Omar, Y. J. Chyan, A. C. Andorn, B. Poeggeler, N. K. Robakis and M. A. Pappolla, *J. Alzheimers Dis.*, 1999, **1**, 139–145.

94. M. A. Cater, K. T. McInnes, Q. X. Li, I. Volitakis, S. La Fontaine, J. F. Mercer and A. I. Bush, *Biochem. J.*, 2008, **412**, 141–152.

95. R. C. Nadal, S. E. Rigby and J. H. Viles, *Biochemistry*, 2008, **47**, 11653–11664.

96. R. Baruch-Suchodolsky and B. Fischer, *Biochemistry*, 2009, **48**, 4354–4370.

97. J. S. Whitson, D. J. Selkoe and C. W. Cotman, *Science*, 1989, **243**, 1488–1490.

98. X. Huang, C. S. Atwood, M. A. Hartshorn, G. Multhaup, L. E. Goldstein, R. C. Scarpa, M. P. Cuajungco, D. N. Gray, J. Lim, R. D. Moir, R. E. Tanzi and A. I. Bush, *Biochemistry*, 1999, **38**, 7609–7616.

99. X. Huang, M. P. Cuajungco, C. S. Atwood, M. A. Hartshorn, J. D. Tyndall, G. R. Hanson, K. C. Stokes, M. Leopold, G. Multhaup, L. E. Goldstein, R. C. Scarpa, A. J. Saunders, J. Lim, R. D. Moir, C. Glabe, E. F. Bowden, C. L. Masters, D. P. Fairlie, R. E. Tanzi and A. I. Bush, *J. Biol. Chem.*, 1999, **274**, 37111–37116.

100. C. Hureau and P. Faller, *Biochimie*, 2009, **91**, 1212–1217.

101. L. Puglielli, A. L. Friedlich, K. D. Setchell, S. Nagano, C. Opazo, R. A. Cherny, K. J. Barnham, J. D. Wade, S. Melov, D. M. Kovacs and A. I. Bush, *J. Clin. Invest.*, 2005, **115**, 2556–2563.

102. B. J. Tabner, S. Turnbull, O. M. El-Agnaf and D. Allsop, *Free Radic. Biol. Med.*, 2002, **32**, 1076–1083.

103. S. I. Dikalov, M. P. Vitek and R. P. Mason, *Free Radic. Bio. Med.*, 2004, **36**, 340–347.

104. D. P. Smith, D. G. Smith, C. C. Curtain, J. F. Boas, J. R. Pilbrow, G. D. Ciccotosto, T. L. Lau, D. J. Tew, K. Perez, J. D. Wade, A. I. Bush, S. C.

Drew, F. Separovic, C. L. Masters, R. Cappai and K. J. Barnham, *J. Biol. Chem.*, 2006, **281**, 15145–15154.

105. T. J. Nelson and D. L. Alkon, *J. Biol. Chem.*, 2005, **280**, 7377–7387.

106. I. V. Murray, M. E. Sindoni and P. H. Axelsen, *Biochemistry*, 2005, **44**, 12606–12613.

107. F. Haeffner, D. G. Smith, K. J. Barnham and A. I. Bush, *J. Inorg. Biochem.*, 2005, **99**, 2403–2422.

108. D. Jiang, L. Men, J. Wang, Y. Zhang, S. Chickenyen, Y. Wang and F. Zhou, *Biochemistry*, 2007, **46**, 9270–9282.

109. N. Yoshimoto, M. Tasaki, T. Shimanouchi H. Umakoshi and R. Kuboi, *J. Biosci. Bioeng.*, 2005, **100**, 455–459.

110. D. G. Smith, R. Cappai and K. J. Barnham, *Biochim. Biophys. Acta*, 2007, **1768**, 1976–1990.

111. F. E. Ali, F. Separovic, C. J. Barrow, R. A. Cherny, F. Fraser, A. I. Bush, C. L. Masters and K. J. Barnham, *J. Pept. Sci.*, 2005, **11**, 353–360.

112. C. G. Diccotosto, D. Tew, C. C. Curtain, D. Smith, D. Carrington, C. L. Masters, A. I. Bush, R. A. Cherny, R. Cappai and K. J. Barnham, *J. Biol. Chem.*, 2004, **279**, 42528–42534.

113. L. Hou and M. G. Zagorski, *J. Am. Chem. Soc.*, 2006, **128**, 9260–9261.

114. G. F. da Silva, V. Lykourinou, A. Angerhofer and L. J. Ming, *Biochim. Biophys. Acta*, 2009, **1792**, 49–55.

115. K. Chen, M. Kazachkov and P. H. Yu, *J. Neural. Transm.*, 2007, **114**, 835–839.

116. K. J. Barnham, F. Haeffner, G. D. Ciccotosto, C. C. Curtain, D. Tew, C. Mavros, K. Beyreuther, D. Carrington, C. L. Masters, R. A. Cherny, R. Cappai and A. I. Bush, *FASEB J.*, 2004, **18**, 1427–1429.

117. C. S. Atwood, G. Perry, H. Zeng, Y. Kato, W. D. Jones, K. Q. Ling, X. Huang, R. D. Moir, D. Wang, L. M. Sayre, M. A. Smith, S. G. Chen and A. I. Bush, *Biochemistry*, 2004, **43**, 560–568.

118. K. Hensley, M. L. Maidt, Z. Yu, H. Sang, W. R. Markesbery and R. A. Floyd, *J. Neurosci.*, 1998, **18**, 8126–8132.

119. S. Nagano, X. Huang, R. D. Moir, S. M. Payton, R. E. Tanzi and A. I. Bush, *J. Biol. Chem.*, 2004, **279**, 14673–14678.

120. D. Metodiewa, *Amino Acids*, 1998, **14**, 181–187.

121. J. Maeda, B. Ji, T. Irie, T. Tomiyama, M. Maruyama, T. Okauchi, M. Staufenbiel, N. Iwata, M. Ono, T. C. Saido, K. Suzuki, H. Mori, M. Higuchi and T. Suhara, *J. Neurosci.*, 2007, **27**, 10957–10968.

122. I. V. Murray, L. Liu, H. Komatsu, K. Uryu, G. Xiao, J. A. Lawson and P. H. Axelsen, *J. Biol. Chem.*, 2007, **282**, 9335–9345.

123. W. R. Markesbery and M. A. Lovell, *Arch. Neurol.*, 2007, **64**, 954–956.

124. M. A. Smith, P. L. Richey Harris, L. M. Sayre, J. S. Beckman and G. Perry, *J. Neurosci.*, 1997, **17**, 2653–2657.

125. M. A. Smith, G. Perry, P. L. Richey, L. M. Sayre, V. E. Anderson, M. F. Beal and N. Kowall, *Nature*, 1996, **382**, 120–121.

126. E. A. Ostrakhovitch, M. R. Lordnejad, F. Schliess, H. Sies and L. O. Klotz, *Arch. Biochem. Biophys.*, 2002, **397**, 232–239.

127. W. Wu, L. M. Graves, I. Jaspers, R. B. Devlin, W. Reed and J. M. Samet, *Am. J. Physiol.*, 1999, **277**, L924–931.

128. B. L. Vallee, J. E. Coleman and D. S. Auld, *Proc. Natl. Acad. Sci. USA*, 1991, **88**, 999–1003.

129. Z. Szallasi, K. Bogi, S. Gohari, T. Biro, P. Acs and P. M. Blumberg, *J. Biol. Chem.*, 1996, **271**, 18299–18301.

130. M. Ebadi, M. A. Elsayed and M. H. Aly, *Biol. Signals*, 1994, **3**, 123–126.

131. C. J. Frederickson, *Int. Rev. Neurobiol.*, 1989, **31**, 145–238.

132. C. J. Frederickson, J. Y. Koh and A. I. Bush, *Nat. Rev. Neurosci.*, 2005, **6**, 449–462.

133. T. G. Smart, A. M. Hosie and P. S. Miller, *Neuroscientist*, 2004, **10**, 432–442.

134. S. L. Sensi, P. Paoletti, A. I. Bush and I. Sekler, *Nat. Rev. Neurosci.*, 2009, **10**, 780–791.

135. T. B. Cole, H. J. Wenzel, K. E. Kafer, P. A. Schwartzkroin and R. D. Palmiter, *Proc. Natl. Acad. Sci. USA*, 1999, **96**, 1716–1721.

136. D. H. Linkous, J. M. Flinn, J. Y. Koh, A. Lanzirotti, P. M. Bertsch, B. F. Jones, L. J. Giblin and C. J. Frederickson, *J. Histochem. Cytochem.*, 2008, **56**, 3–6.

137. R. D. Palmiter, T. B. Cole, C. J. Quaife and S. D. Findley, *Proc. Natl. Acad. Sci. USA*, 1996, **93**, 14934–14939.

138. G. Ravaglia, P. Forti, F. Maioli, B. Nesi, L. Pratelli, L. Savarino, D. Cucinotta and G. Cavalli, *J. Clin. Endocrinol. Metab.*, 2000, **85**, 2260–2265.

139. H. N. Munro, P. M. Suter and R. M. Russell, *Annu. Rev. Nutr.*, 1987, **7**, 23–49.

140. A. L. Monget, P. Galan, P. Preziosi, H. Keller, C. Bourgeois, J. Arnaud, A. Favier and S. Hercberg, *Int. J. Vitam. Nutr. Res.*, 1996, **66**, 71–76.

141. E. Martinez Lista, J. Sole, L. Arola and A. Mas, *Biol. Neonate*, 1993, **64**, 47–52.

142. V. W. Bunker, L. J. Hinks, M. F. Stansfield, M. S. Lawson and B. E. Clayton, *Am. J. Clin. Nutr.*, 1987, **46**, 353–359.

143. L. Baum, I. H. Chan, S. K. Cheung, W. B. Goggins, V. Mok, L. Lam, V. Leung, E. Hui, C. Ng, J. Woo, H. F. Chiu, B. C. Zee, W. Cheng, M. H. Chan, S. Szeto, V. Lui, J. Tsoh, A. I. Bush, C. W. Lam and T. Kwok, *Biometals*, 2010, **23**, 173–179.

144. J. A. Molina, F. J. Jimenez-Jimenez, M. V. Aguilar, I. Meseguer, C. J. Mateos-Vega, M. J. Gonzalez-Munoz, F. de Bustos, J. Porta, M. Orti-Pareja, M. Zurdo, E. Barrios and M. C. Martinez-Para, *J. Neural. Transm.*, 1998, **105**, 479–488.

145. S. W. Suh, K. B. Jensen, M. S. Jensen, D. S. Silva, P. J. Kesslak, G. Danscher and C. J. Frederickson, *Brain Res.*, 2000, **852**, 274–278.

146. A. L. Friedlich, J. Y. Lee, T. van Groen, R. A. Cherny, I. Volitakis, T. B. Cole, R. D. Palmiter, J. Y. Koh and A. I. Bush, *J. Neurosci.*, 2004, **24**, 3453–3459.

147. J. L. Smith, S. Xiong and M. A. Lovell, *Neurotoxicology*, 2006, **27**, 1–5.

148. G. Lyubartseva, J. L. Smith, W. R. Markesbery and M. A. Lovell, *Brain Pathol.*, 2009, **39**, 221–228.

149. P. A. Adlard, J. M. Parncutt, D. I. Finkelstein and A. I. Bush, *J. Neurosci.*, 2010, **30**, 1631–1636.

150. A. I. Bush, W. H. Pettingell, Jr. M. D. Paradis and R. E. Tanzi, *J. Biol. Chem.*, 1994, **269**, 12152–12158.

151. A. I. Bush, W. H. Pettingell, Jr. M. de Paradis, R. E. Tanzi and W. Wasco, *J. Biol. Chem.*, 1994, **269**, 26618–26621.

152. A. I. Bush, G. Multhaup, R. D. Moir, T. G. Williamson, D. H. Small, B. Rumble, P. Pollwein, K. Beyreuther and C. L. Masters, *J. Biol. Chem.*, 1993, **268**, 16109–16112.

153. E. D. Ciuculescu, Y. Mekmouche and P. Faller, *Chemistry*, 2005, **11**, 903–909.

154. S. Scheuermann, B. Hambsch, L. Hesse, J. Stumm, C. Schmidt, D. Beher, T. A. Bayer, K. Beyreuther and G. Multhaup, *J. Biol. Chem.*, 2001, **276**, 33923–33929.

155. I. H. Park, M. W. Jung, H. Mori and I. Mook-Jung, *Biochem. Biophys. Res. Commun.*, 2001, **285**, 680–688.

156. D. E. Hoke, J. L. Tan, N. T. Ilaya, J. G. Culvenor, S. J. Smith, A. R. White, C. L. Masters and G. M. Evin, *FEBS J.*, 2005, **272**, 5544–5557.

157. R. A. Cherny, J. T. Legg, C. A. McLean, D. P. Fairlie, X. Huang, C. S. Atwood, K. Beyreuther, R. E. Tanzi, C. L. Masters and A. I. Bush, *J. Biol. Chem.*, 1999, **274**, 23223–23228.

158. X. Huang, C. S. Atwood, R. D. Moir, M. A. Hartshorn, J. P. Vonsattel, R. E. Tanzi and A. I. Bush, *J. Biol. Chem.*, 1997, **272**, 26464–26470.

159. J. Y. Lee, T. B. Cole, R. D. Palmiter, S. W. Suh and J. Y. Koh, *Proc. Natl. Acad. Sci. USA*, 2002, **99**, 7705–7710.

160. R. D. Terry, *J. Neuropathol. Exp. Neurol.*, 1996, **55**, 1023–1025.

161. D. H. Linkous, P. A. Adlard, P. B. Wanschura, K. M. Conko and J. M. Flinn, *J. Alzheimers Dis.*, 2009, **18**, 565–579.

162. M. Stoltenberg, A. I. Bush, G. Bach, K. Smidt, A. Larsen, J. Rungby, S. Lund, P. Doering and G. Danscher, *Neuroscience*, 2007, **150**, 357–369.

163. M. E. Quinta-Ferreira and C. M. Matias, *Brain Res.*, 2005, **1047**, 1–9.

164. J. A. Carson and A. J. Turner, *J. Neurochem.*, 2002, **81**, 1–8.

165. N. Iwata, S. Tsubuki, Y. Takaki, K. Shirotani, B. Lu, N. P. Gerard, C. Gerard, E. Hama, H. J. Lee and T. C. Saido, *Science*, 2001, **292**, 1550–1552.

166. W. Farris, S. Mansourian, Y. Chang, L. Lindsley, E. A. Eckman, M. P. Frosch, C. B. Eckman, R. E. Tanzi, D. J. Selkoe and S. Guenette, *Proc. Natl. Acad. Sci. USA*, 2003, **100**, 4162–4167.

167. H. Kanemitsu, T. Tomiyama and H. Mori, *Neurosci. Lett.*, 2003, **350**, 113–116.

168. M. A. Leissring, W. Farris, A. Y. Chang, D. M. Walsh, X. Wu, X. Sun, M. P. Frosch and D. J. Selkoe, *Neuron*, 2003, **40**, 1087–1093.

169. F. X. Gomis-Ruth, *Mol. Biotechnol.*, 2003, **24**, 157–202.

170. G. Grasso, A. I. Bush, R. D'Agata, E. Rizzarelli and G. Spoto, *Eur. Biophys. J.*, 2009, **38**, 407–414.

171. A. Caccamo, S. Oddo, M. C. Sugarman, Y. Akbari and F. M. LaFerla, *Neurobiol. Aging*, 2005, **26**, 645–654.

172. D. S. Wang, N. Iwata, E. Hama, T. C. Saido and D. W. Dickson, *Biochem. Biophys. Res. Commun.*, 2003, **310**, 236–241.

173. H. Shinall, E. S. Song and L. B. Hersh, *Biochemistry*, 2005, **44**, 15345–15350.

174. S. E. Lind, J. R. McDonagh and C. J. Smith, *Blood*, 1993, **82**, 1522–1531.

175. P. J. Crouch, D. J. Tew, T. Du, D. N. Nguyen, A. Caragounis, G. Filiz, R. E. Blake, I. A. Trounce, C. P. Soon, K. Laughton, K. A. Perez, Q. X. Li, R. A. Cherny, C. L. Masters, K. J. Barnham and A. R. White, *J. Neurochem.*, 2009, **108**, 1198–1207.

176. K. B. Pierson and M. A. Evenson, *Biochem. Biophys. Res. Commun.*, 1988, **152**, 598–604.

177. C. E. King, P. A. Adlard, T. C. Dickson and J. C. Vickers, *Clin. Exp. Pharmacol. Physiol.*, 2000, **27**, 548–552.

178. J. H. Suh, R. Moreau, S. H. Heath and T. M. Hagen, *Redox Rep.*, 2005, **10**, 52–60.

179. C. Bjorkdahl, M. J. Sjogren, B. Winblad and J. J. Pei, *Neuroreport*, 2005, **16**, 591–595.

180. W. L. An, C. Bjorkdahl, R. Liu, R. F. Cowburn, B. Winblad and J. J. Pei, *J. Neurochem.*, 2005, **92**, 1104–1115.

181. A. S. Zhang and C. A. Enns, *J. Biol. Chem.*, 2009, **284**, 711–715.

182. A. Takeda, *Yakugaku Zasshi*, 2004, **124**, 577–585.

183. W. R. Martin, F. Q. Ye and P. S. Allen, *Mov. Disord.*, 1998, **13**, 281–286.

184. G. Bartzokis, M. Beckson, D. B. Hance, P. Marx, J. A. Foster and S. R. Marder, *Magn. Reson. Imaging*, 1997, **15**, 29–35.

185. C. M. Morris, A. B. Keith, J. A. Edwardson and R. G. Pullen, *J. Neurochem.*, 1992, **59**, 300–306.

186. L. Goodman, *J. Nerv. Ment. Dis.*, 1953, **118**, 97–130.

187. J. R. Connor, S. L. Menzies, S. M. St Martin and E. J. Mufson, *J. Neurosci. Res.*, 1992, **31**, 75–83.

188. J. R. Connor, B. S. Snyder, J. L. Beard, R. E. Fine and E. J. Mufson, *J. Neurosci. Res.*, 1992, **31**, 327–335.

189. J. F. Collingwood, A. Mikhaylova, M. Davidson, C. Batich, W. J. Streit, J. Terry and J. Dobson, *J. Alzheimers Dis.*, 2005, **7**, 267–272.

190. J. Collingwood and J. Dobson, *J. Alzheimers Dis.*, 2006, **10**, 215–222.

191. G. M. Bishop, S. R. Robinson, Q. Liu, G. Perry, C. S. Atwood and M. A. Smith, *Dev. Neurosci.*, 2002, **24**, 184–187.

192. X. Zhu, B. Su, X. Wang, M. A. Smith and G. Perry, *Cell Mol. Life Sci.*, 2007, **64**, 2202–2210.

193. Y. Ke and Z. M. Qian, *Prog. Neurobiol.*, 2007, **83**, 149–173.

194. I. Grundke-Iqbal, J. Fleming, Y. C. Tung, H. Lassmann, K. Iqbal and J. G. Joshi, *Acta Neuropathol.*, 1990, **81**, 105–110.

195. C. M. Morris, J. M. Kerwin and J. A. Edwardson, *Neurodegeneration*, 1994, **3**, 267–275.

196. S. M. LeVine, *Brain Res.*, 1997, **760**, 298–303.

197. C. Bouras, P. Giannakopoulos, P. F. Good, A. Hsu, P. R. Hof and D. P. Perl, *Eur. Neurol.*, 1997, **38**, 53–58.
198. R. J. Castellani, P. I. Moreira, G. Liu, J. Dobson, G. Perry, M. A. Smith and X. Zhu, *Neurochem. Res.*, 2007, **32**, 1640–1645.
199. D. J. Pinero, J. Hu and J. R. Connor, *Cell Mol. Biol. (Noisy-le-grand)*, 2000, **46**, 761–776.
200. R. Danzeisen, T. Achsel, U. Bederke, M. Cozzolino, C. Crosio, A. Ferri, M. Frenzel, E. B. Gralla, L. Huber, A. Ludolph, M. Nencini, G. Rotilio, J. S. Valentine and M. T. Carri, *J. Biol. Inorg. Chem.*, 2006, **11**, 489–498.
201. V. C. Culotta, M. Yang and T. V. O'Halloran, *Biochim. Biophys. Acta*, 2006, **1763**, 747–758.
202. M. Sacharczuk, T. Zagulski, B. Sadowski, M. Barcikowska and R. Pluta, *Neurol. Neurochir. Pol.*, 2005, **39**, 482–489.
203. K. J. Robson, D. J. Lehmann, V. L. Wimhurst, K. J. Livesey, M. Combrinck, A. T. Merryweather-Clarke, D. R. Warden and A. D. Smith, *J. Med. Genet.*, 2004, **41**, 261–265.
204. L. Bertram, M. B. McQueen, K. Mullin, D. Blacker and R. E. Tanzi, *Nat. Genet.*, 2007, **39**, 17–23.
205. O. Ghribi, M. Y. Golovko, B. Larsen, M. Schrag and E. J. Murphy, *J. Neurochem.*, 2006, **99**, 438–449.
206. A. G. Mainous 3rd, S. L. Eschenbach, B. J. Wells, C. J. Everett and J. M. Gill, *Fam. Med.*, 2005, **37**, 36–42.
207. G. Bartzokis, T. A. Tishler, P. H. Lu, P. Villablanca, L. L. Altshuler, M. Carter, D. Huang, N. Edwards and J. Mintz, *Neurobiol. Aging*, 2007, **28**, 414–423.
208. M. Ujiie, D. L. Dickstein and W. A. Jefferies, *Front. Biosci.*, 2002, **7**, e42–47.
209. W. A. Jefferies, D. L. Dickstein and M. Ujiie, *J. Alzheimers Dis.*, 2001, **3**, 339–344.
210. A. Venti, T. Giordano, P. Eder, A. I. Bush, D. K. Lahiri, N. H. Greig and J. T. Rogers, *Ann. N. Y. Acad. Sci.*, 2004, **1035**, 34–48.
211. J. T. Rogers, J. D. Randall, C. M. Cahill, P. S. Eder, X. Huang, H. Gunshin, L. Leiter, J. McPhee, S. S. Sarang, T. Utsuki, N. H. Greig, D. K. Lahiri, R. E. Tanzi, A. I. Bush, T. Giordano and S. R. Gullans, *J. Biol. Chem.*, 2002, **277**, 45518–45528.
212. J. A. Duce, A. Tsatsanis, M. A. Cater, S. A. James, E. Robb, K. Wikhe, S. L. Leong, K. Perez, M. A. Greenough, H.-H. Cho, D. Cappai, R. D. Moir, C. L. Masters, R. E. Tanzi, R. Cappai, K. J. Barnham, G. D. Ciccotosto, J. T. Rogers and A. I. Bush, *Cell*, 2010, in press.
213. A. Yamamoto, R. W. Shin, K. Hasegawa, H. Naiki, H. Sato, F. Yoshimasu and T. Kitamoto, *J. Neurochem.*, 2002, **82**, 1137–1147.
214. J. T. Egana, C. Zambrano, M. T. Nunez, C. Gonzalez-Billault and R. B. Maccioni, *Biometals*, 2003, **16**, 215–223.

CHAPTER 9

The Role of Iron in Neurodegeneration

F. A. ZUCCA, F. A. CUPAIOLI AND L. ZECCA[*]

Institute of Biomedical Technologies – National Research Council of Italy, 20090 Segrate (Milano), Italy

9.1 The Biological Importance of Iron

Iron is a fundamental nutrient that plays an important role in primary life functions for all eukaryotes. The importance of iron in biological systems is due to its capacity as electron donor and acceptor.

Iron is a transition element with different oxidation states, and the most abundant and stable are ferrous (Fe^{2+}) and ferric (Fe^{3+}) states, which are easily convertible. In aqueous solution the presence of Fe^{2+} or Fe^{3+} states depends on the redox potential of the solution: in oxidant solutions or in the presence of oxygen, as in biological systems, iron tends to oxidize to Fe^{3+}, while in reducing solutions or in the presence of reducing agents iron is reduced to Fe^{2+}. Iron solubility also depends on the pH of the solution: acidic solutions promote Fe^{3+} solubilization, in alkaline medium Fe^{2+} is more soluble, while in neutral conditions Fe^{2+} is more soluble than Fe^{3+} (0.1 M vs 10^{-18} M at pH 7). In addition Fe^{3+} can be water-soluble, forming stable complexes with strong chelators.[1] Fe^{3+} prefers oxygen species as ligands, while Fe^{2+} prefers nitrogen and sulfur species. The redox potential of the Fe^{3+}/Fe^{2+} couple in H_2O is $+0.77$ V (eqn 9.1).

$$Fe^{3+}(aq) + e^- \rightleftharpoons Fe^{2+}(aq) \quad E = 0.77\,V \tag{9.1}$$

RSC Drug Discovery Series No. 7
Neurodegeneration: Metallostasis and Proteostasis
Edited by Danilo Milardi and Enrico Rizzarelli
© Royal Society of Chemistry 2011
Published by the Royal Society of Chemistry, www.rsc.org

The redox potential of this reaction falls into the range of biological systems,[2] and makes iron redox reactions essential for a variety of biological processes. Iron complexes can display a great variety of stability constants, depending on the oxidation state and type of ligand: many biological ligands chelated to Fe^{2+} or Fe^{3+} are able to satisfy almost the entire biological range of redox potential reactions (from -0.5 V to 0.6 V).[1] For these reasons, iron is absolutely required for life.

On the other hand, iron can be extremely toxic in aerobic conditions, for example in biological systems. A great deal of experimental evidence shows the ability of free iron or iron chelated to small molecules to catalyze formation of reactive oxygen species (ROS) through the Fenton's reaction (eqn 9.2). Fe^{2+} reacts with oxygen intermediates such as H_2O_2 to produce highly reactive free radical species. The reduction of the resulting Fe^{3+} by ascorbate or other reductants such as O^{2-} regenerates the active form Fe^{2+} (eqn 9.3), which can re-enter the redox cycling with the continuous production of toxic free radicals.[3–5]

$$Fe^{2+} + H_2O_2 \rightarrow Fe^{3+} + OH^\bullet + OH^- \tag{9.2}$$

$$Fe^{3+} + O_2^- \rightarrow Fe^{2+} + O_2 \tag{9.3}$$

Hydroxyl radicals generated in this reaction exert devastating effects on living systems, since they can interact with cellular components resulting in tissue damage. In biological systems these ROS can promote lipid peroxidation, affecting membranes permeability, protein oxidation and DNA double strand breaks.[4,6] Neurons, hepatocytes and kidney cells are the most susceptible to iron toxicity, which leads to neurodegenerative disorders, hepatic damage and renal carcinogenesis.[4]

Biological systems have developed highly regulated mechanisms to maintain adequate levels of iron for normal metabolism, keeping its redox potential under control. In normal conditions iron is bound to functional molecules, transport or storage proteins, avoiding its excess and toxicity that could cause cellular damage.

The total amount of iron in the body of an adult human male is about 3.5–4.0 g (50 mg/kg). The major part of the body iron (65%) is contained in hemoglobin, about 10% is distributed in myoglobin and in iron-containing proteins (enzymes and cytochromes), whereas the remaining is redistributed among transporting and storage proteins in several cell types.[7] Humans on a normal diet swallow 15–20 mg/day of iron, but only 1–2 mg/day of iron is absorbed in the duodenum.[8] The amount of iron is maintained in equilibrium between dietary iron intake and its physiological loss (1 mg/day in men, 1.5 mg/day in women).

Iron is incorporated into protein as an electron carrier or biocatalyst in several forms. It forms the heme group in proteins such as hemoglobin and myoglobin, for oxygen transport/storage/release, and in proteins involved in electron transport such as cytochromes. Moreover iron forms the Fe–S cluster of proteins involved in several pathways including electron transport in the respiratory chain, anaerobic metabolism (dehydrogenases and reductases) and

DNA synthesis (ribonucleotide reductase). Iron is also important during the immune response since it is a key regulator of host–pathogen interaction.[9]

In the brain iron is required for neurotransmitter synthesis: it is a cofactor for tyrosine hydroxylase, tryptophan hydroxylase and for enzymes involved in the formation of catecholamines and 5-hydroxytryptamine.[10] Iron in the brain is also essential in the myelination process and cell division.[11]

9.2 Iron Homeostasis: An Overview

9.2.1 Iron Absorption in the Intestinal Tract

Iron enters the body through the diet and its homeostasis depends exclusively on the regulation of its intestinal absorption, since its excretion is essentially unregulated through sloughing of mucosal cells, blood loss and urinary excretion.

Normally iron, in the form of inorganic non-heme iron, is absorbed in the proximal portion of the duodenum. This part of the intestinal tract is characterized by the presence of highly polarized cells (enterocytes) arranged in finger-like villous epithelium. The enterocytes present a microvillous apical surface and are connected by tight junctions that form a barrier between the lumen and blood vessels. In order to reach the blood circulation iron needs to pass through both apical and basolateral membranes, requiring specific transporters and enzymes that change its oxidation state (Figure 9.1.A).

Dietary iron is normally in the Fe^{3+} state and remains soluble in the intestinal tract because of the acidified microenvironment of the microvilli. Fe^{3+} is reduced to Fe^{2+} by the enterocytic luminal surface ferrireductase Dcytb (duodenal cytochrome B), which uses ascorbate as cofactor.[12] The expression of Dcytb ferrireductase is stimulated in conditions of iron deficiency and hypoxia, but other mechanisms or enzymes could be involved in iron reduction since mice lacking Dcytb do not show iron deficiency.[13]

Subsequently, Fe^{2+} is the substrate for the divalent metal ion transporter 1 (DMT1) that is located in the apical membrane of enterocytes and is responsible for iron uptake.[14,15] DMT1 is a proton symporter that cotransports iron and protons provided by the flow of gastric acid from the stomach to the duodenum; iron uptake is supported by the Fe^{2+} concentration gradient created by ferrireductase activity on the apical surface. The transporter can also import other divalent ions such as Mn^{2+}, Co^{2+}, Zn^{2+}, Cu^{2+} and Pb^{2+}.[16] *DMT1* mutants and knockout mice have important deficits in intestinal iron absorption (and also in its assimilation in erythroblasts as discussed below), confirming that DMT1 is the primary transmembrane dietary iron transporter into intestinal epithelial cells.[14,17] However, other cell types (in liver, placenta and brain) do not strictly require DMT1 for iron uptake, having other metal transporters that are still under investigation. The expression of *DMT1* is induced in conditions of iron deficiency and is probably controlled at post-transcriptional levels.[14,18]

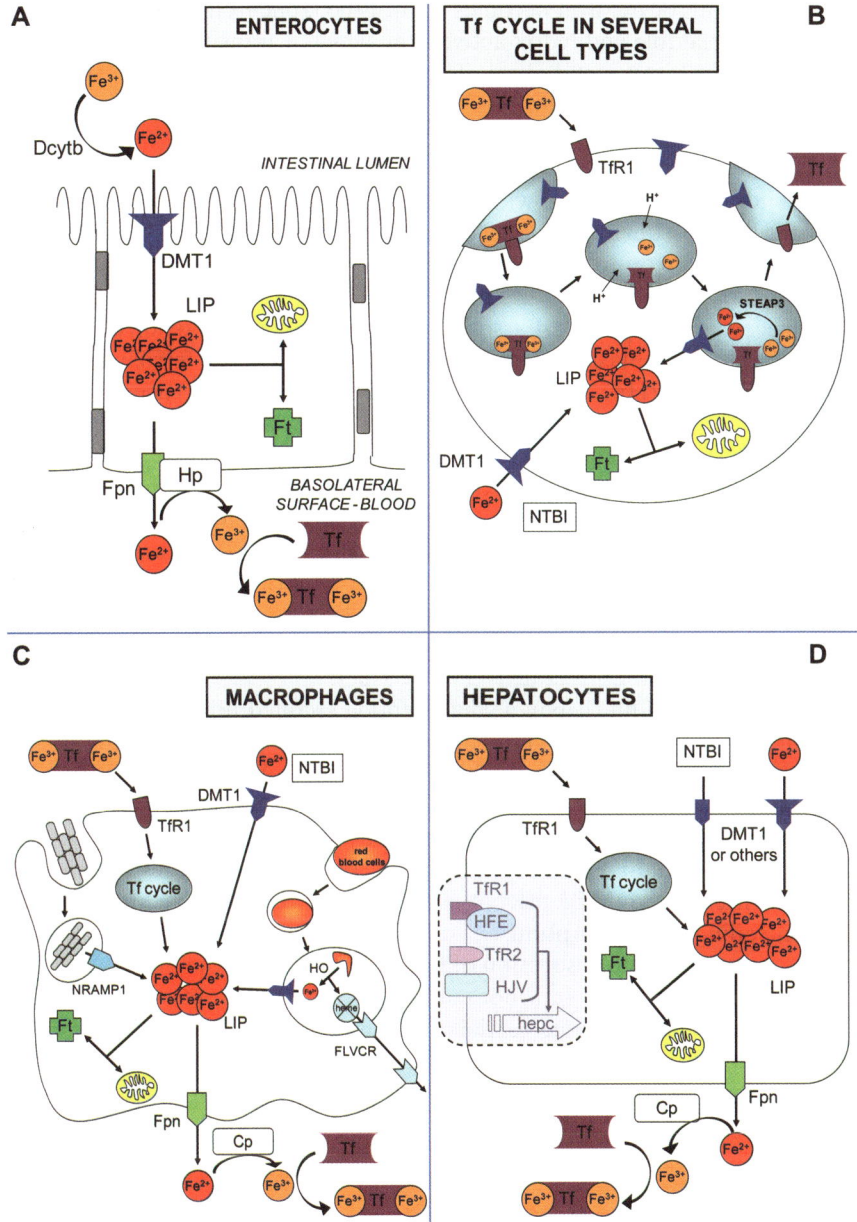

Figure 9.1 Overview of general iron homeostasis.
In this figure major actors in iron homeostasis are summarized at several levels. (A) Absorption of dietary iron by duodenal enterocytes. (B) The Tf cycle that is typical of erythroid cells and many other cellular types. (C) Iron recycling in macrophages. (D) Iron storage in hepatocytes; the shadowed area illustrates the main signal proteins that collaborate in the regulation of hepc expression (for more details see text).

Inside the enterocyte, iron becomes part of the labile iron pool (LIP). This is the pool of chelatable and redox-active iron, which is transitory and serves as a crossroads in cell iron metabolism. The iron can remain in the cell for energy production and storage or can be exported across the basolateral membrane. How iron passes through enterocyes is not totally clear: it has been suggested to occur via chaperone proteins and/or vesicular transcytosis, starting with DMT1 endocytosis, together with normal iron influx via the DMT1 proton symporter.[19] Subsequently, iron exits through the basolateral membrane using the transporter ferroportin (Fpn).[20] The key role of this protein in exporting iron was supported in studies with mice lacking Fpn.[21] Cellular iron export necessarily requires an associated ferroxidase which oxidizes exported Fe^{2+} to Fe^{3+}; ceruloplasmin (Cp), a circulating multicopper ferroxidase,[22] and mainly hephaestin (Hp), a membrane-bound Cp analogue,[23] accomplish this function.

The Fe^{3+} is rapidly chelated by blood circulating transferrin (Tf): this is a high-affinity iron binding protein that immobilizes two Fe^{3+} (holo-transferrin or diferric Tf–Fe) and is present in plasma, the lymphatic system and CSF. Serum Tf is mainly secreted by hepatocytes and is in high excess in plasma, where it is only 30% saturated under normal conditions, although it can be totally saturated in conditions of iron overload. In fact, its main function is to sequester iron and keep it in a non-reactive form, soluble in the aqueous environment to allow correct delivery to tissues.

At this level hepcidin (hepc), a small peptide secreted primarily by the liver, depending on iron loading and inflammatory status,[24] controls the quantity of iron translocated by Fpn into the blood circulation. When this protein binds to Fpn, it induces its phosphorylation and internalization followed by degradation in lysosomes,[25,26] leading to decreased export of cellular iron, which is retained in the intestinal epithelium. Moreover it has been suggested that hepc might downregulate DMT1 mRNA and protein production in enterocytes,[27] thus controlling iron absorption.

In the enterocytes the absorption of heme iron derived from meat in the form of hemoglobin or myoglobin is not completely clear. Attempts to demonstrate the presence of specific heme carrier proteins have not been completely confirmed,[28,29] thus it seems that heme iron is released from the protoporphyrin group by heme oxygenase 1 (HO-1) and can then enter the above mentioned absorption pathway for non-heme iron.

9.2.2 Iron Trafficking, Recycling and Use

Once iron is loaded in the circulatory system (mainly bound to Tf) from the intestine, it is then directed to the major site of utilization, the erythroid bone marrow, where it is used for the formation of circulating erythrocytes.

Erythroid and other cells express on their surface the transferrin receptor-1 (TfR1) which recognizes and binds the circulating Tf–Fe, initiating the Tf cycle.[30] The importance of this mechanism has been confirmed by depletion of

TFR1 gene in mice[31] and by the discovery of spontaneous mutations affecting the Tf gene both in animals and humans,[32,33] which lead to abnormal phenotypes. However, these mutational studies also revealed that, while erythroid cells strictly depend on the Tf cycle, probably as a result of their massive need of iron for hemoglobin synthesis, some cell types can use alternative iron delivery mechanisms.

The DMT1 on the cell surface of erythroblasts can directly import non Tf-bound iron (NTBI), since it is present in the plasma membrane in close proximity to TfR1 and both will invaginate for vesicle formation. The complex Tf–Fe–TfR1 invaginates in a process of clathrin-mediated endocytosis: these early endosomes are acidified by a proton pump and Fe^{3+} is released from Tf-TfR1 at a specific pH of ≈ 5.5.[34] Here it is reduced again by the STEAP3 ferrireductase[35] to Fe^{2+} in order to be cotransported with protons outside the endosomes by DMT1,[36] using the low pH of the endosome as a driving force.

This DMT1 is a different isoform of DMT1 than that of the apical membrane of enterocytes,[18] and its importance in the Tf cycle in erythroid cells was confirmed by disruption of the *DMT1* gene in mice.[14] Finally, the Tf–TfR complex is relocated to the plasma membrane and at the pH of this environment the apo-Tf (iron-free Tf) is released and remains available for a new Tf cycle (Figure 9.1.B).

In erythroblasts, Fe^{2+} released from endosomes can enter the mitochondria, mainly for incorporation into protoporphyrin IX by ferrochetalase, leading to heme formation.[37] Most heme in these cells is used for hemoglobin production. Iron can reach this destination from endosomes in two ways: directly through the contact between endosomes and mitochondria or indirectly using chaperon proteins or by becoming part of LIP.[37] In erythroblasts and in other cell types, mitochondrial iron is also used for biosynthesis of the Fe–S clusters that form prosthetic groups of many enzymes in the respiratory chain.[38] Frataxin, a mitochondrial matrix protein, plays a key role in mitochondrial iron homeostasis and biogenesis of Fe–S clusters since it binds to the Fe–S cluster assembly machinery, supporting its complexation.[39] Mitoferrin is essential for the import of Fe^{2+} into the mitochondria of erythroblasts and other cell types, since animals lacking it are unable to incorporate iron into heme.[40] Heme can be exported from the mitochondria via unknown carriers and its concentration in the erythroblast cytoplasm needs to be controlled by a heme exporter, FLVCR,[41] that acts like a valve to push out extra heme from cytoplasm, avoiding its toxicity. The importance of this heme exporter in normal erythropoiesis was demonstrated in knockout mice for its gene.[42]

Most circulating iron derives from a recycling system for the iron already present in the body; the most important source of recycled iron derives from old and damaged erythrocytes. These are phagocytosed by tissue macrophages, particularly in the spleen, lysed, and the iron (derived from the degradation of heme by HO-1) is released from the phagosomes via DMT1 and then from the cell through Fpn with the assistance of Cp.[21,43] Meanwhile the derived heme group can be exported via FLVCR, preventing its cellular over-accumulation and toxicity,[42] and providing recycled heme to developing erythroid cells. Macrophages can uptake iron via the Tf cycle using TfR1 and other receptors[44] and NTBI via DMT1. Here hepc, whose liver secretion is increased during inflammation in an interleukin-mediated pathway,[45] regulates Fpn, causing its internalization and degradation[43] inside macrophages. The result is a retention of iron recovered from senescent red blood cells inside macrophages with a decrease of circulating iron that could be used by microbes for reproduction. Moreover a DMT1 homologue, NRAMP1, is expressed in the phagosome of macrophages: it depletes iron and manganese from the phagosome, where phagocytosed microorganisms could replicate.[46] Iron overload in macrophages during inflammation is also sustained by the induction of macrophage DMT1 levels by several cytokines.[47] Fe^{2+} export from macrophages via Fpn requires an associated ferroxidase activity: macrophages and hepatocytes use exclusively Cp to convert Fe^{2+} to Fe^{3+} for Tf loading (Figure 9.1.C).[48]

Hepatocytes constitute the major deposit of iron. They store iron mainly in the form of ferritin (Ft) when serum Tf is replete, however it can be mobilized if necessary to other body districts when Tf is inadequately loaded. They can absorb iron through the Tf-Fe/TfR system via TfR1 and TfR2, a lower affinity TfR isotype highly expressed in hepatocytes,[49] but they also have an NTBI uptake system that is efficient when serum iron levels exceed Tf storage capacity.[50,51] The involvement of DMT1 in NTBI uptake in hepatocytes is still under debate;[14,52] some evidence has identified L-type Ca channels, other metal transporters and also TfR2 isotypes in mediating NTBI uptake.[53,54] However, the importance of TfR2 resides mostly in its function as an iron sensor, rather than in iron uptake via the Tf cycle. The levels of the TfR2 protein, whose expression is not under IRE/IRP control (see below), have been reported to increase in response to increased Tf–Fe levels,[55] thus playing a role in the regulation of hepc expression. The exit of iron from hepatocytes seems to occur via Fpn, with the assistance of Cp (Figure 9.1.D).

The circulating hormone hepc, which is normally expressed at high levels in the liver, is the principal actor in systemic iron regulation at several levels: during intestinal absorption of iron, it is utilized for erythropoiesis, macrophage iron recycling and hepatocyte iron storage/mobilization.[56] The importance of this hepatic hormone was suggested by several studies in which animals with iron overload showed an increased expression of hepc,[57] or others in which the inactivation of the hepc gene led to a severe iron overload phenotype,[58,59] or the forced expression of hepc caused profound iron deficiency by blocking its intestinal absorption and retention in macrophages.[60,61]

The expression of hepc is related to different stimuli that influence the regulation of systemic iron: it is largely produced when iron in serum is increased, during iron overload and inflammation, while its expression is diminished during iron request in erythroid cells, iron deficiency and hypoxia.[24,57] Once secreted, hepc binds to Fpn, causing its internalization and lysosomal degradation,[26] thus blocking cellular iron export mainly from intestinal enterocytes and macrophages.

The regulation of hepc expression seems to occur mainly at the level of transcription. However, the mechanism by which iron levels and the inflammatory status can signal via hepc to regulate systemic iron homeostasis is still under investigation: several models, involving Tf saturation sensors such as TfR2, NTBI sensors, transcriptional regulators and transductional signals, have been proposed (for review see reference 62). Mutations that cause decreased hepc production or activity are associated with hemochromatosis, an inherited disease causing severe parenchymal iron deposition in the liver, heart and endocrine tissues accompanied by a paucity of iron in intestinal epithelial cells and tissue macrophages. This disease can be caused by three types of mutation: i) mutations affecting the hepc gene that prevent the formation of the active protein; ii) mutations at the level of Fpn, inhibiting its interaction/regulation by hepc; iii) mutations in other genes, such as *HFE*, *TFR2* and *HJV* that encode for signal proteins highly expressed on hepatocyte surfaces that normally collaborate in regulation of hepc expression (Figure 9.1.D).[63,64]

9.2.3 Intracellular Iron Homeostasis

Each cell needs to maintain intracellular iron homeostasis, buffering its concentration for normal metabolism but simultaneously avoiding its overload, which could be highly toxic. Moreover not all the iron that enters the cell has to exit, and the amount of iron not used for cellular functions needs to be buffered.

For this reason cells produce the most important cellular iron storage protein, in which large amounts of iron are stored in soluble, non-reactive and bio-available form. Ft is a 24-subunits protein arranged to form a hollow shell, with an 80 Å cavity capable of storing up to 4500 atoms of Fe^{3+}, and is present in several organisms.[65] Mammalian Ft are heteropolymers composed of two types of subunit: H- (heavy) and L-chain (light).[66] The ratio between H- and L- subunits is tissue-specific: in general L-rich Ft are characteristic of serum and organs that store iron (*i.e.* liver, spleen) and contain high levels of iron (> 1500 atoms per molecule), while H-rich Ft are typical of brain and heart and contain a lower iron content (<1000 atoms per molecule). If iron stored in Ft is needed by the cells, the degradation of Ft is necessary for iron release: both lysosomal and proteasomal pathways have been suggested for this mechanism. The two chains are encoded by different chromosomes, have different amino acid sequences and have different functions. H-chains have a ferroxidase activity, binding Fe^{2+} and converting it to Fe^{3+} before storage inside the core. L-chains lack ferroxidase activity but are important in the nucleation of the metal core inside the protein.[65] At sub-cellular level, a new type of Ft has been recently discovered in mitochondria of some cell types (testis, neuronal cells and islets of Langerhans): this H-type Ft is synthesized and targeted to mitochondria where through its ferroxidase activity it can sequester potentially harmful free iron.[67]

Intracellular iron homeostasis can be controlled through a post-transcriptional regulatory mechanism. L- and H-Ft mRNA have 5′-untranslated regions (UTRs) necessary for translational control of mRNA Ft pools.[68,69] These can form stable RNA hairpins with a defined secondary structure, the so-called iron responsive elements (IREs).[70] Cytoplasmic proteins, known as iron regulatory proteins (IRPs), have been reported to specifically bind IRE.[71,72] Two types of

IRP have been identified: IRP1 and IRP2, which have a similar sequence but different functions (Figure 9.2).

IRP1 has an iron–sulfur cluster (4Fe·4S) and can operate as a regulatory protein or as cytoplasmic aconitase.[73] When iron is abundant the Fe–S cluster is formed, thus switching the function of IRP1 to aconitase; on the other hand when iron is scarce, the Fe–S cluster is incomplete and IRP1 acts as a RNA binding protein on IREs.[74] IRP2 does not have an Fe–S cluster and iron availability acts on its stability: when iron is abundant it cause IRP2 ubiquitination and degradation, but in conditions of iron depletion IRP2 accumulates in iron deficient cells, allowing its interaction with IREs.[75] Mutational studies have revealed that in most tissues, especially in erythroid cells, IRP2 plays the major role in iron dependent translational regulation.[76] When iron is low, the IRPs bind to the IREs, which in the case of Ft are located upstream of the start codon; this interaction blocks the formation of the initiation complex and the translation of Ft protein.[77] When iron is abundant, IRPs do not interfere with Ft translation, leading to an increase in the iron storage capacity through over-expression of Ft.

Other proteins are controlled by this excellent regulatory system. TfR1 mRNA has multiple IREs on its $3'$ end, and these $3'$ IREs act in mRNA stability control. During iron deficiency the IRPs bind to IREs of its mRNA and avoid degradation via endonucleases, leading to the production of TfR1 necessary for iron uptake. On the other hand, if iron is abundant, IRPs have no effect on TfR1 IREs and nucleases destroy TfR1 mRNA: this leads to a downregulation of TfR1.[78]

In addition Fpn, one enzyme for heme biosynthesis, and DMT1 (two of the four DMT1 isoforms) have been reported to have IREs (respectively at $5'$ for the first two proteins and at $3'$ for the last one) that can be controlled by IRPs in response to iron conditions.[18,79] When iron is abundant, IRPs do not block Fpn translation but the DMT1 mRNA is degraded; on the other hand during iron deficiency, IRPs block the translation of Fpn and prevent the degradation of DMT1 mRNA, enhancing its expression.[80] In addition, hemoglobinization is controlled by IRE/IRP regulatory network (Figure 9.2).

The importance of the IRP/IRE system in the control of iron homeostasis was supported by several studies in which the *IRP1* and *-2* genes were depleted in mice,[76] and the IRE of Ft or Fpn were inactivated by mutations in animals and humans,[81–83] leading to abnormal phenotypes. Although the IRE/IRP system is adequate for the majority of cells, it can be modulated or adapted for cells that need to maintain high levels of iron influx, such as erythroid cells (*i.e.* those maintaining high levels of TfR1).

9.3   Iron in the Brain

9.3.1   Iron Transport from the Blood to Brain Interstitial Fluid: Review of Proposed Mechanisms

After the liver, the brain contains the second highest iron concentration: here in particular it is essential for neuronal development, myelin formation and

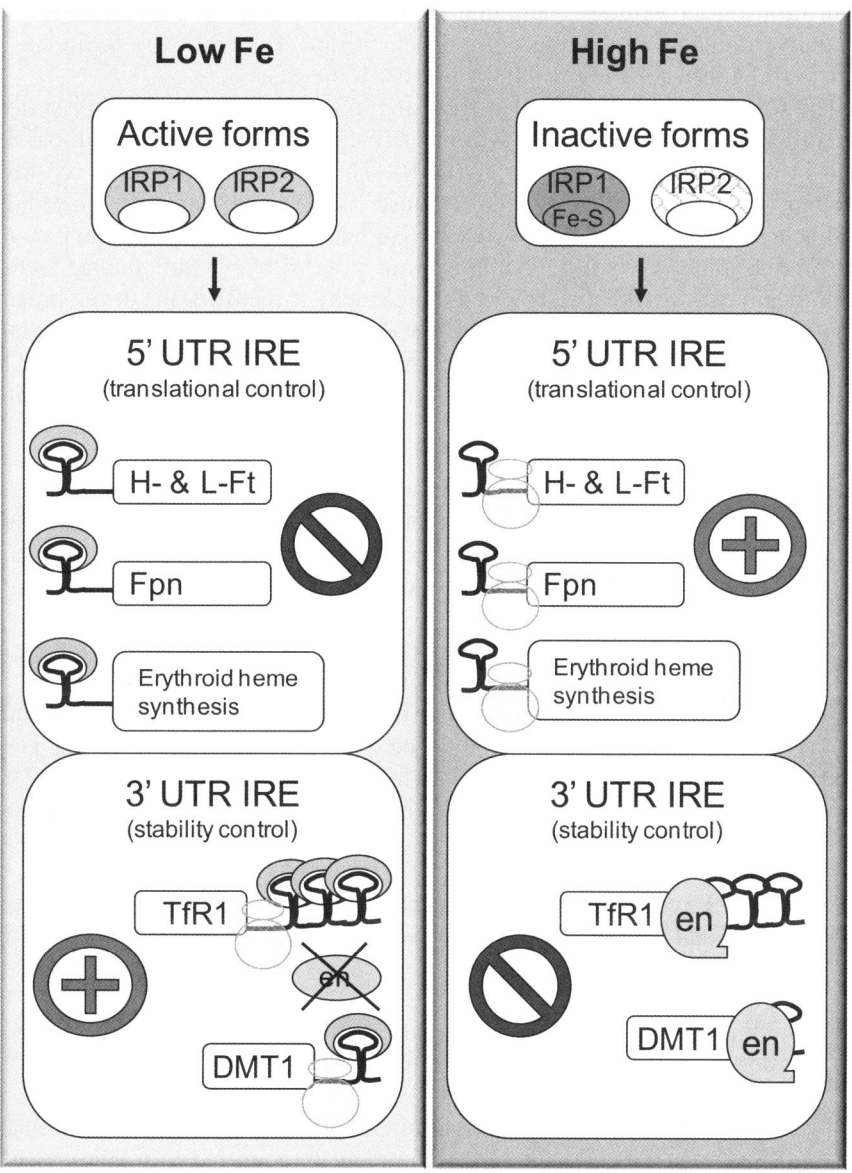

Figure 9.2   IRE/IRP regulatory system of proteins involved in iron metabolism.
IRP1 and IRP2 are sensitive to iron conditions. During iron deficiency these
proteins are active and can bind to the 5′ UTR of proteins involved in iron
storage, export and utilization for heme synthesis, inhibiting their translation.
They also bind to the 3′ UTR of other proteins for iron uptake, protecting
their mRNA from endonucleases and thus promoting protein translation. On
the other hand, when iron is in excess, IRP1 switches its activity to aconitase
and IRP2 is degraded: IRPs can no longer bind to 5′ UTR and 3′ UTR iron
responsive elements (IRE). In this case the translational complex is free to
synthesize proteins involved in iron storage, export and utilization, while the
mRNA of iron uptake proteins is degraded by endonucleases.

neurotransmitter synthesis and metabolism, but iron excess could lead to oxidative damage underlying neurodegeneration. Therefore the brain has a strict mechanism for the regulation of iron homeostasis.

The CNS is relatively excluded from the liver-dependent regulatory system because it is separated from the systemic circulation by a tight epithelial barrier known as the blood–brain barrier (BBB). This is confirmed by the reported absence of brain iron overload in mouse models and humans affected by hemochromatosis.[84] The BBB is a selective barrier formed by capillary endo-thelial cells joined by tight junctions and surrounded by a basal lamina:[85] while endothelial cells are the principal cellular element of the BBB, the development, regulation and functioning of the BBB requires the interaction of capillary endothelial cells with other cell types such as astrocytes and pericytes. The other important barrier that separates the CNS from the systemic circulation is the blood–cerebrospinal fluid (CSF) barrier formed by the epithelial cells of the choroid plexus. The choroid plexus consists of tuft-like fenestrated capillaries wrapped by epithelial cells joined by tight junctions, and it protrudes into brain ventricles into which it secretes the CSF.[85,86] The CSF and brain interstitial fluid, which is derived in part by secretion across the capillary endothelium of the BBB, are free to communicate in several locations by diffusion.[85]

Iron arrives in the blood flow mainly in the Tf–Fe form and can enter the CNS through the TfR1 that is highly expressed in the luminal membrane of the BBB endothelial cells, especially during brain development.[87] The importance of TfR1 in brain iron uptake was confirmed in fetal mice not expressing TfR1 that showed a lethal phenotype with failure of CNS development.[31] Two models of Tf–Fe transport across the BBB have been suggested and debated.[84,88]

i) The Tf–Fe complex could be transported by TfR1 in vesicles directly from the luminal to the abluminal side where endosomes release iron into the interstitial fluid,[84,89] without any release of iron from endosomes to endothelial cytosol, and from there to the brain interstitium. After Tf–Fe binding to the TfR1 at the luminal side and internalization, in acidified endosomes Fe^{3+} is released from Tf and, when vesicles fuse to the abluminal side, this detachment is also supported by the acidic micro-environment of the brain interstitial fluid, leading to the complete release of Fe^{3+}. At this pH Tf remains attached to its receptor and is then recycled to the luminal surface, returning to the blood circulation. Released Fe^{3+} can bind to brain interstitial Tf or to other low molecular weight molecules such as adenosine triphosphate (ATP), citrate and ascorbic acid that are secreted by astrocytic endfeet, accumulate greatly in this microenvironment,[90,91] and are finally transported like NTBI.

The direct transcytosis of Tf–Fe from luminal to abluminal side, which was the first mechanism thought to occur, was not completely confirmed by several studies demonstrating that serum Tf does not cross endothelial cells to enter the interstitial fluid directly.[88,92]

ii) The Tf–Fe is absorbed via TfR1-mediated endocytosis: in acidified endosomes of brain endothelial cells, Fe^{3+} released from Tf is reduced to Fe^{2+} by endosomal reductases and then exported into the cytoplasm, probably through DMT1, although there is some disagreement about DMT1 expression in BBB endothelium.[93] Subsequently it is exported into the interstitial fluid using Fpn, which has been detected immunohisto-chemically in the BBB by some authors[94] but not confirmed by others who detected Fpn only in neurons and oligodendrocytes.[95] Although the presence of Hp in several brain areas has been detected,[96] its role in the brain during export of iron across Fpn is far from understood. Once released by Fpn, Fe^{2+} is rapidly oxidized to Fe^{3+} by the enzyme Cp that is highly expressed at the endfeet of astrocytes: this Cp is anchored to the astrocyte membrane in the form of glycosylphosphatidylinositol-linked Cp.[97] The importance of this membrane-linked Cp was confirmed in patients with aceruloplasminemia: the absence of this ferroxidase prevents the oxidation of Fe^{2+} to Fe^{3+} for Tf binding and transport. This results in large accumulations of Fe^{2+} near astrocytic processes and neurons: the presence of a specific metal importer that is not Tf-dependent in these cells can cause an abnormal iron overload both in astrocytic foot processes[98,99] and in neurons. Finally, iron oxidized by Cp can rapidly bind to Tf of the brain interstitial fluid. There is an alternative pathway for Fe^{2+} released by Fpn in the interstitium: it can bind directly, or after its conversion to Fe^{3+} by Cp, to ATP or citrate released from astrocytes and can be transported in the brain interstitium.[91]

Although the major portion of iron transport across the BBB is normally Tf mediated, NTBI also readily crosses the BBB directly: this was confirmed in hypotransferrinemic mice with very low levels of circulating Tf[100] that showed normal brain levels, thus suggesting the presence of other proteins for iron transport across the BBB, such as melanotransferrin (MTf).[101] MTf has one high affinity binding site for Fe^{3+} and can exist in a soluble form that acts as a transporter and probably delivers iron to the second type, a membrane-anchored form, which acts as a membrane receptor mediating iron internalization in a poorly understood mechanism. The localization of these MTf in normal BBB endothelium[102] has suggested an alternative route for non Tf-dependent iron uptake in physiological conditions and in some neurological diseases, considering its over-expression in CSF and microglia in pathological conditions.[103] However, the MTf contribution might not be essential for iron uptake in normal conditions since mice deficient in MTf did not show any alteration in iron levels.[104]

The existence of NTBI in brain interstitial fluid and CSF was confirmed by several experiments revealing that iron concentrations exceed those of the binding capacity of Tf, which here is fully saturated in normal conditions.[90,91] This NTBI includes citrate–Fe^{3+} or –Fe^{2+}, ascorbate–Fe^{2+}, albumin–Fe^{3+} or –Fe^{2+}, lactoferrin (Lf)–Fe^{3+} and MTf–Fe^{3+} and Ft-bound iron. The pH of the interstitial fluid, which is slightly lower than that of blood, can favor the detachment of iron from Tf for subsequent binding to these molecules.[91] This NTBI can be absorbed mainly by oligodendrocytes and astrocytes that do not express a large amount TfR1, unlike neurons.[105,106]

However, the importance of Tf in iron binding and transport in the interstitial fluid should not be underestimated: interstitial Tf is the main source of iron for neurons and is synthesized and secreted mainly by oligodendrocytes in order to be transported in the brain interstitium to the cells of the CNS. The other main source of Tf is its slow diffusion from the ventricles to the brain interstitium via the CSF circulation, as demonstrated after injection of radiolabeled Tf–Fe into the rat lateral ventricle.[107] Tf can be synthesized by choroid plexus cells,[108] or imported by transcytosis through the choroid plexus from the blood.[92]

Lf has been suggested to be another potential extracellular iron carrier in the brain, considering its similarity to Tf and high affinity for two Fe^{3+}, normally associated with its main antibacterial activity in inflammatory reactions: it is present in CSF and increases during cerebral hemorrhage.[109] Receptors for Lf (LfR) were detected in brain vessels, suggesting that Lf can be imported to the brain from the blood circulation, as demonstrated in *in vitro* studies.[110] It seems that Lf scavenges iron released from degenerating cells, considering that Lf and LfR have been shown to be highly expressed in neurons or glial cells in pathological lesions of several neurological disorders.[111,112] It seems that the contribution of Lf and LfR to iron uptake in neurons and glia is not quantitatively significant under normal conditions.

Recently, the role of the choroid plexus in brain iron homeostasis has been reevaluated, considering recent measurements of its high surface area[113] and measurements of its high blood flow rate,[114] which were both underestimated in the past. For this reason the expression patterns of several proteins involved in iron transport/homeostasis, such as TfR1, DMT1, Fpn, Dcytb, Cp and Ft, were analyzed using *in situ* hybridization: high levels of expression of all these proteins were detected in mouse choroid plexus. These data support the hypothesis that the choroid plexus can mediate a significant amount of iron transport in the CNS.[86]

9.3.2   Iron inside Neurons

Although neurons can potentially take up NTBI present in the brain interstitium, they express high levels of TfR1 and DMT1,[105] suggesting that they mainly take up the Tf–Fe complex in the endosomes, from where iron is exported in the cytoplasm via DMT1 in collaboration with a ferroreductase. When inside the neurons, iron levels need to be efficiently buffered: IRPs respond to iron levels and through the mediation of IREs control the translation of proteins involved in iron homeostasis (TfR1, Ft, DMT1 and Fpn), as in other body regions. In conditions of iron deficiency there is an overexpression of TfR1 in neurons and not in BBB endothelial cells,[88] where only an increase in the cycling rate of endosomes without TfR1 over-expression has been suggested.[115]

Inside neurons, iron can be used for metabolic pathways, stored or secreted if in excess. As described above for other cell types, mitochondria of neurons import iron using the transporter mitoferrin and, with the support of the chaperon protein frataxin, iron is used for the biosynthesis of Fe–S clusters, the prosthetic group of many enzymes of the respiratory chain.

In the cytosol Ft, which is the most important iron binding protein of the brain, plays an important role in iron detoxification against oxidative damage and as iron reserve; moreover it can carry iron to the synapses. Here Fpn has been detected in presynaptic vesicles;[94] this discovery implies that iron may be transported from the cytosol into synaptic vesicles and then released into the synaptic cleft upon vesicle fusion. Here it can be taken up by astrocytes or post-synaptic neurons, by highly expressed DMT1.

Inside neurons neuromelanin (NM), a pigment that accumulates mainly in the substantia nigra and in several other brain areas, strongly chelates iron forming stable complexes thus protecting neurons against its toxicity.[116,117] The other important molecular form of iron is hemosiderin,[118] a degradation product of Ft which stores iron in its insoluble form together with lysosomal components and other degradation products: its levels may exceed iron stored in Ft inside the cells but its mobilization from hemosiderin is not well understood.[119,120] Inside nerve cells iron can exist also in the form of LIP.

The detection of Fpn in soma, axons and dendrites of neurons suggests that excess ferrous iron can be exported from neurons via this transporter:[95] this release may be associated with a Cp activity that oxidizes Fe^{2+} to Fe^{3+}, otherwise Fe^{2+} could be bound by ascorbic acid in the interstitium. In conditions of iron overload, the expression of Fpn is induced in neurons in order to reduce the intracellular iron concentration.[121]

9.3.3 Iron in Glia

The expression of TfR1 in astrocytes, oligodendrocytes and microglia is still under debate,[105,122–124] even if it should be considered that proteins expressed at low levels may be detected only with difficulty. The absence of TfR1 in astrocytes suggests that iron can be absorbed, like NTBI, by a mechanism that does not involve the TfR1; in addition the presence of DMT1 is still under debate.[125] Inside astrocytes, iron is efficiently buffered by Ft. HO-1, which is primarily expressed at low levels in astroglia and only in very few neurons,[126] seems to be involved in iron export from the cell and accumulation in mitochondria; since this protein is over-expressed under stress conditions,[127] its physiological role in iron trafficking is still under debate. The export of iron from astrocytes probably involves Fpn, which was detected in these cells and also induced in conditions of iron overload,[121] and Cp that is anchored to astrocyte membranes.[128] A lack of Cp, in mouse models or in aceruplasminemic patients, caused an accumulation of iron in astrocytes because iron export from the astrocyte is blocked by the absence of Cp.[48,97,129] Moreover the absence of Cp-linked astrocytes reduces the availability of Fe^{3+} for Tf binding, transport in the extracellular fluid and redistribution to neurons, and causes an excessive accumulation of extracellular Fe^{2+}, as explained in the previous paragraph.

In addition, oligodendrocytes seem to take up iron by a mechanism that does not involve the TfR1 since its presence has not been conclusively demonstrated. These cells can absorb NTBI iron (*i.e.* iron–citrate) and then chelate it by Tf, which is synthesized in large amounts by these cells. In oligodendrocytes Tf may have an additional important role in intracellular iron transport along their processes, which can reach long distances in the CNS wrapping around axons. Ft is the other iron binding protein inside oligodendrocytes, since these proteins are mainly expressed in these cells.[105,130] The export of iron from oligodendrocytes takes place via Fpn, which has been demonstrated in these cells.[95]

Microglia can take up NTBI, as demonstrated *in vitro*,[99] and maintains iron buffered mainly by Ft.[105] In resting microglia iron related proteins are detected in low levels.

9.3.4 Iron Export from the Brain

It seems that iron is efficiently maintained in the brain, because it does not diminish in animals with severe systemic iron deficiency as a result of dietary privation.[131] Some iron must be exported, but little is known about possible mechanisms by which iron can exit the brain in normal conditions. The major exit pathway is via the CSF circulation, which drains the interstitial fluid of the brain, and its successive reabsorption into the venous drainage system from the subarachnoid space by crossing the arachnoid membrane of granulations. This could be a regulated process as indicated by the presence of tight junctions.[90] This way of exiting was demonstrated by injecting radiolabeled Tf–Fe into the ventricles of mice; it was in part retained in the brain and in part detected in peripheral blood after injection.[107] It seems that Tf can diffuse in the reverse direction and carry iron released by cells via Fpn to the ventricles and subarachnoid space for exit from the brain.

In the CSF, the Tf binding capacity of iron is low because here the Tf concentration is low and iron chelation by Tf is limited by lower pH, high levels of ascorbic acid and low levels of Cp;[132] the capacity for iron export by Tf is limited in a such line of exit. In the CSF Lf, Ft and NTBI can contribute to iron export from the brain.[91] Moreover iron exit from the brain could be mediated by microglia and phagocytic cells in cases of neuronal death or hemorrhage.

9.4 Iron in Brain Aging: Health and Pathology

The brain contains 2% of the total body iron, corresponding to about 60 mg of non-heme iron.[119,133] Iron in different brain areas is not equally distributed. In normal adult human brain the highest iron concentration is localized in motor areas, especially in the extrapyramidal system: the red nucleus, substantia nigra, globus pallidus, nucleus caudatus and putamen.[133,134] High iron levels are also detected in the nucleus dentatus of the cerebellum, the hippocampus and the nucleus accumbens.[135] In contrast, the iron content in the cortices, thalamus and frontal white matter is significantly lower. Some studies showed a similarity between iron distribution and gamma-aminobutyric acid (GABA) receiving brain areas, suggesting a functional association between iron and neurotransmission.[135]

Under physiological conditions iron distribution in different brain cells is sensitive to neurodevelopment, metabolic activity and the neurodegeneration that occurs with normal aging. Iron accumulates in glial cells and neurons, but one-third to three-quarters of brain iron is stored in glial cells, both microglia and macroglia.[136] The most abundant iron- and Ft-containing brain cells in mouse, rat, monkey and human are oligodendrocytes.[130,135,137] These cells are necessary for myelination; in particular they are responsible for the synthesis of fatty acids and cholesterol required for myelin formation and this process requires iron-dependent enzymes. Oligodendrocytes accumulate iron during neonatal myelination and maintain high iron levels during the total lifespan.[105] The importance of iron in this process is proved by evidence from studies in which myelination was impaired in iron-deficient animals.[138,139]

Moreover, iron acts as a fundamental cofactor for enzymes that synthesize neurotransmitters, such as tyrosine hydroxylase, which catalyzes the hydroxylation of tyrosine to form dihydroxylphenylalanine, the limiting reaction in the synthesis of dopamine, noradrenaline (norepinephrine) and adrenaline (epinephrine). Iron is also contained in tryptophan hydroxylase, the enzyme involved in the first step of the conversion of tryptophan to serotonin;[140] it is implicated in the degradation of neurotransmitters as a cofactor of monoamineoxidase (MAO), the enzyme responsible for the degradation via oxidative deamination of dopamine, adrenaline and noradrenaline.[141]

In the brain up to 90% of non-heme iron is stored in Ft.[133,135] Immunostaining for H- and L-Ft revealed that these proteins are expressed both in neurons and in glia, but the relative abundance of H- and L- forms depends on cell type. Neurons contain H-Ft, microglia and astrocytes contain mainly L-Ft, while oligodendrocytes contain large amounts of both isoforms. In neurons the total amount of Ft is much lower than in oligodendrocytes, and in astrocytes the expression of Ft is very low.[105,142]

Iron accumulates in the brain as a function of age. At birth, in particular in the extrapyramidal area, iron is present in low concentration while a large amount of iron is required for brain development, in particular during the myelination process.[143–145] Non-heme iron accumulates linearly in the whole brain during the first three decades of life, than a plateau is observed until 60 years, when iron progressively increases again.[133] This process seems to be quite specific for different brain regions, and iron is stored mostly in certain types of cell.

In oligodendrocytes, which contain the largest amount of iron bound to Ft and Tf, the iron level during aging remains stable, but it increases in microglia and in astrocytes.[146] In older brain morphologically abnormal microglia were observed to be positive for Ft staining more frequently than normal ones.

This positive reaction indicates that iron overload or excessive iron exposure in microglial cells may be dangerous and might lead to degeneration.[147] Different observations support the hypothesis that iron overload in microglia can stimulate their activation, increasing the inflammatory process that occurs in neurodegenerative diseases.[148] Activated microglia and macrophages involved in inflammatory processes can also influence iron balance.

An age-dependent increase of non-heme iron concentrations is characteristic of brain areas involved in motor function, such as the substantia nigra, putamen, globus pallidus and motor cortex. In the frontal cortex, nucleus caudatus, putamen, substantia nigra and globus pallidus of human subjects at 67–88 years old, H-Ft levels were higher than in young individuals (27–66 years old), while L-Ft levels were increased only in the substantia nigra and globus pallidus.[149,150] During the entire lifespan the expression of Ft and HO-1 increases in the parieto-occipital cortex, while in the hippocampus only the expression of HO-1 seems to be augmented. In these brain areas HO-1 is localized both in glia and in neurons, while Ft has been detected only in glia.[151]

All these observations on the amount of iron during aging come from postmortem tissue, but the amount of iron in the brain can also be detected *in vivo* by non-invasive methods using magnetic resonance imaging (MRI). With the *in vivo* MRI method, called field-dependent R_2, increased iron can be detected for its magnetic properties and because iron bound to Ft interacts with the magnetic resonance scanner environment, exerting a strong magnetic effect. The total amount of iron contained in ferric oxy-hydroxide particles, which form the mineral core of Ft, increases relaxivity R_2 linearly with the field-strength, producing a highly specific and reproducible measurement of iron stores in tissue.[152] These authors showed a robust significant age-related increase in the iron level in the substantia nigra, nucleus caudatus, putamen, globus pallidus and hippocampus, while this increase was significantly lower in white matter.[153,154]

The greatest increase of iron deposition during the lifespan was observed in the substantia nigra. Different studies show that total iron levels in the substantia nigra in young individuals are very low (20 $ng\,mg^{-1}$ tissue in the first years of life)[150] but gradually increase between 16 and 90 years of age, reaching high concentrations such as 200–250 $ng\,mg^{-1}$ tissue.[155] In the substantia nigra the only well-known neuronal iron storage system is NM. NM is a dark brown pigment that accumulates linearly during aging especially in pigmented dopaminergic neurons of the substantia nigra and noradrenergic neurons of the locus coeruleus (Figure 9.3). NM is synthesized by the oxidation of excess catechols that are not accumulated in synaptic vesicles by vesicular monoamine transporter-2.[156] Different studies have demonstrated that NM is able to chelate iron with strong affinity, but also other transition metals and toxic compounds.[117,157] NM forms with iron an octahedral complex that contains a high-spin oxy-hydroxide iron (III) cluster. NM has a complex structure composed of a multimeric system, where each layer is composed of a melanic group bound to aliphatic and peptide chains.[117,158] The melanic structure contains benzothiazine and dihydroxyindole units and the dihydroxyindole

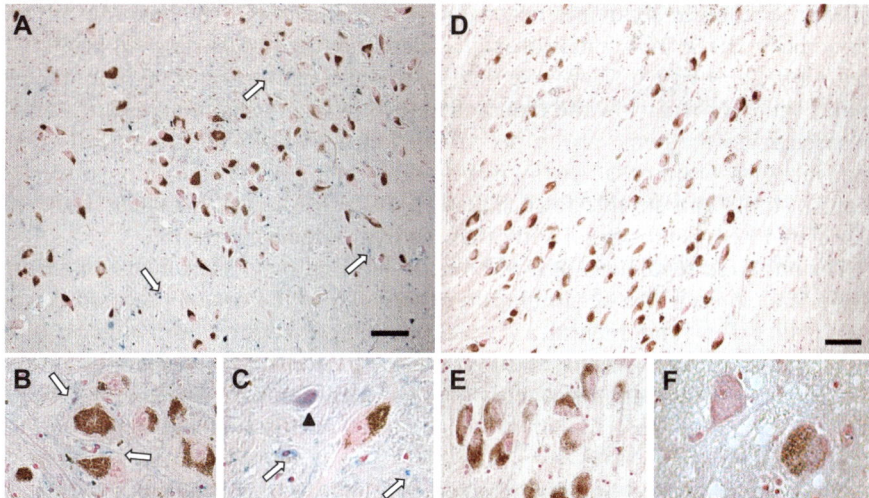

Figure 9.3 Iron and NM in aged brain.

Iron histochemistry with modified Perl's staining of human substantia nigra (A–C) and locus coeruleus (D–F) of normal 88- and 80-year-old subjects. Neuromelanin (NM) present in dopaminergic neurons of the substantia nigra and noradrenergic neurons of the locus coeruleus appears as brown granules and iron (III) deposits are stained blue. In the substantia nigra (A) many iron deposits are present and principally contained in glial cells (arrows). As shown in the two panels at higher magnification, iron deposits are absent in NM-containing neurons (B) but non-pigmented neurons show abundant cytoplasmic deposits (C; arrowhead). This confirms the ability of NM to scavenge iron to form stable complexes. In the locus coeruleus (D) a light staining was detected in very few glial cells and was completely absent from pigmented (E) and non-pigmented neurons (F), as shown at higher magnification. Reactive iron deposits are clearly abundant in the whole parenchyma of the substantia nigra while they are are undetectable in the locus coeruleus; this could explain the higher vulnerability of substantia nigra neurons when compared with the locus coeruleus. Scale bar = 100 µM.

units are responsible for the ability of NM to strongly chelate iron and other metals.[159] In substantia nigra neurons NM immobilizes high quantities of the total iron contained in the tissue, playing a key role in preventing oxidative stress induced by reactive iron, while the remaining iron is mainly stored in glial cells in the form of Ft.[155]

Iron accumulation during aging may not be due only to an incorrect intake but rather to a failure in the regulation of trafficking, storage and cellular iron metabolism. With aging, the vascular system tends to be compromised and the BBB is not always fully functional in the regulation of iron uptake.[160–162] The altered vascularization may lead to increased iron levels in certain brain regions, because moderate or significant hemorrhages occur relatively frequently in subarachnoid or intracerebral areas. During these hemorrhages blood cells leave the circulatory vessel, pouring out into the extravascular space where red blood cells can interact with CNS cells; hemoglobin released from red blood cells can be highly neurotoxic, as demonstrated by *in vitro* studies.[163,164] Additionally *in vivo* studies have demonstrated that intracerebral hemorrhage induce lipid peroxidation and free radical formation;[165] this could be due to iron released from the heme group propelling Fenton's chemistry, with fatal consequences for brain cells. Again, during brain aging iron is partially converted from a stable and soluble form linked to Ft to a form with higher reactivity associated with hemosiderin and other oxyhydroxide derivates.[145]

The impaired regulation of iron homeostasis that may occur in aged neurons could lead to excessive accumulation of free cytosolic iron and decreased iron availability for fundamental enzymes. For these reasons a failure in iron homeostasis/metabolism/distribution and in the control of its redox potential represents an important risk factor for neurodegeneration. We can argue that the age-related increase in iron levels, if not efficiently buffered, might induce neuronal damage.

Iron toxicity is based on Fenton's chemistry and the hydroxyl radical produced during this reaction can induce damage in DNA, lipids and proteins with a devastating effect on neurons. This oxidative damage affects lysosomes and mitochondria in particular, leading to abnormalities in regular degradative pathways and ATP production, with severe consequence for cellular functions. Antioxidants such as superoxide dismutase, catalase and glutathione work in combination to fight against free radical toxicity: superoxide dismutase catalyzes the dismutation of superoxide into oxygen and H_2O_2, then catalase decomposes H_2O_2 to water and oxygen, while glutathione acts as a reducing agent.

Neurons and glia respond in different ways to iron toxicity. Experimental evidence shows that astrocytes are more resistant than neurons to toxicity induced by an excess of iron.[166] Astrocytes have a high level of cellular metal-regulating proteins and in these cells the reduction of glutathione is much more pronounced than in neurons. Moreover neurons have very small quantities of catalase, so they have relatively low antioxidant defenses compared to the highly toxic action of iron overload. This difference makes astrocytes more able than neurons to remove free radical species and to maintain intracellular redox status. Iron overload together with low antioxidant defense makes neurons of specific brain areas very vulnerable to stress factors such as toxins or pathological processes like inflammation, and this condition can easily evolve into neurodegeneration. There is increasing evidence that abnormal iron accumulation in the brain may be involved in the pathogenesis of major neurodegenerative diseases such as Parkinson's (PD) and Alzheimer's disease (AD), and in other neurodegenerative conditions associated with genetic mutations, including aceruloplasminemia, neurodegeneration with brain iron accumulation, and neuroferritinopathy.

9.5   Iron in Parkinson's Disease

Parkinson's disease (PD) is a neurodegenerative disorder characterized by the degeneration of brain areas involved in the control of motor functions. In this disease the substantia nigra degenerates, leading to striatal dopamine deficiency, which is responsible for the major symptoms of PD: bradykinesia, dyskinesia, rigidity and tremor. The main pathological feature is the selective loss of dopaminergic neurons containing NM, while the unpigmented dopaminergic neurons are relatively spared.[167] The pigmented neurons of the locus coeruleus are only secondarily affected. The etiology of the disease is yet unknown, but only rare cases depend certainly on genetic factors: aging remains the main risk factor for PD. Different sources of evidence suggest that in the substantia nigra of PD patients oxidative stress plays an important role in neurodegeneration, and this is probably due to abnormal accumulation of reactive iron.[148,168]

In PD and in other parkinsonian syndromes an increase of the total iron concentration has been observed in the substantia nigra: the increase ranged from 33% to up to 176%.[169–171] The discrepancy in the rate of increase of iron is probably be due to the severity of the disease at the moment of death. Several studies have reported a significant increase in the total iron concentration in the substantia nigra in the most severe cases of PD, but not in milder ones.[172–174]

Microscopic studies demonstrate that in the substantia nigra pars compacta of PD patients there is an increase in reactive iron levels.[168] Moreover a recent paper confirms only an increase of the LIP concentration in the substantia nigra of PD patients, rather than an increase in total iron levels. This excess of redox active iron can explain the oxidative stress damage in parkinsonian substantia nigra.[175] In the locus coeruleus iron levels are very low, and during the entire lifespan they remain constant, suggesting that locus coeruleus neurons are less exposed than those of the substantia nigra to iron toxicity.[155]

In normal subjects the Fe^{3+} : Fe^{2+} ratio is 2 : 1, while in PD patients Fe^{2+} increases, modifying the Fe^{3+} : Fe^{2+} ratio to 1 : 2.[169,172,176,177] The presence of highly reactive iron promotes the conversion of H_2O_2 to hydroxyl radicals via Fenton's reaction, with great amplification of oxidative stress. This stress condition may enhance the release of iron from storage molecules such as Ft, heme proteins such as cytochromes, Fe–S proteins,[178] and eventually from NM.[116,179]

In PD patients, iron content increases also in the lateral globus pallidus, while in the medial globus pallidus the increase was not observed, and this suggests the retrograde degeneration of dopaminergic neurons of the substantia nigra in the disease.[180] There is also an inverse relationship between dopamine concentration and iron content in the putamen of patients with PD, but not in the substantia nigra, supporting the hypothesis of a retrograde degenerative process in the progression of PD.[181] Fe^{3+} deposits are localized in microglia, oligodendrocytes and astrocytes close to neurons of the substantia nigra and in neurons and glia in the putamen and globus pallidus of PD patients.

During normal aging several factors interact to influence the vulnerability of subtantia nigra neurons: the presence of dopamine, whose metabolism can produce reactive species, low levels of antioxidants such as glutathione, and high iron levels. This poorly controlled oxidative condition may be the major cause of death of dopaminergic neurons in the substantia nigra during PD.[182] Dopamine is unstable and it is subject to autoxidation, resulting in the formation of dopamine–quinone with production of superoxide anion radicals. This reaction is catalyzed and accelerated by high iron concentrations.[183] Moreover dopamine autoxidation is further increased in the early stage of the disease when dopaminergic neurons die and the remaining ones increase dopamine production to compensate for the deficit. In neurons and glial cells, dopamine is metabolized by MAO with H_2O_2 formation. The amount of MAO-B isoform of the enzyme increases with age,[184] and its levels seems to be highest in the substantia nigra. In this region large numbers of MAO-B positive astrocytes are present and this may contribute to a local increase in oxidative stress.[185] The cell membrane is very permeable to the H_2O_2 produced in glial cells by MAO-B, and H_2O_2 can cross into neighboring dopaminergic neurons where it may react with Fe^{2+} to produce toxic hydroxyl radicals with further dopamine oxidation. The glial cells are protected from toxic levels of H_2O_2 by the presence of high levels of glutathione and glutathione peroxidase, while neurons do not have such high levels of these protective agents and are more susceptible to the redox dopamine cycle.

A pathological characteristic of PD and other neurodegenerative disorders such as "neurodegeneration with brain iron accumulation" and "Lewy body disease" is the presence of Lewy bodies. They are eosinophilic inclusions of the cytoplasm formed by aggregated proteins with a dense core made of granular material, surrounded by widely radiating filaments. Lewy bodies are mainly contained in dopaminergic neurons of the substantia nigra of PD patients, but recently they have been detected in several brain regions degenerating in different stages of the pathology.[186] Lewy bodies are mostly composed of α-synuclein aggregates, and the oxidation of methionine residues of α-synuclein plays an important role in the formation of soluble α-synuclein oligomers. Moreover iron and free radical producers, such as dopamine and H_2O_2, promote the aggregation of α-synuclein. It seems that Fe^{2+} acts in concert with H_2O_2 to promote α-synuclein aggregation and this hypothesis is confirmed by the inhibition of α-synuclein aggregation in the presence of the iron chelator desferrioxamine.[187] Moreover the aggregation of α-synuclein induces the accumulation of redox reactive iron in neurons of the substantia nigra and this iron may catalyze the formation of free radicals by Fenton's reaction, exacerbating the oxidative stress cycle.[188]

In the substantia nigra of PD patients Ft are more heavily iron-loaded compared to those of normal subjects, indicating that in PD the quantities of Ft are inadequate to sustain heavy iron loading.[149] In PD patients the iron accumulation is not associated with up-regulation of Ft, thus contributing to the increase of oxidative stress in the substantia nigra without changes in IRP1 activity, the predominant form of IRPs in the human mesencephalon.[189] During PD progression the numbers of Ft-loaded microglia are increased in the substantia nigra, and reactive microglia are often associated with degenerating NM-containing neurons.[190]

In NM-containing neurons of healthy substantia nigra and locus coeruleus, iron is mainly sequestered by NM, since Ft and reactive forms of iron such as hemosiderin were not detected in pigmented neurons of both areas.[155] Other authors detected only a light positivity for H-Ft in some pigmented neurons of the substantia nigra.[105] NM is able to bind iron to a high degree in a saturation-dependent manner. Intracytoplasmatic iron deposits are absent from NM-containing neurons of the substantia nigra, probably due to the ability of NM to chelate iron (see Figure 9.3). In pigmented neurons NM scavenges free iron, forming stable complexes, reducing the risk of iron mobilization and the toxic effects of free iron.[159] Pathological increase of iron levels may saturate the iron-chelating sites of NM and the pigment may release iron from low affinity binding sites, which may result in an increased, rather than decreased, production of hydroxyl radicals.[191] In the substantia nigra of PD patients an increase of redox activity was observed in iron–NM aggregates released by dying neurons. The redox activity is higher in patients with the most severe neuronal loss, but it is not detected near pigmented neurons, denoting the key role of released NM in the modulation of iron redox toxicity.[168] The NM concentrations in normal substantia nigra and locus coeruleus are similar, but the iron content in NM of the substantia nigra is higher than in that of the locus coeruleus. These findings, together with the low levels of total iron and Ft in the locus coeruleus, suggest that in this area iron mobilization and toxicity is lower than in the substantia nigra. This may explain the different neuronal vulnerability of substantia nigra and locus coeruleus.[155]

In PD, NM released from degenerating neurons of the substantia nigra remains in the extracellular environment for long periods and in large amounts because of its insolubility.[157,159] *In vitro* studies showed that the NM–iron complex activates microglia, stimulating the release of pro-inflammatory factors such as tumor necrosis factor-α, interleukin-6 and nitric oxide, and causes neurodegeneration in primary ventral midbrain cultures in a microglia-dependent manner.[192,193] In animal models the injection of NM into the substantia nigra, mimicking the extra-neuronal NM, induces microglia activation and neuronal death.[193,194] This evidence suggests that NM released from dying neurons can induce chronic inflammation in the substantia nigra of PD patients, promoting neurodegenerative processes.

PD is also associated with increased HO-1 immunoreactivity in the neuropil and Lewy bodies of affected dopaminergic neurons;[195] in the same neurons an increase in the expression of DMT1 was also observed.[103]

Elevated iron levels in the substantia nigra during PD may be due to impaired gene expression or mutations in genes relevant to iron transport through vessels and cell membranes, or in iron binding. High iron concentration seems to be independent from the Tf/TfR system, in fact there is a significant decrease in TfR density on melanized neurons of the substantia nigra of PD patients.[196] Moreover during PD, Lf and LfR density is augmented in substantia nigra neurons and microvessels.[111,112]

Whether iron accumulation in PD is an early or late event and whether increased iron and iron-induced free radical reactions are the cause of the neurodegeneration or the consequence of a pathological process is still under debate.

9.6   Iron in Alzheimer's Disease

AD is another severe age-related neurodegenerative disorder and is the most frequent cause of a decline in cognitive ability and dementia in the elderly. It is characterized by the progressive degeneration of synapses and neurons of the brain cortices; the degeneration involves primarily the enthorinal cortex and the neurons of the hippocampus. The main symptoms of AD are the significant cognitive and memory loss, difficulty in performing familiar tasks, problems with language, disorientation and personality changes. AD is related to abnormal accumulation of extracellular plaques and intraneuronal tangles. In this disease the extracellular deposition of amyloid-β peptide (Aβ)1–40 and Aβ1–42 aggregates occurs in the brain cortex and hippocampus, resulting in senile plaques; this disease is also characterized by the development of neuro-fibrillary tangles, intracellular aggregates composed of paired helical filaments of hyperphosphorylated cytoskeletal tau protein.[197] In AD an increased expression of apoptotic proteins, an impaired ubiquitin–proteasome system and oxidative stress are also observed.[198]

Brains of AD patients accumulate iron within specific regions, in particular the hippocampus and cerebral cortex, displaying a selective susceptibility to neurodegeneration. The surplus of iron involved in the pathology is a possible source of oxidative stress generated by ROS, explaining the high oxidative load observed in neurons of AD brains. In AD neurons two enzymes involved in oxidative stress are activated: HO-1[199] and NADPH oxidase.[200] It has been demonstrated that HO-1 is significantly over-expressed in cerebral vasculature, in astrocytes and neurons of the hippocampus and temporal cortex, and is co-localized in senile plaques and neuronal neurofibrillary tangles in the same brain areas of AD subjects.[201] Oxidative stress is probably one of the early changes in AD and it can be attributed to an altered redox iron state that occurs also in the preclinical phase of AD.[202] In such patients, redox active iron is associated with both neurofibrillary tangles and senile plaques. Iron and Ft have been observed in AD brains with histochemical and immunohistochemical methods especially in senile plaques, in astrocytes and in microglia surrounding the senile plaques.[105,203] Moreover other histochemical studies in AD patients have demonstrated the presence of Fe^{3+} deposits, co-localized with Fe^{2+}, in senile plaques, neurofibrillary tangles, and neuropils.[204]

The formation of Aβ aggregates is directly influenced by the presence of iron; in fact, in the presence of divalent metals such as Fe^{2+}, Aβ spontaneously self-aggregates into neurotoxic fibrils.[205] In addition, Fe^{3+} is able to influence the formation of amyloid fibrils. The interaction of iron with Aβ fibrils reduces the fibrils and promotes the production of ROS.[206] ROS in turn autoxidize Aβ to create a protease-resistant soluble form and cross-linked Aβ, and have effects on lipid membrane peroxidation, protein oxidation and DNA breakdown. In the pathogenesis of AD, iron is not efficiently buffered by storage proteins, thus iron tends to accumulate in the brain cortices where it interacts with Aβ fibrils. Iron is also able to influence the production of Aβ through the modulation of furin, which is a ubiquitous endoprotease required to cleave many precursor proteins such as those of α- and γ-secretase. Aβ is generated by proteolysis of amyloid-β precursor protein, which is promoted by α-, β- and γ-secretases. The amyloid-β precursor protein is processed into the amyloidogenic or non-amyloidogenic pathway. In the non-amyloidogenic pathway α-secretase cleaves the amyloid-β precursor protein to generate the soluble fragment, precluding the formation of Aβ aggregates. In the amyloidogenic pathway the precursor protein is sequentially cleaved by β- and γ-secretase, resulting in the production of insoluble Aβ aggregates. High iron levels down-modulate furin expression which in turn reduces the ability of α-secretase to produce the soluble fragment and promotes β-secretase activity with consequent enhancement of the production of Aβ insoluble aggregates.[207] In fact the furin mRNA levels are reduced in the brains of AD patients.[208] Iron is also involved in the production of amyloid precursor protein that is a transmembrane metalloprotein cleaved by β- and γ-secretase into 40–42 amino-acid prefibrillar amyloid. In non-amyloidogenic pathway amyloid precursor protein is cleaved by α-secretase in Aβ to release the 90 kDa ectodomain of amyloid precursor protein. On the mRNA encoding the amyloid precursor protein there is a functional IRE in the 5′-UTR, and *in vitro* systems have demonstrated that intracellular iron levels are able to modulate the synthesis of amyloid precursor protein.[209]

Iron, especially Fe^{3+}, is localized also in neurons with neurofibrillary tangles and it is probably involved in their formation. *In vitro* experiments show that Fe^{3+} is able to bind hyperphosphorylated tau protein and promotes its aggregation, inducing the formation of neurofibrillary tangles.[198]

In AD brains an abnormal level of iron-accumulating proteins such as Ft and Tf has been observed, in particular in the hippocampus and cerebral cortex.[210] In AD patients iron is complexed with the Ft that is mainly localized in neuritic processes associated with amyloid plaques.[148] Abnormal Ft has also been detected in AD brain in oligodendrocytes and in dystrophic myelinated neuropils. Compared to physiological Ft, the pathological form has different crystalline structures: in pathological Ft two cubic mixed Fe^{2+}–Fe^{3+} oxides are the major component, whereas ferrihydrite, a hexagonal Fe^{3+} oxide, is the major component of physiological Ft. This situation is probably caused by a dysfunction in ferroxidase activity of pathological Ft.[211] Iron in oligodendrocytes is necessary for myelination, and here the different mineralization of iron inside Ft could affect myelination processes, explaining the decline in myelin content in affected brain regions of AD patients: this loss of myelin can be associated with disturbances in the information flow between neurons.

In hippocampus neurons containing senile plaques hemosiderin has been detected, especially in sulfur-rich dense bodies of these dystrophic neurites. Hemosiderin in this brain region is also localized in lysosomes and siderosomes of glial cells.[212]

In AD patients Tf is significantly decreased compared with aged normal subjects, especially in the white matter of the cerebral cortex. This Tf decrease could indicate reduced iron mobility, utilization and availability. This condition could play a significant role in increased oxidative damage and neuronal degeneration during the course of the disease.[213]

There is also a link between iron transport and the iron overload observed in AD patients. In post-mortem AD brain two isoforms of DMT1 co-localized with Aβ fibrils in the senile plaques. In the cortex and hippocampus of animal models (transgenic mice expressing amyloid precursor protein and human presenilin-1) both isoforms of DMT1 are increased if compared to wild-type control.[214] This finding suggests that iron overload in AD is not only attributed to an imbalance of iron accumulating protein but also to a deregulation of metal transporters.

In AD brains Cp levels may also be altered: Cp tends to increase in many brain regions, with the largest increase in the CA1 region of the hippocampus. Cp is localized also in senile plaques and occasionally in neurofibrillary tangles.[215,216]

There is also a link between AD and conditions of congenital iron overload, for example hemochromatosis, which is mainly associated with mutations in the *HFE* genes. The HFE proteins in the brain are found in blood vessels and in the choroid plexus. In AD patients HFE is found on glial cells in association with neuritic plaques, in astrocytes and on neurons in the vicinity of the plaques. HFE can be observed both in neurons containing neurofibrillary tangles and in neurons lacking tangles. The pattern of neuronal staining for HFE in AD might indicate that HFE can be induced by stress, and neurons that stain for tau in the periplaque area are also HFE-positive.[217] *HFE* mutations cause one of the most common autosomal recessive genetic disorders of the Caucasian population: the most frequent mutations are H63D and C282Y. The C282Y mutation occurs less frequently than the H63D mutation (1.9% *vs* 8.9%) and is more frequently associated with clinical hemochromatosis,[218] but each of these variants alone has relatively little effect on iron status. Probably only the combination of either C282Y or H63D and Tf C2 alleles (see below) might contribute to higher iron loads during the preclinical phase of AD.[219] The *HFE* mutation alone is only a risk factor for AD, since its principal effect is in regulation of systemic iron homeostasis. Several reports indicate an increase in *HFE* mutations in other neurodegenerative disorders such as PD, multiple sclerosis, and amyotrophic lateral sclerosis.[220,221] The *HFE* mutation is not the only mutation in a gene encoding a protein participating in iron management that may have an impact on AD: Tf subtype C2 was also found with increased frequency in AD patients when compared with age-matched controls.[222] AD patients with homozygosity for the apolipoprotein E isoform E4 allele are twice as likely to have a Tf C2 allele than those without or with only a single copy of apolipoprotein E isoform E4.[222,223] As cited before, the presence of the C2 variant in association with an *HFE* mutation can raise the risk of pathogenesis of AD five-fold, and this risk is increased again in presence of apolipoprotein E isoform E4.

These findings suggest that genetic alterations in specific iron homeostasis proteins can increase the risk, anticipate the onset of AD, and contribute to the pathogenesis, but are not able alone to induce the pathology.

9.7   Iron in other Neurological Diseases

PD and AD are not the only common neurodegenerative disorders in which iron homeostasis is involved. Iron, especially iron overload, is also implicated in several other neurological diseases. There is a group of disorders known as "neurodegeneration with brain iron accumulation" (formerly Hallervorden–Spatz syndrome) in which there is a progressive extrapyramidal disease with brain iron accumulation usually in the basal ganglia. The subgroup of "neurodegeneration with brain iron accumulation disease" includes the autosomal recessive disorders aceruloplasminemia, the autosomal dominant disorder neuroferritinopathy (*L-Ft* mutation), and the recessive disorders pantothenate kinase 2-associated neurodegeneration (*PANK2* mutation).

Mutation in the *Cp* gene causes the complete absence of Cp ferroxidase activity with serious consequences. Aceruloplasminemia is inherited in an autosomal recessive manner that results in the inability to oxidize Fe^{2+} and remove its excess: as a consequence iron accumulates in cells inducing biomolecule peroxidation and ROS formation. Congenital aceruloplasminemia is characterized by iron accumulation in the brain and viscera. Patients with aceruloplasminemia are characterized by diabetes mellitus, progressive CNS and retinal degeneration with typical neurologic symptoms of movement disorder (blepharospasm, grimacing, facial and neck dystonia, tremors, chorea) and ataxia (gait ataxia, dysarthria). Patients with aceruloplasminemia also present with psychiatric disturbance, including depression and cognitive dysfunction.[224]

Neuroferritinopathy or dominant adult-onset basal ganglia disease is an autosomal dominant disease characterized by a mutation in the *L-Ft* gene. To date two mutations linked to the disease have been studied: an adenine insertion at position 460–461 in exon 4 of the *L-Ft* gene (FTL460–461InsA mutation) and the insertion of two nucleotides (TC) at position 498–499 (FTL498–499InsTC mutation).[225] The FTL460–461InsA mutation alters the reading frame of the gene and the last 22 amino acids of L-Ft are replaced by 26 different amino acids. The symptoms of this *L-Ft* mutation arise in the fourth decade of life and the disease is characterized by motor dysfunction, manifested by bradykinesia and abnormal posturing of the arms, sometimes with chorea, dystonia, and an akinetic-rigid Parkinsonian syndrome predominating, while the cognitive functions are well preserved.[226] In patients with the FTL460–461InsA mutation abundant spherical inclusions are present in the globus pallidus which contain Ft and iron and produce cyst formations in the basal ganglia. Abnormal aggregates of Ft and iron are also found in the substantia nigra of patients, while in the white matter there are signs of neurodegeneration such as axonal swelling.[227] MRI shows cavitation of the globus pallidus. These patients have biochemical and hematological parameters in the normal range, with the exception of low Ft levels in serum.[226]

The FTL498–499InsTC mutation causes a shift of the reading frame and the last nine amino acids are substituted by a new sequence of 25 amino acids. In this case the symptoms appear at the age of 20 and progress over four decades.

The FTL498–499InsTC mutation is characterized by tremor progressively complicated by cerebellar signs, dyskinesias, rigidity, pyramidal signs and, unlike the FTL460 461InsΛ mutation, also cognitive dysfunction. The brain shows mild cerebral and cerebellar atrophy and cavitation of the putamen. In astrocytes and oligodendroglia in the gray and white matter of affected brain regions intra-nuclear and intra-cytoplasmic bodies containing Ft and iron were observed. These inclusion bodies in the glia are more abundant in the nucleus caudatus, putamen and globus pallidus, where nerve loss is more severe.[228] Both mutations introduce sequences of amino acids that have almost the same size (25 and 26 amino acids), with the same basic properties and a similar number of hydrophobic residues, but no sequence homology. The mutated L-subunit co-assembles with normal H- and L-subunits to form abnormal Ft shells with atypical conformation. This can favor the aggregation of the iron-containing Ft molecules observed in the brain of the patients. Moreover *in vitro* models have demonstrated that the abnormal Ft is degraded faster than the normal form, with consequent release inside the cells of large amounts of LIP.[225]

PANK2 associated neurodegeneration is an autosomal recessive disease with abnormal iron accumulation in the brain, particularly in the globus pallidus, and currently *PANK2* is the only gene associated with this pathology. The symptoms of PANK2 associated neurodegeneration usually occur before 10 years of age. This neurodegenerative disease is characterized in children by progressive dystonia, dysarthria, rigidity, pigmentary retinopathy and basal ganglia iron deposition; adults have "atypical" presentation, prominent speech defects, psychiatric disturbances, and more gradual progression of disease. MRI shows a characteristic pattern in the globus pallidus called "the eye of the tiger" because of its appearance.[229,230] PANK2 is a mitochondrial enzyme that catalyzes the first regulatory step of coenzyme A synthesis. Recently it has been demonstrated that *PANK2* mutation can influence Fpn expression in some cell types and this might alter iron transfer and redistribution into the brain.[231]

Another important neuropathology linked to iron dyshomeostasis is Friedreich's ataxia. This is the most common form of hereditary ataxia with autosomal recessive transmission and usually appears before 25 years of age. Friedreich's ataxia is characterized by the lack of frataxin. Frataxin is a mitochondrial protein localized in the inner mitochondrial membrane that is involved in iron metabolism, and it is necessary for the maintenance of respiratory function, Fe–S cluster assembly and mitochondrial DNA.[232,233] The genetic defect in frataxin is associated with a GAA triplet expansion in the first intron of the gene, resulting in a reduction in frataxin expression by transcriptional inhibition.[234] Patients with Friedreich's ataxia have a deficit in mitochondrial respiration and impairment of Fe–S proteins such as complexes I, II and III and aconitase.[235,236] During the pathology iron accumulates in the mitochondria and this underlines the role of frataxin in iron homeostasis.[237] The main symptoms of Friedreich's ataxia include spinocerebellar and sensory ataxia with absence of deep tendon reflexes, dysarthria, hypertrophic cardio-myopathy and scoliosis, additionally diabetes mellitus, *pes cavus*, hypoacusia and optic atrophy.[238]

Other neurological diseases have been associated with alterations in iron status: restless leg syndrome associated with decreased iron in *substantia nigra* and defects in IRP1 and Huntington's disease with increased iron in basal ganglia.[148]

References

1. J. L. Pierre, M. Fontecave and R. R. Crichton, *Biometals*, 2002, **15**, 341.
2. P. Aisen, C. Enns and M. Wessling-Resnick, *Int. J. Biochem. Cell. Biol.*, 2001, **33**, 940.
3. B. Halliwell and J. M. Gutteridge, *FEBS Lett.*, 1992, **307**, 108.
4. S. J. Stohs and D. Bagchi, *Free Radic. Biol. Med.*, 1995, **18**, 321.
5. C. C. Winterbourn, *Toxicol. Lett.*, 1995, **82–83**, 969.
6. T. P. Ryan and S. D. Aust, *Crit. Rev. Toxicol.*, 1992, **22**, 119.
7. M. Muñoz, I. Villar and J. A. García-Erce, *World J. Gastroenterol.*, 2009, **15**, 4617.
8. J. Chung and M. Wessling-Resnick, *Crit. Rev. Clin. Lab. Sci.*, 2003, **40**, 151.
9. U. E. Schaible and S. H. Kaufmann, *Nat. Rev. Microbiol.*, 2004, **2**, 946.
10. M. B. Youdim and A. R. Green, *Proc. Nutr. Soc.*, 1978, **37**, 173.
11. J. R. Connor and S. L. Menzies, *Glia*, 1996, **17**, 83.
12. A. T. McKie, D. Barrow, G. O. Latunde-Dada, A. Rolfs, G. Sager, E. Mudaly, M. Mudaly, C. Richardson, D. Barlow, A. Bomford, T. J. Peters, K. B. Raja, S. Shirali, M. A. Hediger, F. Farzaneh and R. J. Simpson, *Science*, 2001, **291**, 1755.
13. H. Gunshin, C. N. Starr, C. Direnzo, M. D. Fleming, J. Jin, E. L. Greer, V. M. Sellers, S. M. Galica and N. C. Andrews, *Blood*, 2005, **106**, 2879.
14. H. Gunshin, Y. Fujiwara, A. O. Custodio, C. Direnzo, S. Robine and N. C. Andrews, *J. Clin. Invest.*, 2005, **115**, 1258.
15. B. Mackenzie, H. Takanaga, N. Hubert, A. Rolfs and M. A. Hediger, *Biochem. J.*, 2007, **403**, 59.
16. H. Gunshin, B. Mackenzie, U. V. Berger, Y. Gunshin, M. F. Romero, W. F. Boron, S. Nussberger, J. L. Gollan and M. A. Hediger, *Nature*, 1997, **388**, 482.
17. M. D. Fleming, C. C. Trenor 3rd, M. A. Su, D. Foernzler, D. R. Beier, W. F. Dietrich and N. C. Andrews, *Nat. Genet.*, 1997, **16**, 383.
18. N. Hubert and M. W. Hentze, *Proc. Natl. Acad. Sci. USA*, 2002, **99**, 12345.
19. Y. Ma, M. Yeh, K. Y. Yeh and J. Glass, *Am. J. Physiol. Gastrointest. Liver Physiol.*, 2006, **290**, G417.
20. A. T. McKie, P. Marciani, A. Rolfs, K. Brennan, K. Wehr, D. Barrow, S. Miret, A. Bomford, T. J. Peters, F. Farzaneh, M. A. Hediger, M. W. Hentze and R. J. Simpson, *Mol. Cell*, 2000, **5**, 299.
21. A. Donovan, C. A. Lima, J. L. Pinkus, G. S. Pinkus, L. I. Zon, S. Robine and N. C. Andrews, *Cell. Metab.*, 2005, **1**, 191.
22. S. Cherukuri, R. Potla, J. Sarkar, S. Nurko, Z. L .Harris and P. L. Fox, *Cell. Metab.*, 2005, **2**, 309.

23. C. D. Vulpe, Y. M. Kuo, T. L. Murphy, L. Cowley, C. Askwith, N. Libina, J. Gitschier and G. J. Anderson, *Nat. Genet.*, 1999, **21**, 195.

24. G. Nicolas, C. Chauvet, L. Viatte, J. L. Danan, X. Bigard, I. Devaux, C. Beaumont, A. Kahn and S. Vaulont, *J. Clin. Invest.*, 2002, **110**, 1037.

25. I. De Domenico, D. M. Ward, C. Langelier, M. B. Vaughn, E. Nemeth, W. I. Sundquist, T. Ganz, G. Musci and J. Kaplan, *Mol. Biol. Cell*, 2007, **18**, 2569.

26. E. Nemeth, M. S. Tuttle, J. Powelson, M. B. Vaughn, A. Donovan, D. M. Ward, T. Ganz and J. Kaplan, *Science*, 2004, **306**, 2090.

27. N. P. Mena, A. Esparza, V. Tapia, P. Valdés and M. T. Núñez, *Am. J. Physiol. Gastrointest. Liver Physiol.*, 2008, **294**, G192.

28. M. Shayeghi, G. O. Latunde-Dada, J. S. Oakhill, A. H. Laftah, K. Takeuchi, N. Halliday, Y. Khan, A. Warley, F. E. McCann, R. C. Hider, D. M. Frazer, G. J. Anderson, C. D. Vulpe, R. J. Simpson and A. T. McKie, *Cell*, 2005, **122**, 789.

29. A. Qiu, M. Jansen, A. Sakaris, S. H. Min, S. Chattopadhyay, E. Tsai, C. Sandoval, R. Zhao, M. H. Akabas and I. D. Goldman, *Cell*, 2006, **127**, 917.

30. M. W. Hentze, M. U. Muckenthaler and N. C. Andrews, *Cell*, 2004, **117**, 285.

31. J. E. Levy, O. Jin, Y. Fujiwara, F. Kuo and N. C. Andrews, *Nat. Genet.*, 1999, **21**, 396.

32. C. C. Trenor 3rd, D. R. Campagna, V. M. Sellers, N. C. Andrews and M. D. Fleming, *Blood*, 2000, **96**, 1113.

33. M. Asada-Senju, T. Maeda, T. Sakata, A. Hayashi and T. Suzuki, *J. Hum. Genet.*, 2002, **47**, 355.

34. D. M. Sipe and R. F. Murphy, *J. Biol. Chem.*, 1991, **266**, 8002.

35. R. S. Ohgami, D. R. Campagna, E. L. Greer, B. Antiochos, A. McDonald, J. Chen, J. J. Sharp, Y. Fujiwara, J. E. Barker and M. D. Fleming, *Nat. Genet.*, 2005, **37**, 1264.

36. N. C. Andrews, *Int. J. Biochem. Cell. Biol.*, 1999, **31**, 991.

37. P. Ponka, *Blood*, 1997, **89**, 1.

38. R. Lill and U. Mühlenhoff, *Annu. Rev. Biochem.*, 2008, **77**, 669.

39. J. Gerber, U. Mühlenhoff and R. Lill, *EMBO Rep.*, 2003, **4**, 906.

40. G. C. Shaw, J. J. Cope, L. Li, K. Corson, C. Hersey, G. E. Ackermann, B. Gwynn, A. J. Lambert, R. A. Wingert, D. Traver, N. S. Trede, B. A. Barut, Y. Zhou, E. Minet, A. Donovan, A. Brownlie, R. Balzan, M. J. Weiss, L. L. Peters, J. Kaplan, L. I. Zon and B. H. Paw, *Nature*, 2006, **440**, 96.

41. J. G. Quigley, Z. Yang, M. T. Worthington, J. D. Phillips, K. M. Sabo, D. E. Sabath, C. L. Berg, S. Sassa, B. L. Wood and J. L. Abkowitz, *Cell*, 2004, **118**, 757.

42. S. B. Keel, R. T. Doty, Z. Yang, J. G. Quigley, J. Chen, S. Knoblaugh, P. D. Kingsley, I. De Domenico, M. B. Vaughn, J. Kaplan, J. Palis and J. L. Abkowitz, *Science*, 2008, **319**, 825.

43. C. Delaby, N. Pilard, A. S. Gonçalves, C. Beaumont and F. Canonne-Hergaux, *Blood*, 2005, **106**, 3979.

44. C. I. Raje, S. Kumar, A. Harle, J. S. Nanda and M. Raje, *J. Biol. Chem.*, 2007, **282**, 3252.
45. P. Lee, H. Peng, T. Gelbart, L. Wang and E. Beutler, *Proc. Natl. Acad. Sci. USA*, 2005, **102**, 1906.
46. S. Gruenheid, E. Pinner, M. Desjardins and P. Gros, *J. Exp. Med.*, 1997, **185**, 717.
47. G. Weiss and L. T. Goodnough, *N. Engl. J. Med.*, 2005, **352**, 1011.
48. Z. L. Harris, A. P. Durley, T. K. Man and J. D. Gitlin, *Proc. Natl. Acad. Sci. USA*, 1999, **96**, 10812.
49. H. Kawabata, R. Yang, T. Hirama, P. T. Vuong, S. Kawano, A. F. Gombart and H. P. Koeffler, *J. Biol. Chem.*, 1999, **274**, 20826.
50. M. A. Page, E. Baker and E. H. Morgan, *Am. J. Physiol.*, 1984, **246**, G26.
51. D. Barisani, C. L. Berg, M. Wessling-Resnick and J. L. Gollan, *Am. J. Physiol.*, 1995, **269**, G570.
52. M. Shindo, Y. Torimoto, H. Saito, W. Motomura, K. Ikuta, K. Sato, Y. Fujimoto and Y. Kohgo, *Hepatol. Res.*, 2006, **35**, 152.
53. G. Y. Oudit, H. Sun, M. G. Trivieri, S. E. Koch, F. Dawood, C. Ackerley, M. Yazdanpanah, G. J. Wilson, A. Schwartz, P. P. Liu and P. H. Backx, *Nat. Med.*, 2003, **9**, 1187.
54. R. M. Graham, G. M. Reutens, C. E. Herbison, R. D. Delima, A. C. Chua, J. K. Olynyk and D. Trinder, *J. Hepatol.*, 2008, **48**, 327.
55. M. B. Johnson and C. A. Enns, *Blood*, 2004, **104**, 4287.
56. C. H. Park, E. V. Valore, A. J. Waring and T. Ganz, *J. Biol. Chem.*, 2001, **276**, 7806.
57. C. Pigeon, G. Ilyin, B. Courselaud, P. Leroyer, B. Turlin, P. Brissot and O. Loréal, *J. Biol. Chem.*, 2001, **276**, 7811.
58. G. Nicolas, M. Bennoun, I. Devaux, C. Beaumont, B. Grandchamp, A. Kahn and S. Vaulont, *Proc. Natl. Acad. Sci. USA*, 2001, **98**, 8780.
59. L. Viatte, J. C. Lesbordes-Brion, D. Q. Lou, M. Bennoun, G. Nicolas, A. Kahn, F. Canonne-Hergaux and S. Vaulont, *Blood*, 2005, **105**, 4861.
60. G. Nicolas, M. Bennoun, A. Porteu, S. Mativet, C. Beaumont, B. Grandchamp, M. Sirito, M. Sawadogo, A. Kahn and S. Vaulont, *Proc. Natl. Acad. Sci. USA*, 2002, **99**, 4596.
61. L. Viatte, G. Nicolas, D. Q. Lou, M. Bennoun, J. C. Lesbordes-Brion, F. Canonne-Hergaux, K. Schönig, H. Bujard, A. Kahn, N. C. Andrews and S. Vaulont, *Blood*, 2006, **107**, 2952.
62. M. D. Garrick and L. M. Garrick, *Biochim. Biophys. Acta.*, 2009, **1790**, 309.
63. N. C. Andrews and P. J. Schmidt, *Annu. Rev. Physiol.*, 2007, **69**, 69.
64. N. C. Andrews, *Blood*, 2008, **112**, 219.
65. P. M. Harrison and P. Arosio, *Biochim. Biophys. Acta*, 1996, **1275**, 161.
66. E. C. Theil, *J. Nutr.*, 2003, **133**, 1549S.
67. S. Levi and P. Arosio, *Int. J. Biochem. Cell. Biol.*, 2004, **36**, 1887.
68. N. Aziz and H. N. Munro, *Proc. Natl. Acad. Sci. USA*, 1987, **84**, 8478.
69. M. W. Hentze, T. A. Rouault, S. W. Caughman, A. Dancis, J. B. Harford and R. D. Klausner, *Proc. Natl. Acad. Sci. USA*, 1987, **84**, 6730.

70. J. L. Casey, M. W. Hentze, D. M. Koeller, S. W. Caughman, T. A. Rouault, R. D. Klausner and J. B. Harford, *Science*, 1988, **240**, 924.
71. T. A. Rouault, M. W. Hentze, S. W. Caughman, J. B. Harford and R. D. Klausner, *Science*, 1988, **241**, 1207.
72. M. W. Hentze, S. W. Caughman, T. A. Rouault, J. G. Barriocanal, A. Dancis, J. B. Harford and R. D. Klausner, *Science*, 1987, **238**, 1570.
73. S. Kaptain, W. E. Downey, C. Tang, C. Philpott, D. Haile, D. G. Orloff, J. B. Harford, T. A. Rouault and R. D. Klausner, *Proc. Natl. Acad. Sci. USA*, 1991, **88**, 10109.
74. D. J. Haile, T. A. Rouault, C. K. Tang, J. Chin, J. B. Harford and R. D. Klausner, *Proc. Natl. Acad. Sci. USA*, 1992, **89**, 7536.
75. K. Iwai, R. D. Klausner and T. A. Rouault, *EMBO J.*, 1995, **14**, 5350.
76. E. G. Meyron-Holtz, M. C. Ghosh, K. Iwai, T. LaVaute, X. Brazzolotto, U. V. Berger, W. Land, H. Ollivierre-Wilson, A. Grinberg, P. Love and T. A. Rouault, *EMBO J.*, 2004, **23**, 386.
77. M. Muckenthaler, N. K. Gray and M. W. Hentze, *Mol. Cell*, 1998, **2**, 383.
78. J. L. Casey, D. M. Koeller, V. C. Ramin, R. D. Klausner and J. B. Harford, *EMBO J.*, 1989, **8**, 3693.
79. R. Leipuviene and E. C. Theil, *Cell. Mol. Life Sci.*, 2007, **64**, 2945.
80. M. U. Muckenthaler, B. Galy and M. W. Hentze, *Annu. Rev. Nutr.*, 2008, **28**, 197.
81. C. Beaumont, P. Leneuve, I. Devaux, J. Y. Scoazec, M. Berthier, M. N. Loiseau, B. Grandchamp and D. Bonneau, *Nat. Genet.*, 1995, **11**, 444.
82. J. Kato, K. Fujikawa, M. Kanda, N. Fukuda, K. Sasaki, T. Takayama, M. Kobune, K. Takada, R. Takimoto, H. Hamada, T. Ikeda and Y. Niitsu, *Am. J. Hum. Genet.*, 2001, **69**, 191.
83. H. Mok, J. Jelinek, S. Pai, B. M. Cattanach, J. T. Prchal, H. Youssoufian and A. Schumacher, *Development*, 2004, **131**, 1859.
84. T. Moos, T. Rosengren Nielsen, T. Skjørringe and E. H. Morgan, *J. Neurochem.*, 2007, **103**, 1730.
85. N. J. Abbott, A. A. Patabendige, D. E. Dolman, S. R. Yusof and D. J. Begley, *Neurobiol. Dis.*, 2010, **37**, 13.
86. T. A. Rouault, D. L. Zhang and S. Y. Jeong, *Metab. Brain Dis.*, 2009, **24**, 673.
87. W. A. Jefferies, M. R. Brandon, S. V. Hunt, A. F. Williams, K. C. Gatter and D. Y. Mason, *Nature*, 1984, **312**, 162.
88. T. Moos and E. H. Morgan, *Cell. Mol. Neurobiol.*, 2000, **20**, 77.
89. T. Moos, T. Skjoerringe, S. Gosk and E. H. Morgan, *J. Neurochem.*, 2006, **98**, 1946.
90. M. W. Bradbury, *J. Neurochem.*, 1997, **69**, 443.
91. T. Moos and E. H. Morgan, *J. Neurosci. Res.*, 1998, **54**, 486.
92. A. Crowe and E. H. Morgan, *Brain Res.*, 1992, **592**, 8.
93. J. R. Burdo, S. L. Menzies, I. A. Simpson, L. M. Garrick, M. D. Garrick, K. G. Dolan, D. J. Haile, J. L. Beard and J. R. Connor, *J. Neurosci. Res.*, 2001, **66**, 1198.

94. L. J. Wu, A. G. Leenders, S. Cooperman, E. Meyron-Holtz, S. Smith, W. Land, R. Y. Tsai, U. V. Berger, Z. H. Sheng and T. A. Rouault, *Brain Res.*, 2004, **1001**, 108.

95. T. Moos and T. Rosengren Nielsen, *Semin. Pediatr. Neurol.*, 2006, **13**, 149.

96. Z. M. Qian, Y. Z. Chang, L. Zhu, L. Yang, J. R. Du, K. P. Ho, Q. Wang, L. Z. Li, C. Y. Wang, X. Ge, N. L. Jing, L. Li and Y. Ke, *J. Cell. Biochem.*, 2007, **102**, 1225.

97. S. Y. Jeong and S. David, *J. Biol. Chem.*, 2003, **278**, 27144.

98. T. Oide, K. Yoshida, K. Kaneko, M. Ohta and K. Arima, *Neuropathol. Appl. Neurobiol.*, 2006, **32**, 170.

99. A. Takeda, A. Devenyi and J. R. Connor, *J. Neurosci. Res.*, 1998, **51**, 454.

100. J. L. Beard, J. A. Wiesinger, N. Li and J. R. Connor, *J. Neurosci Res.*, 2005, **79**, 254.

101. I. Moroo, M. Ujiie, B. L. Walker, J. W. Tiong, T. Z. Vitalis, D. Karkan, R. Gabathuler, A. R. Moise and W. A. Jefferies, *Microcirculation*, 2003, **10**, 457.

102. S. Rothenberger, M. R. Food, R. Gabathuler, M. L. Kennard, T. Yamada, O. Yasuhara, P. L. McGeer and W. A. Jefferies, *Brain Res.*, 1996, **712**, 117.

103. Z. M. Qian and Q. Wang, *Brain Res. Brain Res. Rev.*, 1998, **27**, 257.

104. E. O. Sekyere, L. L. Dunn, Y. S. Rahmanto and D. R. Richardson, *Blood*, 2006, **107**, 2599.

105. J. R. Connor and S. L. Menzies, *J. Neurol. Sci.*, 1995, **134**(Suppl, 33).

106. T. K. Dickinson and J. R. Connor, *Brain Res.*, 1998, **801**, 171.

107. T. Moos and E. H. Morgan, *Brain Res.*, 1998, **790**, 115.

108. M. Tsutsumi, M. K. Skinner and E. Sanders-Bush, *J. Biol. Chem.*, 1989, **264**, 9626.

109. A. Terent, R. Hällgren, P. Venge and K. Bergström, *Stroke*, 1981, **12**, 40.

110. C. Fillebeen, L. Descamps, M. P. Dehouck, L. Fenart, M. Benaïssa, G. Spik, R. Cecchelli and A. Pierce, *J. Biol. Chem.*, 1999, **274**, 7011.

111. B. Leveugle, G. Spik, D. P. Perl, C. Bouras, H. M. Fillit and P. R. Hof, *Brain Res.*, 1994, **650**, 20.

112. B. A. Faucheux, N. Nillesse, P. Damier, G. Spik, A. Mouatt-Prigent, A. Pierce, B. Leveugle, N. Kubis, J. J. Hauw and Y. Agid, *Proc. Natl. Acad. Sci. USA*, 1995, **92**, 9603.

113. T. Speake, J. D. Kibble and P. D. Brown, *Am. J. Physiol. Cell. Physiol.*, 2004, **286**, C611.

114. M. A. Maktabi, D. D. Heistad and F. M. Faraci, *Am. J. Physiol.*, 1990, **258**, H414.

115. T. Moos, P. S. Oates and E. H. Morgan, *J. Comp. Neurol.*, 1998, **398**, 420.

116. L. Zecca, L. Casella, A. Albertini, C. Bellei, F. A. Zucca, M. Engelen, A. Zadlo, G. Szewczyk, M. Zareba and T. Sarna, *J. Neurochem.*, 2008, **106**, 1866.

117. L. Zecca, C. Bellei, P. Costi, A. Albertini, E. Monzani, L. Casella, M. Gallorini, L. Bergamaschi, A. Moscatelli, N. J. Turro, M. Eisner,

P. R. Crippa, S. Ito, K. Wakamatsu, W. D. Bush, W. C. Ward, J. D. Simon and F. A. Zucca, *Proc. Natl. Acad. Sci. USA*, 2008, **105**, 17567.

118. R. L. Wixom, L. Prutkin and H. N. Munro, *Int. Rev. Exp. Pathol.*, 1980, **22**, 193.

119. J. F. Schenck and E. A. Zimmerman, *NMR Biomed.*, 2004, **17**, 433.

120. R. J. Ward, M. Ramsey, D. P. Dickson, C. Hunt, T. Douglas, S. Mann, F. Aquad, T. J. Peters and R. R. Crichton, *Eur. J. Biochem.*, 1994, **225**, 187.

121. P. Aguirre, N. Mena, V. Tapia, M. Arredondo and M. T. Núñez, *BMC Neurosci.*, 2005, **6**, 3.

122. T. Moos, *J. Comp. Neurol.*, 1996, **375**, 675.

123. H. H. Hoepken, T. Korten, S. R. Robinson and R. Dringen, *J. Neurochem.*, 2004, **88**, 1194.

124. Z. M. Qian, Y. To, P. L. Tang and Y. M. Feng, *Exp. Brain Res.*, 1999, **129**, 473.

125. E. Huang, W. Y. Ong and J. R. Connor, *Neuroscience*, 2004, **128**, 487.

126. D. E. Barañano and S. H. Snyder, *Proc. Natl. Acad. Sci. USA*, 2001, **98**, 10996.

127. H. M. Schipper, L. Bernier, K. Mehindate and D. Frankel, *J. Neurochem.*, 1999, **72**, 1802.

128. D. R. Richardson, *J. Lab. Clin. Med.*, 1999, **134**, 454.

129. S. Y. Jeong and S. David, *J. Neurosci.*, 2006, **26**, 9810.

130. J. R. Connor, S. L. Menzies, S. M. St Martin and E. J. Mufson, *J. Neurosci. Res.*, 1990, **27**, 595.

131. J. L. Beard, B. Felt, T. Schallert, M. Burhans, J. R. Connor and M. K. Georgieff, *Behav. Brain Res.*, 2006, **170**, 224.

132. J. M. Gutteridge, *Clin. Sci. (Lond).*, 1992, **82**, 315.

133. B. Hallgren and P. Sourander, *J. Neurochem.*, 1958, **3**, 41.

134. S. Aoki, Y. Okada, K. Nishimura, A. J. Barkovich, B. O. Kjos, R. C. Brasch and D. Norman, *Radiology*, 1989, **172**, 381.

135. J. M. Hill, *Brain Iron: Neurochemistry and Behavioural Aspects*, ed. M. B. H Youdim, Taylor & Francis, London, UK, 1988, pp. 1–24.

136. J. A. Gaasch, P. R. Lockman, W. J. Geldenhuys, D. D. Allen and C. J. Van der Schyf, *Neurochem. Res.*, 2007, **32**, 1196.

137. J. R. Connor and R. E. Fine, *J. Neurosci. Res.*, 1987, **17**, 51.

138. J. L. Beard, J. A. Wiesinger and J. R. Connor, *Dev. Neurosci.*, 2003, **25**, 308.

139. E. Ortiz, J. M. Pasquini, K. Thompson, B. Felt, G. Butkus, J. Beard and J. R. Connor, *J. Neurosci. Res.*, 2004, **77**, 681.

140. S. Yehuda and M. B. H. Youdim, *Am. J. Clin. Nutr.*, 1989, **50**, 618.

141. J. S. Richardson, *Int. J. Neurosci.*, 1993, **70**, 75.

142. J. R. Connor, K. L Boeshore, S. A. Benkovic and S. L. Menzies, *J. Neurosci. Res.*, 1994, **37**, 461.

143. J. Daubling, *Prog. Brain Res.*, 1968, **29**, 417.

144. E. M. Taylor and E. H. Morgan, *Dev. Brain Res.*, 1990, **55**, 35.

145. R. R. Crichton, *Inorganic Biochemistry of Iron Metabolism*, Ellis Horwood, New York, 1991.

146. T. A. Rouault, *Nat. Genet.*, 2001, **28**, 299.

147. K. O. Lopes, D. L. Sparks and W. J. Streit, *Glia*, 2008, **56**, 1048.
148. L. Zecca, M. B. Youdim, P. Riederer, J. R. Connor and R. R. Crichton, *Nat. Rev. Neurosci.*, 2004, **5**, 863.
149. J. R. Connor, B. S. Snyder, P. Arosio, D. A. Loeffler and P. LeWitt, *J. Neurochem.*, 1995, **65**, 717.
150. L. Zecca, M. Gallorini, V. Schünemann, A. X. Trautwein, M. Gerlach, P. Riederer, P. Vezzoni and D. Tampellini, *J. Neurochem.*, 2001, **76**, 1766.
151. W. Hirose, K. Ikematsu and R. Tsuda, *Leg. Med. (Tokyo)*, 2003, **5**(Suppl 1), S360.
152. G. Bartzokis, T. A. Tishler, P. H. Lu, P. Villablanca, L. L. Altshuler, M. Carter, D. Huang, N. Edwards and J. Mintz, *Neurobiol. Aging*, 2007, **28**, 414.
153. D. Aquino, A. Bizzi, M. Grisoli, B. Garavaglia, M. G. Bruzzone, N. Nardocci, M. Savoiardo and L. Chiapparini, *Radiology*, 2009, **252**, 165.
154. G. Bartzokis, M. Beckson, D. B. Hance, P. Marx, J. A. Foster and S. R. Marder, *Magn. Reson. Imaging*, 1997, **15**, 29.
155. L. Zecca, A. Stroppolo, A. Gatti, D. Tampellini, M. Toscani, M. Gallorini, G. Giaveri, P. Arosio, P. Santambrogio, R. G. Fariello, E. Karatekin, M. H. Kleinman, N. Turro, O. Hornykiewicz and F. A. Zucca, *Proc. Natl. Acad. Sci. USA*, 2004, **101**, 9843.
156. D. Sulzer, J. Bogulavsky, K. E. Larsen, G. Behr, E. Karatekin, M. H. Kleinman, N. Turro, D. Krantz, R. H. Edwards, L. A. Greene and L. Zecca, *Proc. Natl. Acad. Sci. USA*, 2000, **97**, 11869.
157. L. Zecca, F. A. Zucca, H. Wilms and D. Sulzer, *Trends Neurosci.*, 2003, **26**, 578.
158. L. Zecca, P. Costi, C. Mecacci, S. Ito, M. Terreni and S. Sonnino, *J. Neurochem.*, 2000, **74**, 1758.
159. L. Zecca, F. A. Zucca, A. Albertini, E. Rizzio and R. G. Fariello, *Neurology*, 2006, **67**, S8.
160. A. Brun and E. Englund, *Ann. Neurol.*, 1986, **19**, 253.
161. D. A. Snowdon, *Ann. Intern.*, 2003, **139**(5 Pt 2), 450:
162. B. A. Faucheux, A. M. Bonnet, Y. Agid and E. C. Hirsch, *Lancet*, 1999, **353**, 981.
163. Y. S. Levy, J. Y. Streifler, H. Panet, E. Melamed and D. Offen, *Neurotox. Res.*, 2002, **4**, 609.
164. R. F. Regan and S. S. Panter, *Neurosci. Lett.*, 1993, **153**, 219.
165. K. R. Wagner, F. R. Sharp, T. D. Ardizzone, A. Lu and J. F. Clark, *J. Cereb. Blood Flow Metab.*, 2003, **23**, 629.
166. G. J. Kress, K. E. Dineley and I. J. Reynolds, *J. Neurosci.*, 2002, **22**, 5848.
167. A. Kastner, E. C. Hirsch, O. Lejeune, F. Javoy-Agid, O. Rascol and Y. Agid, *J. Neurochem.*, 1992, **59**, 1080.
168. B. A. Faucheux, M. E. Martin, C. Beaumont C, J. J. Hauw, Y. Agid and E. C. Hirsch, *J. Neurochem.*, 2003, **86**, 1142.
169. E. Sofic, P. Riederer, H. Heinsen, H. Beckmann, G. P. Reynolds, G. Hebenstreit and M. B. Youdim, *J. Neural. Transm.*, 1988, **74**, 199.

170. D. T. Dexter, F. R. Wells, A. J. Lees, F. Agid, Y. Agid, P. Jenner and C. D. Marsden, *J. Neurochem.*, 1989, **52**, 1830.

171. D. A. Loeffler, J. R. Connor, P. L. Juneau, B. S. Snyder, L. Kanaley, A. J. DeMaggio, H. Nguyen, C. M. Brickman and P. A. LeWitt, *J. Neurochem.*, 1995, **65**, 710.

172. P. Riederer, E. Sofic, W. D. Rausch, B. Schmidt, G. P. Reynolds, K. Jellinger and M. B. Youdim, *J. Neurochem.*, 1989, **52**, 515.

173. D. T. Dexter, F. R. Wells, F. Agid, Y. Agid, A. J. Lees, P. Jenner and C. D. Marsden, *Lancet.*, 1987, **2**, 1219.

174. E. C. Hirsch, J.-P. Brandel, P. Galle, F. Javoy-Agid and Y. Agid, *J. Neurochem.*, 1991, **56**, 446.

175. A. Wypijewska, J. Galazka-Friedman, E. R. Bauminger, Z. K. Wszolek, K. J. Schweitzer, D. W. Dickson, A. Jaklewicz, D. Elbaum and A. Friedman, *Parkinsonism Relat. Disord.*, 2010, **16**, 329.

176. P. Riederer, W. D. Rausch, B. Schmidt, P. Kruzik, C. Konradi, E. Sofic, W. Danielczyk, M. Fischer and E. Ogris, *Mt. Sinai J. Med.*, 1988, **55**, 21.

177. D. T. Dexter, J. Sian, P. Jenner and C. D. Marsden, *Adv. Neurol.*, 1993, **60**, 273.

178. D. Kaur and J. Andersen, *Ageing Res. Rev.*, 2004, **3**, 327.

179. M. Fasano, B. Bergamasco and L. Lopiano, *J. Neurochem.*, 2006, **96**, 909.

180. D. T. Dexter, A. Carayon, F. Javoy-Agid, Y. Agid, F. R. Wells, S. E. Daniel, A. J. Lees, P. Jenner and C. D. Marsden, *Brain.*, 1991, **114**(Pt 4), 1953.

181. M. Gerlach, D. Ben-Shachar, P. Riederer and M. B. Youdim, *J. Neurochem.*, 1994, **63**, 793.

182. P. Jenner and C. W. Olanow, *Neurology*, 1996, **47**, S161.

183. D. Sulzer and L. Zecca, *Neurotox. Res.*, 2000, **1**, 181.

184. M. J. Kumar and J. K. Andersen, *Mol. Neurobiol.*, 2004, **30**, 77.

185. K. N. Westlund, R. M. Denney, R. M. Rose and C. W. Abell, *Neuroscience*, 1988, **25**, 439.

186. H. Braak, K. Del Tredici, H. Bratzke, J. Hamm-Clement, D. Sandmann-Keil and U. Rüb, *J. Neurol.*, 2002, **249**(Suppl 3), III/1.

187. M. Hashimoto, A. Takeda, L. J. Hsu, T. Takenouchi and E. Masliah, *J. Biol. Chem.*, 1999, **274**, 28849.

188. S. Altamura and M. U. Muckenthaler, *J. Alzheimers Dis.*, 2009, **16**, 879.

189. B. A. Faucheux, M. E. Martin, C. Beaumont, S. Hunot, J. J. Hauw, Y. Agid and E. C. Hirsch, *J. Neurochem.*, 2002, **83**, 320.

190. K. Jellinger, W. Paulus, I. Grundke-Iqbal, P. Riederer and M. B. Youdim, Brain iron and ferritin in Parkinson's and Alzheimer's diseases. *J. Neural Transm. Park. Dis. Dement. Sect., 1990, 2, 327.*

191. K. L. Double, M. Gerlach, V. Schünemann, A. X. Trautwein, L. Zecca, M. Gallorini, M. B. Youdim, P. Riederer and D. Ben-Shachar, *Biochem. Pharmacol.*, 2003, **66**, 489.

192. H. Wilms, P. Rosenstiel, J. Sievers, G. Deuschl, L. Zecca and R. Lucius, *FASEB J.*, 2003, **17**, 500.

193. W. Zhang, K. Phillips, A. R. Wielgus, J. Liu, A. Albertini, F. A. Zucca, R. Faust, S. Y. Qian, D. S. Miller, C. F. Chignell, B. Wilson, V. Jackson-

Lewis, S. Przedborski, D. Joset, J. Loike, J. S. Hong, D. Sulzer and L. Zecca, *Neurotox. Res.*, 2011, **19**, 63–72.

194. L. Zecca, H. Wilms, S. Geick, J. H. Claasen, L. O. Brandenburg, C. Holzknecht, M. L. Panizza, F. A. Zucca, G. Deuschl, J. Sievers and R. Lucius, *Acta Neuropathol.*, 2008, **116**, 47.

195. H. M. Schipper, A. Liberman and E. G. Stopa, *Exp. Neurol.*, 1998, **150**, 60.

196. B. A. Faucheux, J. J. Hauw, Y. Agid and E. C. Hirsch, *Brain Res.*, 1997, **749**, 170.

197. C. Haass and D. J. Selkoe, *Nat. Rev. Mol. Cell. Biol.*, 2007, **8**, 101.

198. S. Mandel, T. Amit, O. Bar-Am and M. B. Youdim, *Prog. Neurobiol.*, 2007, **82**, 348.

199. A. Takeda, M. A. Smith, J. Avilá, A. Nunomura, S. L. Siedlak, X. Zhu, G. Perry and L. M. Sayre, *J. Neurochem.*, 2000, **75**, 1234.

200. S. Shimohama, H. Tanino, N. Kawakami, N. Okamura, H. Kodama, T. Yamaguchi, T. Hayakawa, A. Nunomura, S. Chiba, G. Perry, M. A. Smith and S. Fujimoto, *Biochem. Biophys. Res. Commun.*, 2000, **273**, 5.

201. H. M. Schipper, *Free Radic. Biol. Med.*, 2004, **37**, 1995.

202. M. A. Smith, X. Zhu, M. Tabaton, G. Liu, D. W. McKeel Jr, M. L. Cohen, X. Wang, S. L. Siedlak, B. E. Dwyer, T. Hayashi, M. Nakamura, A. Nunomura and G. Perry, *J. Alzheimers Dis.*, 2010, **19**, 363.

203. I. Grundke-Iqbal, J. Fleming, Y. C. Tung, H. Lassmann, K. Iqbal and J. G. Joshi, *Acta Neuropathol.*, 1990, **81**, 105.

204. M. A. Smith, P. L. R. Harris, L. M. Sayre and G. Perry, *Proc. Natl. Acad. Sci. USA*, 1997, **94**, 9866.

205. A. I. Bush, *Trends Neurosci.*, 2003, **26**, 207.

206. C. S. Atwood, X. Huang, R. D. Moir, R. E. Tanzi and A. I. Bush, *Met. Ions Biol. Syst.*, 1999, **36**, 309.

207. L. Silvestri and C. Camaschella, *J. Cell. Mol. Med.*, 2008, **12**, 1548.

208. E. M. Hwang, S. K. Kim, J. H. Sohn, J. Y. Lee, Y. Kim, Y. S. Kim and I. Mook-Jung, *Biochem. Biophys. Res. Commun.*, 2006, **349**, 654.

209. J. T. Rogers, A. I. Bush, H. H. Cho, D. H. Smith, A. M. Thomson, A. L. Friedlich, D. K. Lahiri, P. J. Leedman, X. Huang and C. M. Cahill, *Biochem. Soc. Trans.*, 2008, **36**, 1282.

210. J. C. Sipe, P. Lee and E. Beutler., Brain iron metabolism and neurodegenerative disorders. *Dev. Neurosci.*, 2002, **24**, 188.

211. C. Quintana and L. Gutiérrez, *Biochim. Biophys. Acta*, 2010, May 3. Epub ahead of print.

212. C. Quintana, S. Bellefqih, J. Y. Laval, J. L. Guerquin-Kern, T. D. Wu, J. Avila, I. Ferrer, R. Arranz and C. Patiño, *J. Struct. Biol.*, 2006, **153**, 42.

213. J. R. Connor, B. S. Snyder, J. L. Beard, R. E. Fine and E. J. Mufson, *J. Neurosci. Res.*, 1992, **31**, 327.

214. W. Zheng, N. Xin, Z. H. Chi, B. L. Zhao, J. Zhang, J. Y. Li and Z. Y. Wang, *FASEB J.*, 2009, **23**, 4207.

215. R. J. Castellani, M. A. Smith, A. Nunomura, P. L. Harris and G. Perry, *Free Radic. Biol. Med.*, 1999, **26**, 1508.

216. D. A. Loeffler, A. A. Sima and P. A. LeWitt, *Free Radic. Res.*, 2001, **35**, 111.
217. S. Y. Lee and J. R. Connor, *Neurobiol. Aging*, 2005, **26**, 803.
218. P. D. Phatak, D. H. Ryan, J. Cappuccio, D. Oakes, C. Braggins, K. Provenzano, S. Eberly and R. L. Sham, *Blood Cells Mol. Dis.*, 2002, **29**, 41.
219. D. J. Lehmann, M. Worwood, R. Ellis, V. L. J. Wimhurst, A. T. Merryweather-Clarke, D. R. Warden, A. D. Smith and K. J. H. Robson, *J. Med. Genet.*, 2006, **43**, e52.
220. X. S. Wang, S. Lee, Z. Simmons, P. Boyer, K. Scott, W. Liu and J. Connor, *J. Neurol. Sci.*, 2004, **227**, 27.
221. R. J. Guerreiro, J. M. Bras, I. Santana, C. Januario, B. Santiago, A. S. Morgadinho, M. H. Ribeiro, J. Hardy, A. Singleton and C. Oliveira, *BMC Neurol.*, 2006, **6**, 24.
222. K. Namekata, M. Imagawa, A. Terashi, S. Ohta, F. Oyama and Y. Ihara, *Hum. Genet.*, 1997, **101**, 126.
223. R. I. Hussain, C. G. Ballard, J. A. Edwardson and C. M. Morris, *Neurosci. Lett.*, 2002, **317**, 13.
224. A. McNeill, M. Pandolfo, J. Kuhn, H. Shang and H. Miyajima, *Eur. Neurol.*, 2008, **60**, 200.
225. S. Levi, A. Cozzi and P. Arosio, *Best Pract. Res. Clin. Haematol.*, 2005, **18**, 265.
226. A. R. Curtis, C. Fey, C. M. Morris, L. A. Bindoff, P. G. Ince, P. F. Chinnery, A. Coulthard, M. J. Jackson, A. P. Jackson, D. P. McHale, D. Hay, W. A. Barker, A. F. Markham, D. Bates, A. Curtis and J. Burn, *Nature Genetics*, 2001, **28**, 350.
227. D. E. Crompton, P. F. Chinnery, C. Fey, A. R. Curtis, C. M. Morris, J. Kierstan, A. Burt, F. Young, A. Coulthard, A. Curtis, P. G. Ince, D. Bates, M. J. Jackson and J. Burn, *Blood Cells Mol. Dis.*, 2002, **29**, 522.
228. R. Vidal, B. Ghetti, M. Takao, C. Brefel-Courbon, E. Uro-Coste, B. S. Glazier, V. Siani, M. D. Benson, P. Calvas, L. Miravalle, O. Rascol and M. B. Delisle, *J. Neuropathol. Exp. Neurol.*, 2004, **63**, 363.
229. S. J. Hayflick, S. K. Westaway, B. Levinson, B. Zhou, M. A. Johnson, K. H. Ching and J. Gitschier, *N. Engl. J. Med.*, 2003, **348**, 33.
230. A. McNeill, D. Birchall, S. J. Hayflick, A. Gregory, J. F. Schenk, E. A. Zimmerman, H. Shang, H. Miyajima and P. F. Chinnery, *Neurology*, 2008, **70**, 1614.
231. M. Poli, M. Derosas, S. Luscieti, P. Cavadini, A. Campanella, R. Verardi, D. Finazzi and P. Arosio, *Neurobiol. Dis.*, 2010, Apr 23. Epub ahead of print.
232. H. Koutnikova, V. Campuzano, F. Foury, P. Dollé, O. Cazzalini and M. Koenig, *Nat. Genet.*, 1997, **16**, 345.
233. R. B. Wilson and D. M. Roof, *Nat. Genet.*, 1997, **16**, 352.
234. V. Campuzano, L. Montermini, M. D. Moltò, L. Pianese, M. Cossée, F. Cavalcanti, E. Monros, F. Rodius, F. Duclos, A. Monticelli, F. Zara, J. Cañizares, H. Koutnikova, S. I. Bidichandani, C. Gellera, A. Brice, P. Trouillas, G. De Michele, A. Filla, R. De Frutos, F. Palau, P. I. Patel,

S. Di Donato, J. L. Mandel, S. Cocozza, M. Koenig and M. Pandolfo, *Science*, 1996, **271**, 1423.
235. A. Rötig, P. de Lonlay, D. Chretien, F. Foury, M. Koenig, D. Sidi, A. Munnich and P. Rustin, *Nat. Genet.*, 1997, **17**, 215.
236. R. Lodi, J. M. Cooper, J. L. Bradley, D. Manners, P. Styles, D. J. Taylor and A. H. Schapira, *Proc. Natl. Acad. Sci. USA*, 1999, **96**, 11492.
237. M. Babcock, D. de Silva, R. Oaks, S. Davis-Kaplan, S. Jiralerspong, L. Montermini, M. Pandolfo and J. Kaplan, *Science*, 1997, **276**, 1709.
238. A. E. Harding, *Brain*, 1981, **104**, 589.

CHAPTER 10

Aluminium in Neurodegenerative Diseases

S. BOLOGNIN[a] AND P. ZATTA[b*]

[a] Department of Neurochemistry, New York State Institute for Basic Research in Developmental Disabilities, 1050 Forest Hill Road, Staten Island, NY, 10314, USA; [b] CNR-Institute for Biomedical Technologies, Padova unit "Metalloproteins", V.le G. Colombo, 3, 35121, Padova, Italy

10.1 Neurodegenerative Disorders and Metal Ions

The term neurodegenerative disorders (ND) refers to a vast group of heterogeneous diseases often characterized by the deposition of proteins within neurons or brain parenchyma. This detrimental alteration typically occurs because of the failure of several proteins to fold correctly, or to remain correctly folded, giving rise to many different types of biological malfunction.[1] Although the mechanism initiating this molecular pathway is still unclear, accumulating evidence supports the hypothesis that, along with many other etiological factors, metal ion dyshomeostasis generated by various mechanisms could be a major contributing factor to the initiation and promotion of the pathology.[12] Indeed, metal ions such as copper (Cu), zinc (Zn), iron (Fe) and aluminium (Al) have all been proposed as modulators of the aggregation of some specific proteins that are directly linked to these diseases. Moreover, many ND show, among other features, a common impairment of metal ion homeostasis in the brain. These metals, with the exception of Al, are fundamental for correct brain functioning. They need to be strictly regulated, however, to avoid the triggering of detrimental cell processes: depletion as well as accumulation of these metals

RSC Drug Discovery Series No. 7
Neurodegeneration: Metallostasis and Proteostasis
Edited by Danilo Milardi and Enrico Rizzarelli
© Royal Society of Chemistry 2011
Published by the Royal Society of Chemistry, www.rsc.org

can lead both to abnormal interactions with proteins or nucleic acids and to consequent cell damage. The brain therefore strictly regulates the metal ion fluxes across the blood–brain barrier (BBB).

Notably, ageing is considered to be one of the most significant risk factors for ND and there is mounting evidence of a general age-related increase of the above metals in the brain. Brain metal accumulation, especially for redox metals such as Cu and Fe, leads to increased oxidative stress (with the production of excess superoxide and hydroxyl radicals), and is associated with severe neuronal damage in physiological ageing as well as in ND (*e.g.* Alzheimer's disease).[2] Metals may then provide the link between protein misfolding and aggregation, oxidative stress and the cascade of biochemical alterations, eventually leading to neuronal cell death. Their essential role in a variety of general cellular functions is unanimously recognized, as well as the fact that they are required by at least one quarter of all proteins as cofactors.[3] The role of metal ions in neurodegeneration is more controversial:[4,5] for a few and very rare ND (*e.g.* Wilson disease or neuroferritinopathy) metal dysmetabolism (essentially due to genetic or environmental factors) was clearly established as the primary cause of the disease, while for AD and the other most common ND, the multifactorial character of the pathology and the overall modest increase in metal concentrations (though in the presence of evident metal dysmetabolism) makes it difficult to assign clear and conclusive roles to the various agents, notably the uncertain involvement of Al in the etiology of AD.

10.2 Aluminium Toxicity: Important Evaluations

Al is one of the most abundant elements in the earth's crust. Due to its strong interactions with oxygen donors, Al can bind strongly to several biological macromolecules. It is now well established that Al is a neurotoxic agent. Indeed, many papers report that it induces severe toxic effects on the central nervous system (CNS):[6] for example, it has been shown that Al can increase the activity of the acetylcholinesterase (AchE).[7–9] Various animal studies reveal that Al exposure causes important neuropathological and neurobehavioral changes resulting in impaired learning ability.[10,11] Two aspects need to be considered before approaching this issue. Firstly, Al has a complex hydrolysis pH-dependent chemistry in biological systems, which may account for the many inconsistencies reported in the literature on the effects of Al on animal or cellular models. For example, when Al inorganic salts such as chloride, sulfate, hydroxide or perchlorate are dissolved in water at a calculated concentration of 10 mM, the analytical Al concentration in solution is about 50 μM. The use of Al–lactate or Al–aspartate, however, increases the soluble Al concentration to 50–330 μM. Hence, a careful evaluation of the metal bioavailability under physiological conditions has to be taken into account when designing Al studies. Secondly, a distinction has to be made between the concepts of neurotoxicity and neurodegeneration. Indeed, Al has been widely described as a neurotoxic element[13] when it cannot be excreted or it is in direct contact with the brain. Some studies have summarized the effects of occupational exposure to Al, suggesting that it induces relevant neurotoxic effects following acute or subacute exposure.[14] Nevertheless, besides the well-known neurotoxicity of Al at high concentration, the role of the metal at lower concentration in affecting pathways related to neurodegenerative mechanisms has not been adequately investigated.

10.3 Aluminium and AD: Is There a Link?

Since the 1970s it has been hypothesized that exposure to Al may enhance the pathogenesis of AD, mainly in genetically predisposed subjects.[15] Indeed, Alfrey[16] described for the first time a neurological condition resembling AD dementia which was called dialysis encephalopathy (DE). DE consists of an abnormal general accumulation of Al in the brain of uremic patients with renal failure undergoing chronic dialysis, which occurred when tap water, without any further purification, was used in the dialysis process.[17,18] The effects of Al on cognitive functions were reversible since the condition of DE patients greatly improved after removal of Al uptake and following a treatment with desferrioxamine (DFO).[19] In fact, once Al was removed from the "dialysis bath" the DE practically disappeared. These findings gave rise to widespread speculation as to whether AD and Al could be linked, but no conclusive results had yet been established.[20] Early studies using laser microprobe mass analysis (LAMMA) showed high Al concentrations within the AD neurofibrillary tangles.[21,22] More recently, significantly increased levels of Al have been reported in the parietal cortex of the AD brain as compared with controls.[23,24] Nevertheless, the epidemiological results which addressed the problem of increased exposure to Al, for example in drinking water, in connection with the incidence of AD were controversial.[20] It is the case that many nephrologists currently use Al salts to decrease the hyperphosphatemia in uremic subjects with no major incidence of AD among these patients compared to the general population. Thus, Al itself cannot be a sufficient trigger of AD and there must be another factor for the potential AD–Al connection.

In this complex scenario, we have investigated the role of Al in an attempt to elucidate its contribution, if any, to AD pathology/pathogenesis. As β-amyloid peptide (Aβ) overproduction and anomalous oligomerization are thought to be central to the pathogenesis of AD we wonder whether Al could affect its aggregational pathway and eventually modulate its toxicity in a biological environment.

10.4 The Role of Aluminium in the Aggregation and Toxicity of β-Amyloid Peptide

The Aβ peptide is the principal constituent of the senile plaques (SP) and it is believed to play a central role in AD etiopathogenesis. This peptide derives from the endoproteolysis of a large transmembrane precursor protein (APP) which is sequentially cleaved by two secretases, namely γ- and β-secretase.

Aβ can spontaneously and rapidly self-aggregate to form insoluble high-molecular weight aggregates which can arrange into insoluble fibrils that finally precipitate extracellularly forming SP. SP and neurofibrillary tangles, the latter arising from intraneuronal deposition of the τ protein,[25] are the typical hallmarks of AD. Early observations identified the accumulation of Aβ into SP as the main culprit for the widespread neurodegeneration characterizing this pathology ("*amyloid cascade hypothesis*").[26] Nevertheless, later observations indicated that early-stage Aβ aggregates, "*oligomers*", may be more relevant to AD etiology and may better correlate with the severity of dementia than insoluble deposits.[27] Higher levels of soluble oligomers were indeed found in AD brains compared to controls,[28] and were proposed to play a paramount role in triggering the early events causing the disease. Thus, the focus of research has moved towards the investigation of the Aβ oligomerization pathway.

The relevant effect of Aβ oligomers as a cofactor for neuronal impairment, and the still elusive role played by metals, led us to carry out comparative and more detailed investigations into the adducts formed when Aβ reacts with Cu, Fe, Zn, and with Al.[29] Electron microscopy analysis showed that Aβ–Al was characterized by a large population of small oligomers, which could be responsible for the significant toxicity on neuroblastoma cells in terms of alteration of cell morphology, decrease in cell viability, increase in membrane fluidity justified by its high hydrophobicity,[30] and possible increase of late apoptosis (Figure 10.1). No significant toxic effects were observed in neuroblastoma cells after 24 hours of treatment both with Aβ and with the other Aβ–metal complexes. Al seemed somehow to be able to "freeze" the oligomeric state of Aβ, stabilizing this assembly with respect to the conformations obtained for other Aβ–metal complexes.

These findings show a significant involvement of Al, compared to the other metal ions utilized in our study, in promoting a specific Aβ aggregation, which is able to produce marked toxic effects on neuroblastoma cells.

10.5 Aluminium–Amyloid Complex and Cell Membranes

Many studies reported that AD is characterized by abnormal lipid profiles, similar to those of atherosclerosis patients.[31] Several studies indicated that Aβ neurotoxicity might be mediated through direct interaction between the peptide and cellular membranes.[32,33] Both cholesterol and plasma membrane lipids have been shown to have roles in regulating APP metabolism.[34,35] Notably, the amphipathic character of Aβ might explain its detrimental association with the membrane and can explain why several membrane lipids can interact with Aβ and affect the oligomerization process.[36] In this regard, it has been reported that neuronal membranes can increase Aβ conversion into toxic oligomers,[37] and that part of the critical balance between toxic and inert Aβ pools is determined by the relative amounts of lipids in the direct environment of the SP.[38] Therefore, the relationship between Aβ and the cellular membrane could be crucial in the process leading to the pathology. In accordance with this

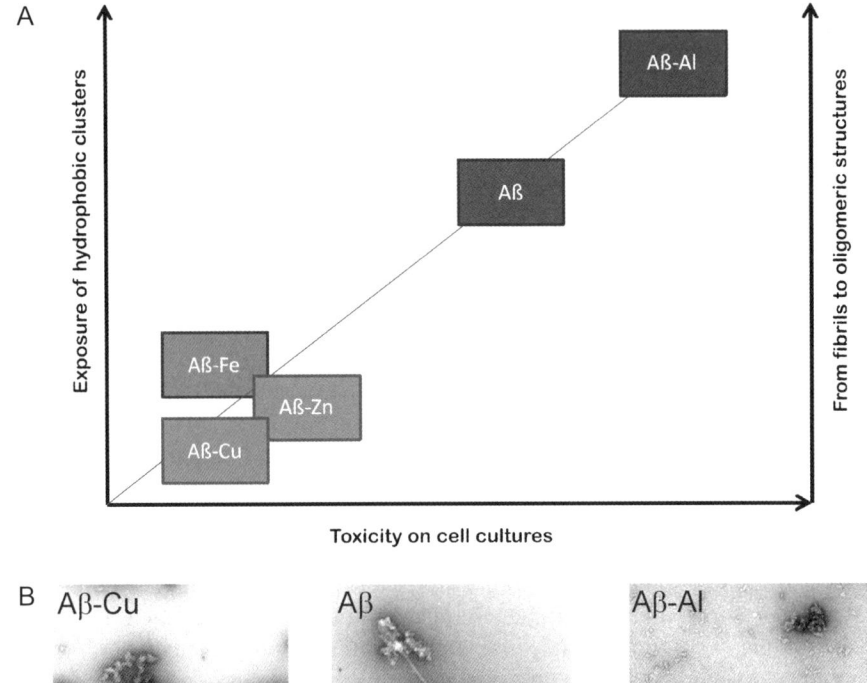

Figure 10.1 A schematic representation of the significant toxicity of Aβ–Al.
The complex triggers the appearance of a large population of small oligomers which could be responsible for the observed decrease in cell viability. A: No significant toxic effect was observed, in our model (neuroblastoma cell after 24 hours of treatment), in the presence of the other Aβ–metal complexes. B: Transmission electron microscopy (TEM) of Aβ–Cu, Aβ and Aβ–Al.

hypothesis, perturbation in the lipid distribution was reported several years ago in many AD patients,[39,40] and hypercholesterolemia is considered an early risk factor for the development of AD.[31] *In vitro* studies indicated that increased cellular cholesterol levels result in the increased production of Aβ peptides.[41] Moreover, it has been reported that proteins relevant to Aβ generation localize in the membrane rafts.[42] Nevertheless, the actual mechanism underlying this interaction remains elusive. Several hypotheses have been proposed including the alteration of the physiological characteristics of the membrane,[39] lipid peroxidation,[43] and the formation of calcium-permeable ion channels which allow excessive Ca^{2+} influx, disrupting physiological homeostasis.[44] This latter event could occur either through the modulation of an existing Ca^{2+} channel or through the formation of a new cation-selective channel.

Curtain *et al.*[37] showed that penetration of the lipid bilayer is closely related to conditions favorable to Aβ oligomerization. Moreover, oligomeric Aβ has been found to cause membrane fluctuation.[45] We therefore studied the effect of Aβ and various Aβ–metal complexes (Aβ–Al, Aβ–Cu, Aβ–Fe, Aβ–Zn) on a lipid model of cellular membrane using the X-ray diffraction technique.[46] Synthetic dimyristoylphosphatidylcholine (DMPC) and dimyristoylphospha-tidylethanolamine (DMPE) bilayers were used because they represent phospholipids located in the outer and in the inner monolayer of the membrane, respectively.[47] Aβ–Al was the most effective complex in perturbing DMPC bilayer when compared with the other metal complexes. It is worth noting that this effect was peculiar to the Aβ–Al complex, since neither the peptide alone nor the Al salt affected DMPC when incubated with concentrations similar to those of the Aβ–Al complex that produced relevant alterations. Considerably less pronounced was the effect of the Aβ–Al complex on DMPE bilayer compared to that induced in DMPC. DMPE molecules pack more tightly than those of DMPC owing to their smaller polar groups and higher effective charge, resulting in a very stable bilayer system that is not significantly affected by water.[48] This property allowed for the incorporation of the Aβ–Al complex into DMPC bilayers, disrupting their arrangement and consequently the whole of the bilayer structure. On the other hand, the Al complexed with Aβ may induce a change in the net charge of the peptide which can promote abnormal lipid–peptide interactions, thus promoting pathological oligomerization of Aβ. Recently, our group has demonstrated that when Al was bound to Aβ, forming a stable metallorganic complex, the surface hydrophobicity of the peptide dramatically increased as a consequence of metal-induced conformational changes, favoring misfolding/aggregation phenomena.[30] Aβ–Al, thanks to its higher lipophilicity compared with the other Aβ–metal complexes, could intercalate with the acyl chain region, altering the bilayer arrangement.

An interesting hypothesis would be to relate the different abilities of the Aβ–metal complexes in perturbing membranes to the different toxic species produced. In support of our hypothesis, Demuro *et al.*[33] proposed that Aβs are responsible for a generalized increase in membrane permeability induced specifically by spherical amyloid oligomers.

Moreover, we determined whether complexing with Al affected the ability of radioactively iodinated Aβ to cross the *in vivo* BBB. We found that the rates of Aβ and Aβ–Al uptake were similar, but Aβ–Al entered the parenchymal space of the brain more readily.[49] This complex also had a longer half-life in blood and increased permeation at the striatum and thalamus, suggesting that it would have more access to brain cells than the peptide alone.

10.6 Aluminium–Amyloid Complex and Calcium

There is much evidence to suggest that Ca^{2+} dyshomeostasis could exert an important role in promoting AD-related neuronal injury.[50,51] It has been demonstrated that elevated concentrations of cytosolic Ca stimulate Aβ aggregation and amyloidogenesis.[52,53] The Ca hypothesis of ND was first proposed by Khachaturian.[54] According to this theory intracellular Ca alteration could be the cause of AD. Up to now, increasing evidence indicates that calcium dysregulation could be relevant to AD pathogenesis[55] in that it seems to occur early in the AD process;[56,57] targeting calcium signaling could thus be a valuable therapeutic approach. Mitochondrial Ca^{2+} overload has also been linked to the increased production of reactive oxygen species (ROS) and the release of pro-apoptotic factors.[58] It has been reported that mitochondria isolated from AD brains showed Aβ accumulation and morphological alterations,[59] and that Aβ may alter the structural properties of the mitochondrial membrane.[60] One proposed mechanism of Aβ toxicity suggests that Aβ can enter the plasma membrane, forming ion-conducting pores. We thus investigated the effects of Aβ, alone and complexed with Al, Zn, Cu or Fe, on cortical neuronal cell culture $[Ca^{2+}]_i$ homeostasis.[29]

We reported that, among the various Aβ–metal complexes, only the Aβ–Al complex altered glutamate-driven $[Ca]_i$ rises and was able to enhance *N*-methyl-D-aspartate (NMDA) receptor-mediated $[Ca^{2+}]_i$ dyshomeostasis. We speculate that some of the enhanced NMDA-triggered $[Ca^{2+}]_i$ rises observed in cortical neurons pre-incubated with Aβ–Al could be due to a partial impairment of mitochondrial Ca^{2+} buffering.

A large body of evidence indicates that the accumulation of Aβ in mitochondria is associated with decreased enzymatic activity of respiratory chain complexes III and IV and a reduced rate of oxygen consumption.[60,61] Thus, we evaluated the effects of Aβ and its metal complexes on the functioning of isolated rat brain mitochondria. Aβ–Al inhibited the oxidative respiration in isolated rat brain mitochondria and it induced a decrease in state 3 respiration.[29] These results appear to be in agreement with a previous study indicating that extracellular treatment with small spherical Aβ oligomers, unlike monomers and fibrils, can cause disruption of $[Ca^{2+}]_i$.[33] Our studies also indicate that Aβ induced a decrease in state 3 respiration, but the phenomenon was strongly exacerbated when the peptide was conjugated with Al.

Changes of membrane fluidity triggered by exposure to Aβ–metal complexes were also evaluated by steady-state fluorescence anisotropy of mitochondria-bound 6-diphenyl-1,3,5-hexatriene (DPH) and *N,N,N*-trimethyl-4-(6-phenyl-1,3,5-hexatrien-1-yl) phenylammonium p-toluene sulfonate (TMA-DPH) but no perturbations of the internal lipid domains (as monitored by DPH) and polar heads group/hydrophobic tail border areas (as monitored by TMA-DPH) in the presence of Aβ–Al complex at 4 µM (the concentration that strongly inhibits state 3 respiration) were registered. It is likely that Aβ–Al exerted its inhibition without affecting the membrane fluidity, or that the low concentration of the peptide used in this study did not produce a change detectable by the anisotropy assay.

Although further studies are clearly necessary to provide a more detailed understanding of the alterations promoted by the Aβ–Al complex in neurons and mitochondria, these data support the idea that such complexes might play an important role in the impairment of neuronal metabolism and Ca^{2+} homeostasis.

10.7 Aluminium–Amyloid Complex and Gene Expression in Human Neuroblastoma Cell Culture

To investigate the potential effects induced by Aβ and Aβ–metal complexes on gene expression in SH-SY5Y cells, we performed a microarray analysis of the total genome, focusing in particular on those genes that the literature considered to be involved in the pathology.[62,89] This powerful technique allowed us to identify the differential gene alteration after treatment with Aβ and Aβ–metal complexes. As expected, each Aβ–metal complex produced different alterations in the gene expression profile (up- or downregulation) of AD-related genes and this effect was much more pronounced for Aβ–metal complexes than for Aβ alone. In this regard, to highlight the specific changes in gene expression produced by Aβ–metal complexes and to exclude the effect of the metal itself, the microarray analysis was performed in parallel with metals (Al, Cu, Fe and Zn) at 10 times higher concentrations, and with Aβ alone under strictly comparable experimental conditions. From the total of 35 129 human neuroblastoma genes explored, Aβ–Al was the only treatment able to induce a significant upregulation of four AD-related genes, including amyloid precursor protein (APP) family members *APLP1* and *APLP2* (amyloid precursor-like protein-1 and -2). APLP1 is primarily expressed in the nervous system, whereas APP and APLP2 are constitutively expressed in most cell types.[63] Both proteins have been found in the amyloid plaques in the brains of AD patients[64] and thus may contribute to pathogenesis.[65,66] Considering the synaptic localization of APLP1 and APLP2 and their homology to APP, these proteins may be involved in synaptogenesis and synaptic plasticity, as has been suggested for APP. In addition, to assess a potential role of APLP1 in AD, preliminary studies of immunohistochemical distribution of APP in human hippocampal formation, a region heavily affected in AD brain, revealed an accumulation of *APLP1* in neuritic plaques.[64]

Another upregulated gene after Aβ–Al treatment was the gene encoding for microtubule-associated protein τ (*MAPT*) which is the main component, in hyperphosphorylated form, of the aberrant paired helical filaments (PHFs) found in AD.[67,68] PHFs contribute to neurofibrillary tangles (NFT), protein aggregates that, in association with SP, are the aberrant structures found in the brains of patients with AD. Moreover, to validate the microarray data, a quantitative real-time polymerase chain reaction (qRT-PCR) was performed under the same experimental conditions used for the microarray analysis. Coherently, a significant increase of mRNA levels was found for all the four genes considered after 24 hours of treatment with Aβ–Al complex, with respect to the control and the other treatment (Aβ, Aβ–Fe, and metals alone).

These findings allow us to speculate about the pathological role of Al complexed with Aβ as one important factor, among many others, that characterize this multifactorial disease.

10.8 Aluminium and Other Neurodegenerative Disorders

The direct involvement of Al in disease etiology has been proposed only for AD. However, elevated levels of several metal ions such as Mn, Cu, Fe, Zn and Al have also been observed in the *substantia nigra* (SN) of Parkinson's disease (PD) patients.[69,70] The molecular mechanisms that lead to degeneration of the dopaminergic neurons in the brain stem and, particularly, in the SN of PD patients are still unknown.

Notably, some surviving nigral dopaminergic neurons were shown to contain cytosolic filamentous inclusions known as Lewy bodies whose major fibrillar material is the presynaptic protein α-synuclein.[71,72] Little is currently known about the effects of elevated metal concentrations on the structural properties and on aggregation behavior of α-synuclein, though several studies indicate a crucial role for metal-induced oxidative damage. However, it was shown that α-synuclein fibril formation is greatly promoted in the presence of metal ions, with Al having the most significant effect, along with Cu, Fe, and manganese (Mn).[73] Several early epidemiological studies have also shown that exposure to metals such as Cu, Zn, Mn, Fe and Al may be associated with increased incidence of PD.[74] Furthermore, post-mortem analysis of PD brain tissues, using various quantitative methods and histochemical techniques,[76] revealed an increased concentration of Fe in nigral neurons[75] and of Al and Fe in Lewy bodies.[69] The study of metal–α-synuclein interactions, besides being associated with oxidative stress, has also been stimulated by the discovery that metals are able to induce structural change in α-synuclein conformation. Many studies have reported that enhanced α-synuclein aggregation in the presence of Cu[77] but also of Al, Fe and Mn[73,78] can effectively cause acceleration in the rate of α-synuclein fibril formation. The clinical significance of this interaction, as well as the involvement of Al in the etiology of the disease, is still unclear and merits further investigation.

10.9 Chelation Therapy

Much evidence suggests that dyshomeostasis of Al but also other metals (*e.g.* Cu, Fe and Zn) occurs in AD and other ND. Thus, the use of chelating agents to scavenge free metals, that are present in excess or are miscompartmentalized in the brain, may represent a very promising therapeutic option. Generally, the treatment with chelating agents has been successful for a few rare diseases (*e.g.* Wilson's disease) in which a dramatic brain metal accumulation takes place, as a result of specific genetic defects. Yet, for the most common sporadic ND, the design of effective non-toxic chelating molecules remains a challenging task. The requirements for a chelating compound to be effective include the ability to cross the BBB as well as a specific but moderate chelation activity to avoid depletion of biometals.

Among the molecules which have been proposed for the chelation of Al, desferrioxamine (DFO) is certainly the most representative example. DFO is a chelating agent currently approved by the Food and Drug Administration (FDA) of the USA for the treatment of Fe overload. DFO therapy, which dramatically increases Fe excretion, has led to significant improvements in the length and quality of life of patients who suffer from β-thalassemia and other kinds of refractory anemia. Initially, some studies proposed the use of DFO for treating AD, in relation to Al, which shares with Fe a number of physio-chemical features such as a similar ionic radius, a similar charge density and similar kinds of protein ligand.[79] A two-year long, single-blind study reported that DFO decreased the progression of the disease by chelating Al.[80] This result was further supported by a pilot study which demonstrated the ability of DFO to lower brain concentrations of Al in AD patients through intramuscular injections.[81] According to Hider[82] this effect was also partly attributed to the affinity of DFO to bind, not only Al, but also Fe, Cu and Zn, while for others the beneficial effect of the treatment could be due to iron removal, since DFO mainly chelates this ion.[83,84] It was also reported that decreased concentrations of Zn and Fe were found in a post-mortem analysis of DFO-treated subjects.[84] Thus, in contrast to the initial promising results, later studies doubt the effectiveness of DFO for treating AD.

A second example is Feralex G (FXG), a chelator of Fe and Al which has been shown to be effective in disaggregation of paired helical filaments in brain cells. Kruck *et al.*[85] demonstrated, *in vitro*, that combinations of antioxidants and metal chelators such as FXG synergistically scavenge ROS and down-regulate ROS-triggered gene expression. These molecules may be of use in ND associated with metal-ion induced toxicity and excessive production of ROS. In particular, Shin *et al.*[86] proposed the potential clinical usefulness of FGX in the Al/Fe chelation therapy for patients with AD.

10.10 Conclusions

The hypothesis of metal involvement in neurodegenerative mechanisms has generally gained considerable acceptance. Nevertheless, unanimous agreement regarding the quantitative/qualitative features of this imbalance and its contribution to the disease progression has not been reached. The Al–AD link is particularly contentious: this metal was first proposed as the major etiological factor in AD, but after 30 years of studies no convincing results have so far appeared. This is due to the fact that contrasting reports on the actual concentration of the metal in the brain have been published and the goal of recognizing and characterizing unambiguously its distribution patterns and its potential accumulation has not yet been accomplished. There may be numerous reasons for the several analytical discrepancies; the use of different experimental and methodological approaches and the constantly improving technologies are two sources of variability.

There are many arguments in the literature supporting or excluding the involvement of Al in AD. We think it could be premature to discard the role of this metal at least as a cofactor in AD aetiology especially considering the role of Al in promoting a peculiar conformational change which stabilizes Aβ in its oligomeric form.[87,88]

References

1. C. Soto and L. D. Estrada, *Arch. Neurol.*, 2008, **65**, 184.
2. D. A. Butterfield, T. Reed, S. F. Newman and R. Sultana, *Free Rad. Biol. Med.*, 2007, **43**, 658.
3. M. Ferrer, O. V. Golyshina, A. Beloqui, P. N. Golyshin and K. N. Timmis, *Nature*, 2007, **445**, 91.
4. P. Zatta and A. Frank, *Brain Res. Rev.*, 2007, **54**, 19.
5. P. Zatta, R. Lucchini, S. J. van Rensburg and A. Taylor, *Brain Res. Bull.*, 2003, **62**, 15.
6. H. Erazi, W. Sansar, S. Ahboucha and H. Gamrani, *Comptes Rend. Biol.*, 2010, **333**, 23.
7. P. Zatta, M. Ibn-Lkhayat-Idrissi, P. Zambenedetti, M. Kilyen and T. Kiss, *Brain Res. Bull.*, 2002, **59**, 41.
8. K. R. Dave, A. R. Syal and S. S. Katyare, *Brain Res. Bull.*, 2002, **58**, 225.
9. R. R. Kaizer, M. C. Corrêa, L. R. Gris, C. S. da Rosa, D. Bohrer, V. M. Morsch and M. R. Schetinger, *Neurochem. Res.*, 2008, **33**, 2294.
10. A. C. Miu, C. E. Andreescu, R. Vasiu and A. I. Olteanu, *Int. J. Neurosci.*, 2003, **113**, 1197.
11. T. Kaur, R. K. Bijarnia and B. Nehru, *Drug Chem. Toxicol.*, 2009, **32**, 215.
12. P. Zatta, *Metal Ions and Neurodegenerative Disorders*, World Scientific Publishing Co., Singapore, 2002, pp. 1–536.
13. P. Zatta, *Coord. Chem. Rev.*, 2002, **228**, 91.
14. D. Krewski, R. A. Yokel, E. Nieboer, D. Borchelt, J. Cohen, J. Harry, S. Kacew, J. Lindsay, A. M. Mahfouz and V. Rondeau, *J. Toxicol. Environ. Health-Part B-Crit. Rev.*, 2007, **10**, 1.
15. A. Campbell, *J. Alzheimers Dis.*, 2006, **10**, 165.
16. A. C. Alfrey, G. R. LeGendre and W. D. Kaehny, *N. Engl. J. Med.*, 1976, **294**, 184.
17. P. Zatta, P. Zambenedetti, E. Reusche, F. Stellmacher, A. Cester, P. Albanese, G. Meneghel and M. Nordio, *Nephrol. Dialysis Transplant.*, 2004, **19**, 2929.
18. P. Zatta, *J. Alzheimers Dis.*, 2006, **10**, 33.
19. R. A. Yokel, *J. Toxicol. Environ. Health*, 1994, **41**, 131.
20. E. Reusche, *Metal Ions and Neurodegenerative Disorders*, World Scientific Publishing Co., Singapore, 2003, pp. 117–138.
21. P. F. Good and D. P. Perl, *Ann. Neurol.*, 1993, **34**, 413.
22. C. Bouras, P. Giannakopoulos, P. F. Good, A. Hsu, P. R. Hof and D. P. Perl, *Eur. Neurol.*, 1997, **38**, 53.

23. R. A. K. Srivastava and J. C. Jain, *J. Neurol. Sci.*, 2002, **196**, 45.
24. S. Yumoto, S. Kakimi, A. Ohsaki and A. Ishikawa, *J. Inorg. Biochem.*, 2009, **103**, 1579.
25. K. Iqbal, F. Liu, C. X. Gong, A. D. Alonso and I. Grundke-Iqbal, *Acta Neuropathol.*, 2009, **118**, 53.
26. J. Hardy and D. J. Selkoe, *Science*, 2002, **297**, 353.
27. D. J. Selkoe, *Behav. Brain Res.*, 2008, **192**, 106.
28. Z. van Helmond, J. S. Miners, P. G. Kehoe and S. Love, *Brain Pathol.*, 2010, **20**, 787.
29. D. Drago, A. Cavaliere, N. Mascetra, D. Ciavardelli, C. de Ilio, P. Zatta and S. L. Sensi, *Rejuv. Res.*, 2008, **11**, 861.
30. F. Ricchelli, D. Drago, B. Filippi, G. Tognon and P. Zatta, *Cell. Molec. Life Sci.*, 2005, **62**, 1724.
31. I. J. Martins, T. Berger, M. J. Sharman, G. Verdile, S. J. Fuller and R. N. Martins, *J. Neurochem.*, 2009, **111**, 1275.
32. R. Kayed, Y. Sokolov, B. Edmonds, T. M. McIntire, S. C. Milton, J. E. Hall and C. G. Glabe, *J. Biol. Chem.*, 2004, **279**, 46363.
33. A. Demuro, E. Mina, R. Kayed, S. C. Milton, I. Parker and C. G. Gabe, *J. Biol. Chem.*, 2005, **280**, 17294.
34. S. Osawa, S. Funamoto, M. Nobuhara, S. Wada-Kakuda, M. Shimojo, S. Yagishita and Y. Ihara, *J. Biol. Chem.*, 2008, **283**, 19283.
35. M. O. W. Grimm and U. Muller, *Nat. Cell Biol.*, 2006, **8**.
36. G. Thakur, M. Micic and R. M. Leblanc, *Colloids Surf. B-Biointerfac.*, 2009, **74**, 436.
37. C. C. Curtain, F. E. Ali, D. G. Smith, A. I. Bush, C. L. Masters and K. J. Barnham, *J. Biol. Chem.*, 2003, **278**, 2977.
38. I. C. Martins, I. Kuperstein, H. Wilkinson, E. Maes, M. Vanbrabant, W. Jonckheere, P. Van Gelder, D. Hartmann, R. D'Hooge, B. De Strooper, J. Schymkowitz and F. Rousseau, *EMBO J.*, 2008, **27**, 224.
39. S. R. Ji, Y. Wu and S. F. Sui, *J. Biol. Chem.*, 2002, **277**, 6273.
40. J. W. Pettegrew, K. Panchalingam, R. L. Hamilton and R. J. McClure, *Neurochem. Res.*, 2001, **26**, 771.
41. K. Fassbender, M. Simons, C. Bergman, M. Stroick, D. Lutjohann, P. Keller, H. Runz, S. Kuhl, T. Bertsch, K. von Bergmann, M. Hennerici, K. Beyreuther and T. Hartmann, *Proc. Natl. Acad. Sci. USA*, 2001, **98**, 5856.
42. P. C. Reid, Y. Urano, T. Kodama and T. Hamakubo, *J. Cell. Molec. Med.*, 2007, **11**, 383.
43. V. Koppaka and P. H. Axelsen, *Biochemistry*, 2000, **39**, 10011.
44. H. Lin, R. Bhatia and R. Lal, *FASEB J.*, 2001, **15**, 2433.
45. M. Morita, K. Osoda, M. Yamazaki, F. Shirai, N. Matsuoka, H. Arakawa and S. Nishimura, *Brain Res.*, 2009, **1295**, 186.
46. M. Suwalsky, S. Bolognin and P. Zatta, *J. Alzheimers Dis.*, 2009, **17**, 81.
47. J. M. Boon and B. D. Smith, *Curr. Opin. Chem. Biol.*, 2002, **6**, 749.
48. M. Suwalsky, C. Schneider, H. D. Mansilla and J. Kiwi, *J. Photochem. Photobiol. B-Biol.*, 2005, **78**, 253.

49. W. A. Banks, M. L. Niehoff, D. Drago and P. Zatta, *Brain Res.*, 2006, **1116**, 215.
50. M. P. Mattson, *Aging Cell*, 2007, **6**, 337.
51. M. Kawahara and Y. Kuroda, *Brain Res. Bull.*, 2000, **53**, 389.
52. A. M. Isaacs, D. B. Senn, M. L. Yuan, J. P. Shine and B. A. Yankner, *J. Biol. Chem.*, 2006, **281**, 27916.
53. N. Pierrot, P. Ghisdal, A. S. Caumont and J. N. Octave, *J. Neurochem.*, 2004, **88**, 1140.
54. Z. S. Khachaturian, *Calcium Hypothesis of Aging and Dementia*, New York Academy of Science, New York, 1994, vol. 747, pp. 1–11.
55. L. Bojarski, J. Herms and J. Kuznicki, *Neurochem. Int.*, 2008, **52**, 621.
56. F. M. LaFerla, K. N. Green and S. Oddo, *Nat. Rev. Neurosci.*, 2007, **8**, 499.
57. J. T. Yu, R. C. C. Chang and L. Tan, *Prog. Neurobiol.*, 2009, **89**, 240.
58. P. Bernardi and A. Rasola, *Subcell. Biochem.*, 2007, **45**, 487.
59. P. Fernández-Vizarra, A. P. Fernández, S. Castro-Blanco, J. M. Encinas, J. Serrano, M. L. Bentura, P. Muñoz, R. Martínez-Murillo and J. Rodrigo, *Neurobiol. Dis.*, 2004, **15**, 287.
60. A. M. Aleardi, G. Benard, O. Augereau, M. Malgat, J. C. Talbot, J. P. Mazat, T. Letellier, J. Dachary-Prigent, G. C. Solaini and R. Rossignol, *J. Bioenergetics Biomemb.*, 2005, **37**, 207.
61. C. Caspersen, N. Wang, J. Yao, A. Sosunov, X. Chen, J. W. Lustbader, H. W. Xu, D. Stern, G. McKhann and S. D. Yan, *FASEB J.*, 2005, **19**, 2040.
62. http:\\www.alzgene.org, The AlzGene Database, Alzheimer Research Forum.
63. L. Adlerz, M. Beckmann, S. Holback, R. Tehranian, V. Cortéz Toro and K. Iverfelt, *Molec. Brain Res.*, 2003, **119**, 62.
64. T. A. Bayer, K. Paliga, S. Weggen, O. D. Wiestler, K. Beyreuther and G. Multhaup, *Acta Neuropathol.*, 1997, **94**, 519.
65. S. Neumann, S. Schöbel, S. Jäger, A. Trautwein, C. Haas, C. U. Pietrzik and S. F. Lichtenhaler, *J. Biol. Chem.*, 2006, **281**, 7583.
66. L. Adlerz, S. Holback, G. Multhaup and K. Iverfeldt, *J. Biol. Chem.*, 2007, **282**, 10203.
67. J. Avila, I. Santa-Maria, M. Perez, F. Hernandez and F. Moreno, *J. Biomed. Biotechnol.*, 2006.
68. G. V. Johnson and C. D. Bailey, *J. Alzheimers Dis.*, 2002, **4**, 375.
69. E. C. Hirsch, J. P. Brandel, P. Galle, F. Javoyagid and Y. Agid, *J. Neurochem.*, 1991, **56**, 446.
70. D. T. Dexter, A. Carayon, F. Javoy-Agid, Y. Agid, F. R. Wells, S. E. Daniel, A. J. Lees, P. Jenner and C. D. Marsden, *Brain*, 1991, **114**, 1953.
71. M. G. Spillantini, M. L. Schmidt, V. M. Lee, J. Q. Trojanowski, R. Jakes and M. Goedert, *Nature*, 1997, **388**, 839.
72. M. G. Spillantini, R. A. Crowther, R. Jakes, N. J. Cairns, P. L. Lantos and M. Goedert, *Neurosci. Lett.*, 1998, **251**, 205.
73. V. N. Uversky, J. Li and A. L. Fink, *J. Biol. Chem.*, 2001, **276**, 44284.
74. K. M. Powers, T. Smith-Weller, G. M. Franklin, W. T. Longstreth Jr., P. D. Swanson and H. Checkoway, *Neurology*, 2003, **60**, 1761.

75. K. A. Jellinger, *Drugs & Aging*, 1999, **14**, 115.
76. L. Zecca, M. B. H. Youdim, P. Riederer, J. R. Connor and R. R. Crichton, *Nat. Rev. Neurosci.*, 2004, **5**, 863.
77. A. Binolfi, G. R. Lamberto, R. Duran, L. Quintanar, C. W. Bertocini, J. M. Souza, C. Cerveñansky, M. Zweckstetter, C. Griesinger and C. O. Fernández, *J. Amer. Chem. Soc.*, 2008, **130**, 11801.
78. S. R. Paik, J. H. Lee, D. H. Kim, C. S. Chang and J. Kim, *Arch. Biochem. Biophys.*, 1997, **344**, 325.
79. P. C. D'Haese and M. E. De Broe, *Aluminium and Alzheimer's Disease*, Elsevier, Amsterdam, 2001, p. 221.
80. M. D. Crapper, A. Chew, R. H. Bradley and D. E. Sykes, *Surf. Interf. Anal.*, 1991, **17**, 177.
81. T. P. Kruck, J. G. Cui, M. E. Percy and W. J. Lukiw, *Cell. Molec. Neurobiol.*, 2004, **24**, 443.
82. R. C. Hider, Y. M. Ma, F. Molina-Holgado, A. Gaeta and S. Roy, *Biochem. Soc. Trans.*, 2008, **36**, 1304.
83. D. R. Richardson and P. Ponka, *J. Lab. Clin. Med.*, 1998, **132**, 351.
84. M. P. Cuajungco, K. Y. Faget, X. D. Huang, R. E. Tanzi and A. I. Bush, in *Molecular Basis of Dementia*, ed. J. H. Growdon, R. J. Wurtman, S. Corkin and R. M. Nitsch, New York Academy of Science, New York, 2000, vol. 920, pp. 292–304.
85. T. R. Kruck, M. E. Percy and W. J. Lukiw, *Neuroreport*, 2008, **19**, 245.
86. R. W. Shin, T. P. A. Kruck, H. Murayama and T. Kitamoto, *Brain Res.*, 2003, **961**, 139.
87. D. Drago, S. Bolognin and P. Zatta, *Curr. Alzheimer Res.*, 2008, **5**, 500.
88. P. Zatta, D. Drago, S. Bolognin and S. L. Sensi, *Trends Pharmacol. Sci.*, 2009, **30**, 346.
89. V. Gatta, D. Drago, K. Fincati, M. T. Valenti, L. Dalle Carbonare, S. L. Sensi and P. Zatta, *PLoS One*, 2011, **27**, 15965.

CHAPTER 11

Metal Toxicity and Metallostasis in Amyotrophic Lateral Sclerosis

H. L. LELIE,[a] J. P. WHITELEGGE,[b] D. R. BORCHELT[c] AND J. S. VALENTINE[a]*

[a] UCLA Department of Chemistry and Biochemistry, Box, 951569, 607 Charles E. Young Drive, Los Angeles, CA 90025-1569, USA; [b] The Pasarow Mass Spectrometry Laboratory, Box 42 NPI-Semel Institute, David Geffen School of Medicine, UCLA, 760 Westwood Plaza, Los Angeles, CA 90024-1759, USA; [c] Department of Neuroscience, McKnight Brain Institute, SantaFe HealthCare Alzheimer's Disease Research Center, University of Florida, 100 Newell Drive, Gainesville, FL 32611, USA

11.1 Introduction

Metals have long been suspected to play a role in the motor neuron disease amyotrophic lateral sclerosis (ALS) (see Figure 11.1). Thus early epidemiological case studies linked ALS with exposure to lead,[1–3] selenium,[4] mercury,[5,6] manganese,[7] aluminium,[7] or cadmium,[8] and more recently it was reported that significantly increased levels of iron,[9] copper,[10] zinc,[9] calcium,[11] aluminium,[12] cadmium,[8] selenium,[4] and bromine;[13] and decreased levels of manganese,[13] magnesium,[13] cesium[9] and sodium[13] were present in tissue, serum, and bone from ALS patients and from mouse models of familial ALS. The earlier studies listed above were based on whole tissue analysis, but more recently, with the advent of more

RSC Drug Discovery Series No. 7
Neurodegeneration: Metallostasis and Proteostasis
Edited by Danilo Milardi and Enrico Rizzarelli
© Royal Society of Chemistry 2011
Published by the Royal Society of Chemistry, www.rsc.org

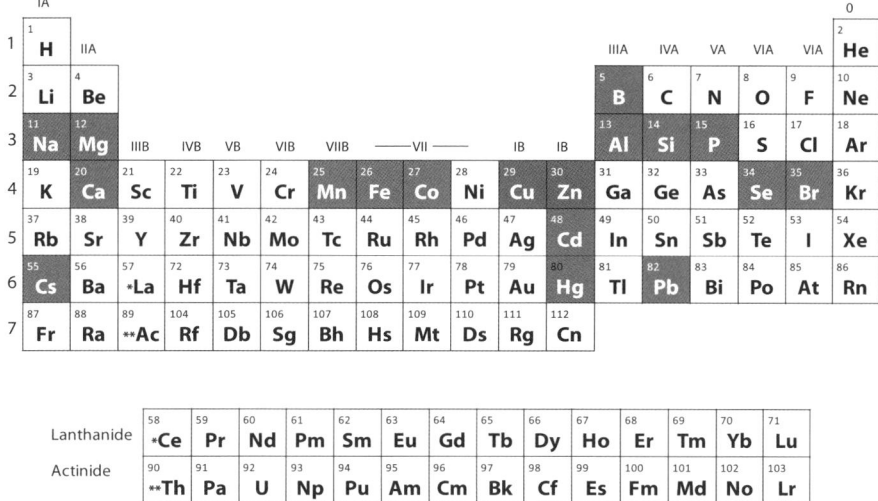

Figure 11.1 Heavy metals and other elements associated with ALS.

powerful and precise instrumentation as well as the availability of mouse models of ALS, it has been determined that subcellular changes are occurring in the distributions of iron, copper and zinc ions in ALS diseased tissue.[14–16]

A major discovery that further linked metals to ALS was the discovery in 1993 that mutations in *SOD1*, the gene coding for the metalloenzyme copper–zinc superoxide dismutase, caused many cases of familial ALS.[17] Subsequent to that discovery, various theories were advanced to explain how copper and/or zinc might play major roles in the SOD1-linked ALS disease. These theories range from pro-oxidant effects of copper bound to mutant SOD1 proteins to protein aggregation caused by abnormal amounts of metal-deficient mutant SOD1 proteins, but no single theory has been entirely ruled in or ruled out at this time.[18,19]

The major purpose of this chapter is to review the evidence for the involvement of metals in ALS. This involves two main areas of research, one where metals in general were shown to have an effect on all forms of ALS and a second area in which specifically copper and zinc are suspected to play a direct role in mutant SOD1 mediated ALS. The implications for the possibility of abnormalities in metallostasis and a role for copper and zinc in the toxic conversion of mutant SOD1 are explored.

11.2 Background

ALS has an incidence rate of about 2 per 100 000 and a lifetime risk of 1 in 1000, making it the most commonly occurring motor neuron disease.[20] The earliest symptoms of ALS are weakness, muscle twitching, and cramping.[21] These symptoms give way to the gradual spreading of muscle weakness and paralysis of all the voluntary muscles. Death typically occurs within 2–3 years after diagnosis, although longer (and shorter) survival times are possible. Both upper motor neurons, which originate from the brain and run down the track of the spinal cord, and lower motor neurons, which originate in the ventral horn of the spinal cord and connect to the specific innervated muscle, are affected. Upper motor neurons transmit signals from the brain to lower motor neurons, which are responsible for sending the impulses that control muscle movement. These motor neurons represent some of the largest cells in the body, with some extending processes up to 1 m long, and they bear a strong electrical load due to the necessity of sending rapid and frequent action potentials.[22]

Motor neurons are not all equally affected by ALS and can be divided into two subtypes on this basis. The ALS-resistant motor neurons, which control sphincter and ocular function, contain lower densities of calcium permeable glutamate receptors and increased calcium buffering capacity as compared to the ALS-vulnerable motor neurons,[23] possibly rendering those cell types more resistant to calcium-mediated excitotoxicity. It is interesting in this regard that riluzole, the only drug that is currently approved by the US FDA for ALS, increases lifespan, although only slightly, by targeting calcium-mediated excitotoxicity.

ALS has a complex etiology, and environmental and genetic factors have both been linked to disease development. Epidemiological studies on ALS have long concluded that external environmental factors must play an important role in the disease, and excessive exercise, military service, geography, age, smoking, spinal trauma, viruses, and exposure to environmental toxins such as heavy metals and chemical agents such as pesticides have all been suggested as possible risk factors for disease.[24–27] Nevertheless, no one specific class of agents has been definitively identified.

Ninety percent of all ALS cases are sporadic and only ten percent are familial, but genetic studies on inherited ALS have nonetheless provided the greatest breakthroughs in the field and have set the framework for under-standing the disease. In a landmark report published in 1993, a genetic linkage between mutations in the SOD1 gene and ALS was reported,[17] and it is now established that 20% of genetic cases are due to mutations in the antioxidant enzyme copper–zinc superoxide dismutase (SOD1). This discovery in 1993 was particularly significant since it led to the creation in 1994 of the first transgenic mouse models of the disease, expressing mutant human SOD1.[28] Studies based upon these mouse models of ALS have increased our understanding of the many pathological abnormalities associated with the disease, including the roles of oxidative stress, apoptosis, excitotoxicity, neurofilament disorganiza-tion, mitochondrial dysfunction, proteasome inhibition, impaired axonal transport, reactive astrocytes, and protein aggregation (Figure 11.2).[19,29]

The normal primary function of SOD1 is thought to be acting as a major antioxidant enzyme in cells by reducing levels of intracellular superoxide radical. However, the targeted deletion of the SOD1 gene in mice does not produce catastrophic abnormalities or cause ALS-like symptoms.[30] Thus SOD1-mediated ALS is not apparently related to a loss in this antioxidant

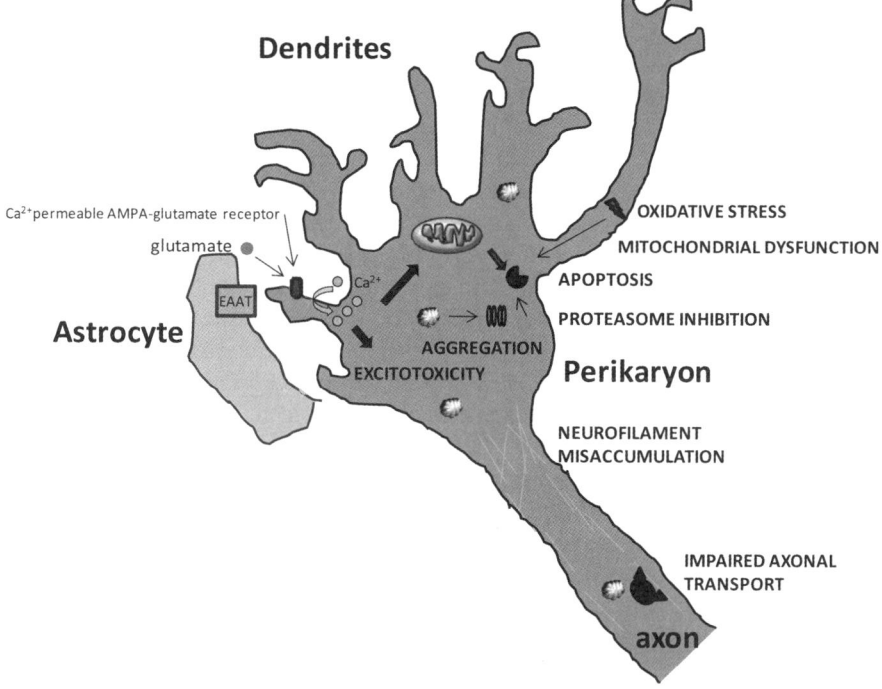

Figure 11.2 Pathogenic mechanisms in ALS.

function, but rather the ALS mutations render the mutant SOD1 proteins toxic. More recently, mutations in the RNA processing proteins, TAR DNA binding protein (TDP-43), and FUS (FUsed in Sarcoma) were discovered to also be associated with ALS.[31–33] Understanding how these DNA/RNA binding proteins can lead to motor neuron death and how those processes are related to those elicited by mutant SOD1 promises to be a rewarding avenue of research in the future.[34]

11.3 Toxic Metals and ALS

Decades before any genes involved in familial ALS were identified, the search for clues concerning ALS came mainly from epidemiological studies. Some of the earliest clues about the causes of ALS originated in the remote western Pacific region of the world in the Micronesian islands, the largest of which is Guam. In the 1950's, it was observed that the frequency of ALS in this region was fifty times higher than in the rest of the world. ALS in these patients was often also associated with a Parkinsonism dementia complex. Initially, it was believed that the extreme prevalence was due to a genetic factor, but shortly after the 1950's, the incidence began to decline dramatically suggesting that a strong environmental factor may have been responsible. One of the many early candidates considered for this environmental factor was excessive exposure to the transition metal manganese. This theory was supported by the observation that volcanic formations in Guam contain high levels of manganese, leading to high manganese content in soil and drinking water as measured by neutron activation analysis (NAA).[12,35,36] Excessive manganese exposure can lead to symptoms similar to those of ALS, and epidemiological studies revealed a cluster of ALS cases among manganese miners in Guam. When spinal nervous tissue from Guam patients was analyzed, an increase in the content of manganese in the areas of degeneration was reported.[12] Interestingly, an increase in other metals including aluminium and calcium was also observed in those same affected tissues. Loss of bone density and further tissue studies revealed dramatic variation in calcium levels in tissue subsections, supporting an involvement of calcium homeostasis in the disease.[4,7] Other indications of universal disturbances in metallostasis were observed in blood samples where lower levels of manganese and selenium were reported for patients versus controls.[4] These early studies appeared hopeful and provided a lot of promise toward understanding ALS pathology.

Over time, however, follow-up studies on patient exposure and tissue metal levels proved inconsistent. For example, one study on the drinking water and soil in Guamanian villages where large clusters of ALS cases existed did not show high levels of manganese present as before but actually indicated the complete opposite, a lack of both manganese and calcium.[37] Moreover new investigations reported difficulty in reproducing the earlier findings concerning tissue metal levels,[38] while others implicated a completely different set of metals.[39] Studies as recent as 2001, where blood metal levels were measured to validate earlier findings of changes associated with motor neuron disease, did not confirm earlier studies with a larger sample size of sporadic ALS cases and in fact led to contradictory findings.[8] Taken together, these studies demonstrate that fluctuations in metal levels are very frequently associated with ALS in Guam, but the specific metals vary and the extent of the fluctuations remains unclear. Currently, the view on the reports of metals in Guam ALS and ALS–Parkinsonism dementia patients is inconclusive. While some level of involvement of metallic elements appears probable, determining which ones are relevant remains elusive due to variation in the methodologies used in the early studies, the relatively small pool of patients, and inconsistent results. Nevertheless, these studies laid the groundwork for future studies on the association of metals and ALS.

The Guam studies inspired several subsequent studies of metals and ALS. In a 1983 clinical survey of metal exposure of ALS patients at the Mayo Clinic, it was discovered that ALS patients had a slightly increased exposure to heavy metals relative to controls.[40] Follow-up investigations with a large sample size of male ALS patients revealed that they were frequently employed in blue collar jobs involving welding and soldering, thus implicating increased lead exposure.[3] Very recent studies showed that sporadic ALS patients had a higher than normal exposure to lead on average, which led to higher levels in blood and bone.[2,41] These findings support a pathological role for increases in exposure to heavy metals leading to increased levels in the body. Ultimately, however, no retrospective study has yet conclusively defined epidemiological and environmental factors that lead to ALS, largely due to inconsistencies in these reports.[5] It should be noted, however, that inconsistencies in the literature may be partially explained by the fact that tissue investigations on ALS are generally limited to post-mortem end-stage samples, rendering it difficult to assess disease-relevant changes. Recent investigations implementing more advanced techniques with the ability to measure more localized microscopic regions of tissue have reinvigorated the possibility that metal ions and other elements such as iron, manganese, copper, zinc, bromine, sodium, magnesium, and calcium homeostasis are affected in neurodegeneration.[13,42–44] Ross *et al.* have proposed a model in which multi-metal toxicity is responsible for ALS pathology.[45] In this model, each metal can contribute to toxicity in its own way, either directly through aberrant redox chemistry, perturbed homeostasis, changes in metalloenzyme activities, or altered signaling, and so forth. As analytical technology advances, we anticipate future successes in this currently inconclusive area of ALS research.

11.4 Metallothioneins and ALS

Metallothioneins are small (6–7 kDa) peptides responsible for scavenging and transporting mono- and divalent metal ions.[46] Four main species exist in humans, MT-I to MT-IV, and in nervous tissue, MT-I and MT-II are localized to astrocytes and MT-III is present in neurons. It is generally believed that these proteins play a protective role against disease by regulating dangerously high levels of metals and scavenging ROS.[47,48] This protective role was confirmed when knockout mice for either the astrocyte specific metallothioneins (I and II) or the neuronal metallothioneins (MT-III) were crossed with G93A SOD1 ALS mice resulting in acceleration of disease.[49] Gene expression profiles found that MT-I and -II were up-regulated prior to symptoms in spinal cord of G93A SOD1 mice but MT-III was up-regulated later, suggestive of an initial glial response.[50,51] In later stages, it appears that MT-I becomes up-regulated in other tissues such as liver, kidney, and skeletal muscle of transgenic mice, indicating that a systemic response to metal homeostasis and oxidative stress exist in ALS.[52,53]

Metallothioneins, which play important roles in normal metallostasis as well as in providing protection from toxic metals in human tissues, have been observed to be elevated in liver, kidney, spinal cord gray matter and astroglia from ALS patients relative to non-ALS controls, indicating a systemic break-down in metallostasis associated with the disease.[54] Metallothioneins have been shown to mediate tissue expression of SOD1 when copper is present,[55] and ALS patients have been shown to have decreased expression of metallothioneins relative to controls.[56] It has further been suggested that one of the mechanisms by which exercise helps patients is through increased expression of metal-lothioneins.[57] Intriguingly, a single nucleotide polymorphism in the metallothionein gene, *MT-1*, was associated with some sporadic ALS patients that had high exposure to environmental toxicants, possibly linking a genetic susceptibility to metal toxicity.[58] The combination of the transgenic mice studies with the reports on human patients suggest that metallothioneins may play a role in countering disease, possibly due to their function in maintaining metallostasis.

11.5   SOD1 and ALS

11.5.1   SOD1 and Familial ALS

SOD1, originally termed hemocuprein, hepatocuprein, or erythrocuprein, was first isolated from red blood cells in 1938 and identified as a major copper-binding protein in erythrocytes. After its superoxide dismutase activity was discovered by McCord and Fridovich in 1969,[59] it was renamed copper–zinc superoxide dismutase. Research in subsequent years led to extensive characterization of the protein, and we now have considerable knowledge of its physical, chemical, and biological properties.[18]

SOD1 is a 153-amino acid homodimeric enzyme that adopts a beta barrel fold. Each subunit contains two large loops that encase a zinc and copper ion, the latter a prerequisite for activity. In the human SOD1 protein, there are four cysteines, two of which are involved in an intramolecular disulfide bond. The fully metallated, disulfide-oxidized, dimeric protein is extremely stable, a property that is apparent from its high melting temperature (above 90 °C), and its extreme resistance to proteolytic degradation and denaturation by 10 M urea or 4% SDS.[18,60]

SOD1 catalyzes a two-step disproportionation reaction of two superoxide anions to produce one molecule of hydrogen peroxide and one molecule of molecular oxygen as the copper ion is alternately reduced and oxidized (eqns. 11.1–3).

$$O_2^- + Cu(II)ZnSOD \rightarrow O_2 + Cu(I)ZnSOD \quad (11.1)$$

$$O_2^- + Cu(I)ZnSOD + 2H^+ \rightarrow H_2O_2 + Cu(II)ZnSOD \quad (11.2)$$

$$Sum : 2O_2^- + 2H^+ \rightarrow H_2O_2 + O_2 \quad (11.3)$$

SOD1-linked ALS is not due to a loss of SOD activity. This conclusion is based on several key pieces of evidence. Firstly, the majority of SOD1 ALS mutations do not result in loss of SOD activity.[61] Additionally, *sod1* knockout mice do not develop the disease.[30] Furthermore, with the exception of northern European families that inherit the D90A mutation, the disease shows a dominant pattern of inheritance. Thus, mutations in SOD1 that are associated with ALS are thought to instill a toxic property to the protein.

Several hypotheses have been put forth for this toxic gain-of-function by mutant SOD1, including a conversion to pro-oxidant activity of enzyme-bound copper,[62] but the currently most widely accepted hypothesis for this gain-of-function is the accumulation of misfolded and abnormally oligomerized mutant SOD1.[60,63] This hypothesis was derived from the presence of prominent cytoplasmic inclusions containing SOD1 that were identified within the motor neurons of mouse models of ALS and in the spinal cords of human ALS patients.[60,64] It is also possible that some combination of oxidative damage and protein aggregation could be involved in SOD1-linked ALS. For example, oxidative modification of histidine residues at the active site was shown to make the protein more prone to aggregation,[65] and the inclusions found in ALS tissues show signs of post-translational modifications such as glycosylation and side chain modification[66] and are also commonly co-localized with regions of lipid peroxidation, suggesting that oxidative damage could play some sort of role in their formation or toxic action.[67]

Understanding how SOD1 aggregates has been an area of intense study. Some important clues were derived from studies on the composition of SOD1 aggregates isolated from transgenic mice spinal cord.[68] Counter to several hypotheses, aggregated SOD1 was not post-translationally modified, nor did it co-aggregate with any prominent binding partners. However, a recent study by Wang *et al.* of mice that express mutant SOD1 fused to yellow fluorescent protein (YFP) reported that aggregates of the mutant fusion protein contain HSP70.[69]

The structure of the large aggregates that form inclusions has been another area of interest, particularly whether the mutant SOD1 produces amyloid or amyloid-like structures. Recent studies of human SOD1-linked ALS have not found convincing evidence for amyloid structures.[70] However, purified apo WT SOD1, even under physiologically relevant conditions, has a propensity to form amyloid-like fibers.[71,72] Amyloid-like fibers have also been seen in crystal structures of some ALS mutant SOD1 proteins,[73,74] and fibrillar structures that bind Thioflavin-S (a dye that selectively binds amyloid fibrils) have been detected in tissues of symptomatic transgenic mice that express mutant SOD1.[75] These findings correlate with the detection of detergent-insoluble and sedimentable forms of mutant SOD1 in spinal cord tissues of symptomatic transgenic mice.[76] Interestingly, in the mouse models, these large aggregates of mutant protein accumulate largely in the interval between the onset of the first noticeable symptoms and paralysis, indicating that forms of the mutant protein other than large aggregates are likely to be involved in initiating the disease. At present, there are many unresolved questions regarding the structure of the aggregates *in vivo* and the relative contribution of different types of aggregates to disease pathogenesis.

It is unclear how aggregates may be toxic, but several hypotheses exist. It has been suggested that aggregates can mediate aberrant chemical reactions following misfolding of mutant SOD1, or they can sequester essential cellular components.[34,77] They may also monopolize chaperones and overburden protein degradation pathways. Aggregates have also been found to interfere with mitochondrial functions,[78] induce endoplasmic reticulum (ER) stress,[79] and inhibit axonal transport.[80]

Despite their associations with many neurodegenerative diseases, it is not known if the formation of the visible aggregates themselves is toxic, correlative, or beneficial; with the latter being attributed to the clearing of otherwise toxic components. For instance, in the case of Alzheimer's disease, it is believed that the smaller oligomers of amyloid beta peptide harbor greater neurotoxicity than the large amyloid plaques that represent large extracellular aggregates of amyloid peptide.[81] A similar scenario is possible for mutant SOD1 with smaller oligomeric assemblies of protein harboring greater toxicity than the larger aggregates that are visible as pathologic inclusions. However, it is entirely possible that the smaller oligomeric assemblies and the larger aggregates impinge on different cellular processes with each exacting some type of toxicity to the motor nervous system.

11.5.2   SOD1 and Sporadic ALS

It is unknown if SOD1 also plays a role in the pathogenesis of sporadic ALS, but several lines of evidence suggest that it may. SOD1-linked familial ALS has a very similar clinical pathology to sporadic ALS, although treatments developed using the G93A-SOD1 ALS mouse model, in which the protein is highly over-expressed, have not been successfully translated to human patients.[82] Inclusions containing SOD1 have recently been identified in sporadic ALS patients.[83] Smaller, soluble, oligomeric forms of WT SOD1 may also be neurotoxic entities.[76] It is interesting to note that wild-type SOD1, in its metal-free form, does have a propensity to oligomerize *in vitro* under physiological conditions[71,72] and thus might be able to take on a toxic form *in vivo* and play a role in sporadic ALS. Interestingly, oxidatively damaged wild-type SOD1 has been reported to misfold and acquire toxic properties similar to those of mutant SOD1 proteins, suggesting the additional possibility that oxidized wild-type SOD1 could be involved in sporadic ALS.[84,85] A 2007 article from Gruzman *et al.* reported the identification of a 32-kDa covalently cross-linked SOD1 dimer associated with both sporadic and familial ALS patient tissues specifically.[86] On a related note, a recent study shows that sporadic ALS patients have increased SOD1 mRNA transcripts in affected tissues thereby linking sporadic ALS, SOD1, and possibly aggregation.[87]

It is also possible that toxic exposure to metallic elements could influence the levels of SOD1 expressed. Exposure of yeast cells to exogenous Cu, Ca, Cd, Co, Cr, Li, Mn, Zn, K, Na, Ni, Rb or Fe led to the dramatic and consistent up-regulation of SOD1. Zinc resulted in the highest SOD1 induction, while other antioxidant proteins, such as thioredoxin peroxidase, cytochrome c peroxidase, and heat shock proteins all were down-regulated.[88] It was concluded that SOD1 plays some role, apart from its role as a superoxide dismutase, in detoxification of heavy metals. Studies using the human HepG2 hepatoma cell model revealed that mammalian systems recapitulate the yeast results; heavy metals were found to induce SOD1 through a metal responsive element (MRE) that is 273 to 276 base pairs upstream of the SOD1 gene.[89] One possible explanation is that SOD1 acts in the buffering and maintenance of metallostasis of these various metals.

Recent investigations further suggest that a strong link may exist between the SOD1 protein and tissue metallostasis.[90] Synchrotron based X-ray fluorescence imaging carried out on SOD1 transgenic mouse spinal cord sections revealed a dramatic effect of over-expression of SOD1 on the levels of metals in these tissues. Over-expression of either ALS-mutant or wild-type SOD1 resulted in a dramatic shift in the localization of copper and iron from the spinal cord white matter to the spinal cord gray matter without affecting the total metal levels.[90] Copper levels reflected the distribution of SOD1, which was higher in the gray matter than the white matter in transgenic mice, and iron levels were correlated with copper levels, possibly reflecting the linked metabolism of these two metals. Interestingly, zinc levels were much higher in the spinal cord white matter of diseased animals (*i.e.*, mutants G93A SOD1 and H46R/H48Q SOD1) compared to the non-transgenic and wild-type SOD1 controls. Deciphering the relationship between SOD1 and both normal metallostasis and toxic metal exposures may provide clues about the disease process in ALS.

11.5.3 Stabilization of the SOD1 Protein by Zinc and Copper

In any discussion of metals in ALS and SOD1, it is pertinent to understand the critical roles played by metals in the maturation, structure and function of SOD1. SOD1 comprises up to 1% of all soluble cytosolic proteins in neurons.[91] Maturation of SOD1 requires removal of the initial methionine, N-terminal acetylation, acquisition of zinc and copper ions, oxidative formation of the disulfide bonds in each subunit, and dimerization, but the timing of these post-translational modifications is not yet known. A recent focus in the SOD1–ALS field has been on investigating these maturation steps, in particular metal acquisition, since metal ion binding contributes significantly to the stability and other characteristics of the mature protein.

The mechanism of zinc acquisition by SOD1 *in vivo* is not well understood. Some have proposed that it can occur through simple diffusion or that zinc is delivered by metallothionein (reviewed by Potter and Valentine, 2003),[92] while others suggest that it may be assisted by the copper chaperone for SOD1, known as CCS.[93] Tight zinc binding is believed to be an early step in maturation and is required for the proper folding of SOD1.[94] Zinc coordinates to SOD1 via His 63, His 71, His 80, and Asp 83, binding in a distorted tetra-hedral geometry. Its binding significantly stabilizes the structure of SOD1.[95]

Copper insertion is better characterized than zinc insertion and is thought to occur along two main pathways in humans: a CCS-dependent and a CCS-independent pathway.[96] In the CCS-dependent pathway, CCS, which is present in less than one tenth the levels of SOD1,[93,97] docks with SOD1, delivering copper and facilitating formation of the intramolecular disulfide bond in an oxygen-dependent mechanism.[91] SOD1 has four cysteines, Cys 6, Cys 57, Cys 111, and Cys 146, with Cys 57 and Cys 146 forming the native intramolecular disulfide bond in each subunit of the mature protein. The formation of the disulfide bond is important in stabilizing the dimer interface by anchoring a loop of the protein (Glu 49 to Asn 53).[98] The free thiols, C6 and C111, serve an unknown role or no role at all, however many mutations have been associated with these sites including C6S, C6F, C6G, C111Y.

Some organisms such as *Saccharomyces cerevisiae*, wild-type yeast cells, require CCS while other organisms such as *Caenorhabditis elegans* do not even have an endogenous copper chaperone for SOD1.[96] Therefore, when the *ccs1* gene is removed from yeast, yeast SOD1 does not become activated. However, human SOD1 can be activated even when expressed in a *ccs1Δ* null yeast cell via a CCS-independent pathway, and this pathway includes the oxidation of the disulfide without a requirement for oxygen.[99] The precise loading mechanism of the CCS-independent pathway has not yet been determined.

It is unclear whether SOD1 dimerization occurs before or after the other post-translational modifications, but we do know that the dimer is quite a stable state for all forms of the protein except the metal-free, disulfide-reduced species.[100–102] Thus dimerization could occur at any point along the maturation pathway, possibly depending on the local steady state concentrations of SOD1, CCS, and glutathione. Fully mature dimeric SOD1, with metals bound and disulfide bonds intact, has an extremely stable quaternary structure with a melting temperature of above 90 °C.[100]

Even without the metal ions bound, the wild-type SOD1 disulfide oxidized apoprotein is relatively stable, with a melting temperature around 52 °C. Nevertheless, apo SOD1, with the disulfide bond intact or reduced, has been found to form amyloid-like species readily under relatively mild conditions,[63,71,103,104] and it has been suggested that the increased propensity toward aggregation of some of the mutant SOD1 proteins may be due to their reduced ability to bind metals *in vivo*. In order to address this question, new analytical methodology was developed to assess the metallation state of soluble and aggregated SOD1 proteins *in vivo* in the SOD1-ALS transgenic mice.[90] Metallation levels of mutant SOD1-containing aggregates isolated from mouse spinal cord were found to be very low. By contrast, the soluble mutant and wild-type SOD1 from the mouse spinal cords was found to be highly metallated in most cases, with a total metallation state of about four metal ions per SOD1 dimer (usually around three zinc ions and about one copper ion per dimer).[90] These results suggest that aggregated SOD1 is derived from immature SOD1 prior to metallation. Several other lines of evidence also suggest that the toxic form of SOD1 is derived from nascent metal-free SOD1 species (see Seetharaman *et al.* for a review of immature SOD1 and ALS).[105]

Some of the ALS-mutant SOD1 proteins are severely destabilized in their apo states while others have stabilities very similar to those of wild-type SOD1 in both the apo and metallated states regardless of the status of the disulfide bond,[106] and ALS-mutant SOD1 proteins that are highly stable when expressed, isolated, and biophysically characterized[106] are nonetheless found to aggregate *in vivo* in the transgenic mice and/or in cell culture assays whereas the wild-type SOD1 protein is not.[107–109] These results suggest that the kinetics of folding and maturation of the mutant SOD1 proteins may be altered relative to wild-type SOD1 and that aggregation may occur from a folding intermediate of the protein. This hypothesis is supported by experiments carried out by Kopito and coworkers who used a cell-free reticulocyte lysate expression system and demonstrated that the kinetics of metal acquisition to a hyperstable state were impaired for ALS-mutant SOD1 relative to wild-type SOD1.[94] Zinc appears to be especially important in this folding process since it is likely to be an early event and its binding is kinetically more labile than that of copper, allowing greater freedom across the energy field landscape of folding.[110,111] From all of these data, it seems possible that the key to mutant SOD1 mediated toxicity may lie in its initial misfolding when coming off the ribosome, possibly exacerbated by slow or improper zinc binding.

11.6   ALS and Some Specific Metals

11.6.1   Copper

Cellular metabolism of copper is tightly regulated, probably owing to its potential for promoting metal-mediated oxidative stress.[112] Copper ion import, transport, and insertion into proteins are tightly regulated, and storage and scavenging is accomplished by both glutathione and the metallothioneins. Copper distribution is believed to follow a gradient whereby increased copper affinity determines its localization, with SOD1 and metallothioneins having the highest affinities for copper,[113] suggesting that SOD1 can play a role in storing copper in a non-toxic form, particularly if intracellular copper levels get too high.

   Disruption of the role of SOD1 in copper buffering has been suggested as a problem that could occur in mutant SOD1-mediated ALS.[114] Rae *et al.* calculated that a normal yeast cell has roughly 390 000 copper ions, about 13% (50 000) of which are bound to SOD1.[112] In mutant SOD1-ALS mice, SOD1 is over-expressed to levels that are at least sixfold and up to tenfold higher than normal. Recent studies have measured the metallation state of SOD1 derived from the spinal cord of such mice.[90] An individual dimer of SOD1 that is fully metallated contains four metal ions, two copper and two zinc, and the tissue metallation state of SOD1 represents the average number of copper and zinc ions bound to each dimer of SOD1 averaged over the whole tissue. The results revealed that, while the copper levels were close to two per dimer for the endogenous mouse SOD1 (with two zincs in the zinc site), the copper metallation state is only half, at about one copper per dimer for wild-type and G93A and G37R SOD1 mutants over-expressed in these mice (each with now about three zincs per dimer) relative to the endogenous mouse SOD1. Since the total tissue metal levels do not change dramatically, these data imply that the mutant and wild-type SOD1 bind 30–40% of total cellular copper. The fact that the wild-type mice do not fall ill despite the obviously large impact on copper metabolism suggests that altered copper metallostasis is unlikely to play a major role in ALS pathology in the transgenic mice.

An early theory concerning the toxic gain-of-function of ALS-mutant SOD1 proteins was aberrant pro-oxidant copper chemistry from copper bound to mutant SOD1 proteins.[18,115,116] However, studies on transgenic mice expressing a mutant SOD1 incapable of binding metals at the normal copper sites have shown that copper is not required for the mouse to fall ill,[107] and investigations using a mouse knockout of the copper chaperone for SOD1 (CCS) revealed that the lack of copper insertion into SOD1 did not change onset or progression of the disease.[117] These and subsequent studies appear to eliminate a major role for copper in SOD1-mediated toxicity.[107,117–119]

At this point, it seems clear that copper chemistry can be ruled out as the major factor causing SOD1-linked ALS. Nevertheless, copper has been implicated in other neurodegenerative disorders, such as Alzheimer's disease and prion encephalopathies, and there are many reports suggesting that copper may be somehow linked to ALS. For example, excess copper was found in erythrocytes of SOD1-associated familial ALS patients,[120–123] and a cross of a copper-deficient mouse with an SOD1-ALS mouse led to delayed onset of disease but did not decrease mortality.[124] Administration of a copper chelator prolonged the lifespan of G93A SOD1 transgenic mice,[16,124] and further studies revealed that mutant SOD1 but not wild-type SOD1 led to a dramatic up-shift in proteins responsible for copper homeostasis in the spinal cords of these mice.[16] These results suggest that while copper chemistry is an unlikely factor in pathology, disruptions in its normal homeostasis may be relevant.

11.6.2 Iron

Iron is a redox-active metal that, even more than copper, has the potential to act as a pro-oxidant and its metabolism is tightly linked to that of copper as iron transport into cells relies on the multi-copper oxidase, ceruloplasmin.[114] Increased iron levels in spinal cord and brain and increased levels of the iron storage protein, ferritin, during later stages of disease have been reproducibly shown in samples from ALS patients.[1,9] Interestingly, increased mitochondrial ferritin levels were detected in a G37R mouse model of ALS, confirming altered iron metabolism, and treatment with an iron chelator prolonged survival for this mouse strain by a very significant five weeks.[14] More studies are needed to determine whether iron homeostasis plays a significant role in ALS pathology.

11.6.3 Zinc

Zinc is the second most abundant metal in biology, after iron, and it is vital for protein stability, DNA replication, regulation of transcription, protein translation, metabolic protein function, signaling pathways, and as a cofactor in many enzymatic reactions. Zinc is also well known to play major roles in the central nervous system, and neurons have been shown to contain three distinct pools of zinc, 80% bound to protein, 5–15% in vesicles of glutaminergic synapses, and 5% free labile zinc.[125] Zinc is not a redox-active metal, but its levels are nonetheless tightly regulated, and alterations in zinc levels could impart adverse effects due to its role in various signaling pathways.[46] Zinc levels within spinal cord and serum were not significantly different in ALS patients versus controls.[120] However, zinc deficiency accelerated disease processes in a G93A hSOD1 mouse.[126] The effects of zinc supplementation on the mice models have been mixed, with low levels appearing beneficial while high levels were toxic.[126] Interestingly, as mentioned earlier, increased levels of the zinc-responsive metallothioneins MT-I and MT-II have been associated with early stages of disease in transgenic mice,[127] suggestive of increased zinc levels. More recently it was discovered that the transgenic mice spinal cords have excess labile zinc localized in neurons and astrocytes when they are undergoing neurodegeneration as measured by colocalization of HNE (an endogenous neurotoxic agent) and the zinc probe TFL-Zn.[15] Increased levels of chelatable zinc are observed in cell cultures of immune cells undergoing apoptosis and in neurons which underwent ischemia or seizure activity.[48] Excessive zinc levels are observed in a number of pathologies associated with neurodegeneration, ranging from oxidative stress and protein aggregation to apoptosis (for a comprehensive review of zinc in neurodegenerative diseases, see Smith *et al.*).[46] In the case of ALS, zinc levels appear to increase in spinal cord white matter as a result of disease but the overall levels remain unchanged.[90] The studies thus far on zinc suggest an important link between zinc and ALS; however no specific role for zinc in pathogenesis has been defined.

Approximately 10% of all zinc in the central nervous system is located in vesicles within the presynaptic cleft of glutaminergic neurons. Release of this zinc into the synaptic cleft is important for signaling as it binds many receptors including N-methyl-D-aspartate (NMDA) and γ-aminobutyric acid (GABA) receptors and voltage-gated calcium channels.[128] It has been postulated that one of the mechanisms of aging involves the aberrant uptake of zinc through voltage-gated calcium channels following glutamate receptor activation and that this zinc can lead progressively to mitochondrial dysfunction and eventually apoptosis.[125] A similar mechanism was proposed in ALS,[15] in which mutant SOD1 could also contribute to a breakdown in zinc homeostasis, leading to mitochondrial dysfunction and apoptosis of motor neurons. This mechanism was modeled in spinal motor neuron culture in which it was found that exposure to zinc led to an acutely toxic effect, which they found to be mediated by the activation of calcium-permeable AMPA/kainite channels.[129] Based on these findings, one would expect the use of a zinc chelator to help counter disease. To test this, $20\,\text{mg}\,\text{kg}^{-1}\,\text{day}^{-1}$ of the zinc chelator, TPEN, was administered to G93A SOD1 mice, resulting in a modest beneficial effect with a five to ten day increase in survival time.[15] These results combined with the link between neuronal zinc and other neurodegenerative diseases indicate a need for additional studies to clarify the role of zinc in ALS.

11.6.4 Calcium

The role of calcium in the molecular pathology of motor neurons in ALS has been very well characterized. Excess calcium influx into the neurons and poor calcium buffering contribute to mitochondrial degeneration and apoptotic cell death. During normal neurotransmission, glutamate is released from a presynaptic neuron and binds to the AMPA receptors of a postsynaptic neuron, leading to a large influx of calcium ions. Normally, the glutamate is promptly removed from the synaptic cleft by the excitatory amino acid transporter (EAAT2) on glial cells. However in ALS, EAAT2 levels are diminished, which in turn leads to over-activation of AMPA receptors and hence excessive calcium influx.[130] Interestingly, a knock down of EAAT2 in rats led to a deteriorating motor syndrome with ALS-like symptoms.[131] Relative to other neurons, motor neurons express lower levels of GluR2, which cause AMPA receptors to be more permeable to calcium.[132] Furthermore, diseased motor neurons have a particularly difficult time recovering to basal levels after AMPA receptor activation, and the excitotoxic pathway is thus enhanced.[11] Beta-*N*-methylamino alanine (BMAA), which is one of the candidates for the neurotoxic ALS-causing agent in Guam, has been shown to activate AMPA receptors, in turn propagating the same calcium influx mechanism.[133] The ALS drug riluzole acts to inhibit tetrodotoxin sodium channels, thereby reducing calcium influx and countering the effects of overstimulation of glutamate receptors.[134] Further contribution to improper calcium handling in the motor neuron may result from decreased expression of the calcium binding proteins calbindin D-28k and parvalbumin.[135] Furthermore, calmodulin has also been found to coaggregate with mutant SOD1, possibly leading to lower levels of this major calcium buffering protein.[68] The studies on calcium provide a clear example of how metal ions and metallostasis function specifically in this neurodegenerative disease.

11.7 Conclusions

Neurodegenerative diseases are frequently associated with the appearance of abnormal protein deposits in nervous tissue, and it has been proposed that this pathological hallmark is indicative of an underlying proteopathy and that failures of normal proteostasis may be the root cause of many of these diseases.[136,137] More recently, it has been appreciated that maintaining the delicate balance of different metal ion concentrations present in the cell may also be important in maintaining normal proteostasis, and it is now clear that failures in metallostasis could also play an important role in disease. In the case of ALS, both sporadic and familial, metals have long been linked to the disease, although in many different ways, and any proposed pathogenic mechanism should certainly address their potential involvement. In the case of SOD1-linked familial ALS, more studies are needed to determine how SOD1 proteins become toxic to motor neurons. As analytical technologies for detection of metals continue to improve, we can expect to learn more about how metallostasis and proteostasis are linked under normal conditions and how dysfunction in either might lead to ALS. It is also clear that involvement of toxic metals in either sporadic or familial ALS cannot be ruled out and should continue to be considered.

Acknowledgements

The authors wish to thank Dr Edith B. Gralla and Dr Sadaf Sehati for helpful discussions and review of this chapter. They also would like to thank the supporting NINDS grant P01NS049. HL was supported by the NIH Chemistry Biology Interface Training Program; Grant number: 5T32GM008496.

References

1. M. Qureshi, R. H. Brown Jr, J. T. Rogers and M. E. Cudkowicz, *Open Neurol. J.*, 2008, **2**, 51–54.
2. F. Kamel, D. M. Umbach, T. L. Munsat, J. M. Shefner, H. Hu and D. P. Sandler, *Epidemiology*, 2002, **13**, 311–319.
3. C. Armon, L. T. Kurland, J. R. Daube and P. C. O'Brien, *Neurology*, 1991, **41**, 1077.
4. H. Nagata, S. Miyata, S. Nakamura, M. Kameyama and Y. Katsui, *J. Neurol. Sci.*, 1985, **67**, 173–178.
5. F. O. Johnson and W. D. Atchison, *NeuroToxicology*, 2009, **30**, 761–765.
6. J. Praline, A.-M. Guennoc, N. Limousin, H. Hallak, B. de Toffol and P. Corcia, *Clin. Neurol. Neurosurg.*, 2007, **109**, 880–883.
7. T. Kihira, M. Mukoyama, K. Ando, Y. Yase and M. Yasui, *J. Neurol. Sci.*, 1990, **98**, 251–258.
8. R. Pamphlett, R. McQuilty and K. Zarkos, *NeuroToxicology*, 2001, **22**, 401–410.
9. W. R. Markesbery, W. D. Ehmann, J. M. Candy, P. G. Ince, P. J. Shaw, L. Tandon and M. A. Deibel, *Neurodegeneration*, 1995, **4**, 383–390.
10. Y. Ihara, K. Nobukuni, H. Takata and T. Hayabara, *Neurol. Res.*, 2005, **27**, 105–108.
11. E. Guatteo, I. Carunchio, M. Pieri, F. Albo, N. Canu, N. B. Mercuri and C. Zona, *Neurobiol. Dis.*, 2007, **28**, 90–100.
12. Y. Fumio, Y. Masayuki, Y. Yoshiro, I. Shiro, D. C. Gajdusek, J. G. Clarence, Jr and C. Kwang-Ming, *Psychiatry Clin. Neurosci.*, 1980, **34**, 75–82.
13. B. Ostachowicz, M. Lankosz, B. Tomik, D. Adamek, P. Wobrauschek, C. Streli and P. Kregsamer, *Spectrochim. Acta, Part B*, 2006, **61**, 1210–1213.
14. S. Y. Jeong, K. I. Rathore, K. Schulz, P. Ponka, P. Arosio and S. David, *J. Neurosci.*, 2009, **29**, 610–619.

15. J. Kim, T.-Y. Kim, J. J. Hwang, J.-Y. Lee, J.-H. Shin, B. J. Gwag and J.-Y. Koh, *Neurobiol. Dis.*, 2009, **34**, 221–229.

16. T. Eiichi, O. Eriko and O. Shin-ichi, *J. Neurochem.*, 2009, **111**, 181–191.

17. D. R. Rosen, T. Siddique, D. Patterson, D. A. Figlewicz, P. Sapp, A. Hentati, D. Donaldson, J. Goto, J. P. O'Regan and H. X. Deng, *Nature*, 1993, **362**, 59–62.

18. J. S. Valentine, P. A. Doucette and S. Zittin Potter, *Annu. Rev. Biochem.*, 2005, **74**, 563–593.

19. H. Ilieva, M. Polymenidou and D. W. Cleveland, *J. Cell Biol.*, 2009, **187**, 761–772.

20. S. Yoshida, D. W. Mulder, L. T. Kurland, C. P. Chu and H. Okazaki, *Neuroepidemiology*, 1986, **5**, 61–70.

21. L. P. Rowland and N. A. Shneider, *N. Engl. J. Med.*, 2001, **344**, 1688–1700.

22. M. Manuel, C. Iglesias, M. Donnet, F. Leroy, C. J. Heckman and D. Zytnicki, *J. Neurosci.*, 2009, **29**, 11246–11256.

23. J. Grosskreutz, L. Van Den Bosch and B. U. Keller, *Cell Calcium*, 2010, **47**, 165–174.

24. F. Kamel, D. M. Umbach, L. Stallone, M. Richards, H. Hu and D. P. Sandler, *Environ. Health Perspect.*, 2008, **116**, 943–947.

25. F. Fang, R. Bellocco, M. A. Hernán and W. Ye, *Neuroepidemiology*, 2006, **27**, 217–221.

26. F. Kamel, D. M. Umbach, H. Hu, T. L. Munsat, J. M. Shefner, J. A. Taylor and D. P. Sandler, *Neurodegener. Dis.*, 2005, **2**, 195–201.

27. L. Migliore and F. Coppedè, *Mutat. Res., Fundam. Mol. Mech. Mutagen.*, 2009, **667**, 82–97.

28. M. E. Gurney, H. Pu, A. Y. Chiu, M. C. D. Canto, C. Y. Polchow, D. D. Alexander, J. Caliendo, A. Hentati, Y. W. Kwon, H.-X. Deng, W. Chen, P. Zhai, R. L. Sufit and T. Siddique, *Science*, 1994, **264**, 1772–1775.

29. D. W. Cleveland and J. D. Rothstein, *Nat. Rev. Neurosci.*, 2001, **2**, 806–819.

30. A. G. Reaume, J. L. Elliott, E. K. Hoffman, N. W. Kowall, R. J. Ferrante, D. F. Siwek, H. M. Wilcox, D. G. Flood, M. F. Beal, R. H. Brown, Jr., R. W. Scott and W. D. Snider, *Nat. Genet.*, 1996, **13**, 43–47.

31. J. Sreedharan, I. P. Blair, V. B. Tripathi, X. Hu, C. Vance, B. Rogelj, S. Ackerley, J. C. Durnall, K. L. Williams, E. Buratti, F. Baralle, J. de Belleroche, J. D. Mitchell, P. N. Leigh, A. Al-Chalabi, C. C. Miller, G. Nicholson and C. E. Shaw, *Science*, 2008, **319**, 1668–1672.

32. C. Vance, B. Rogelj, T. Hortobagyi, K. J. De Vos, A. L. Nishimura, J. Sreedharan, X. Hu, B. Smith, D. Ruddy, P. Wright, J. Ganesalingam, K. L. Williams, V. Tripathi, S. Al-Saraj, A. Al-Chalabi, P. N. Leigh, I. P. Blair, G. Nicholson, J. de Belleroche, J.-M. Gallo, C. C. Miller and C. E. Shaw, *Science*, 2009, **323**, 1208–1211.

33. T. J. Kwiatkowski, Jr., D. A. Bosco, A. L. LeClerc, E. Tamrazian, C. R. Vanderburg, C. Russ, A. Davis, J. Gilchrist, E. J. Kasarskis, T. Munsat, P. Valdmanis, G. A. Rouleau, B. A. Hosler, P. Cortelli,

P. J. de Jong, Y. Yoshinaga, J. L. Haines, M. A. Pericak-Vance, J. Yan, N. Ticozzi, T. Siddique, D. McKenna-Yasek, P. C. Sapp, H. R. Horvitz, J. E. Landers and R. H. Brown, Jr., *Science*, 2009, **323**, 1205–1208.

34. J. D. Rothstein, *Annals of Neurology*, 2009, **65**, S3–S9.

35. D. Reed, C. Plato, T. Elizan and L. T. Kurland, *Am. J. Epidemiol.*, 1966, **83**, 54–73.

36. Y. Yase, T. Kumamoto, F. Yoshimasu and Y. Shinjo, *Neurol. India*, 1968, **16**, 46–50.

37. Y. Fumio, Y. Masayuki, Y. Yoshiro, U. Yushiro, T. Shoji, I. Shiro, S. Kazuhisa, D. C. Gajdusek, J. G. Clarence, Jr. and C. Kwang-Ming, *Psychiatry Clin. Neurosci.*, 1982, **36**, 173–179.

38. M. Yasui, Y. Yase, T. Kihira, K. Adachi and Y. Suzuki, *Eur. Neurol.*, 1992, **32**, 95–98.

39. Y. Richard, M. G. Ralph, D. C. Gajdusek, T. Akio, U. Takashi, K. Yoko, C. Kwang-Ming, S. Itsuro, C. P. Chris and J. G. Clarence, Jr., *Ann. Neurol.*, 1984, **15**, 42–48.

40. R. A. Roelofs-Iverson, D. W. Mulder, L. R. Elveback, L. T. Kurland and C. A. Molgaard, *Neurology*, 1984, **34**, 393–395.

41. F. Fang, L. C. Kwee, K. D. Allen, D. M. Umbach, W. Ye, M. Watson, J. Keller, E. Z. Oddone, D. P. Sandler, S. Schmidt and F. Kamel, *Am. J. Epidemiol.*, **171**, 1126–1133.

42. M. Szczerbowska-Boruchowska, M. Lankosz, J. Ostachowicz, D. Adamek, A. Krygowska-Wajs, B. Tomik, A. Szczudlik, A. Simionovici and S. Bohic, *X-Ray Spectrom.*, 2004, **33**, 3–11.

43. E. Kapaki, C. Zournas, G. Kanias, T. Zambelis, A. Kakami and C. Papageorgiou, *J. Neurol. Sci.*, 1997, **147**, 171–175.

44. E. J. Kasarskis, L. Tandon, M. A. Lovell and W. D. Ehmann, *J. Neurol. Sci.*, 1995, **130**, 203–208.

45. P. M. Roos, O. Vesterberg and M. Nordberg, *Exp. Biol. Med.*, 2006, **231**, 1481–1487.

46. A. P. Smith and N. M. Lee, *Amyotrophic Lateral Scler.*, 2007, **8**, 131–143.

47. M. P. Cuajungco and G. J. Lees, *Neurobiol. Dis.*, 1997, **4**, 137–169.

48. P. Milena, F. Sergi, G. Mercedes, Q. Albert, M. Amalia, C. Javier and H. Juan, *J. Neurosci. Res.*, 2005, **79**, 522–534.

49. K. Puttaparthi, W. L. Gitomer, U. Krishnan, M. Son, B. Rajendran and J. L. Elliott, *J. Neurosci.*, 2002, **22**, 8790–8796.

50. E. Tokuda, S.-I. Ono, K. Ishige, A. Naganuma, Y. Ito and T. Suzuki, *Toxicology*, 2007, **229**, 33–41.

51. I. Sheila, M. Dan H, S. Andrew P and L. E. E. Nancy M, *Ann. N. Y. Acad. Sci.*, 2005, **1053**, 121–136.

52. S.-I. Ono, Y. Endo, E.-I. Tokuda, K. Ishige, K.-i. Tabata, S. Asami, Y. Ito and T. Suzuki, *Biol. Trace Elem. Res.*, 2006, **113**, 93–103.

53. J.-L. Gonzalez de Aguilar, C. Niederhauser-Wiederkehr, B. Halter, M. De Tapia, F. Di Scala, P. Demougin, L. Dupuis, M. Primig, V. Meininger and J.-P. Loeffler, *Physiol. Genomics*, 2008, **32**, 207–218.

54. P. A. E. Sillevis Smitt, T. P. J. Mulder, H. W. Verspaget, H. G. T. Blaauwgeers and D. Troost, J. M. B. V. de Jong, *Neurosignals*, 1994, **3**, 193–197.
55. L. Tapia, M. González-Agüero, M. F. Cisternas, M. Suazo, V. Cambiazo, R. Uauy and M. González, *Biochem. J.*, 2004, **378**, 617–624.
56. I. Hozumi, M. Yamada, Y. Uchida, K. Ozawa, H. Takahashi and T. Inuzuka, *Amyotrophic Lateral Scl.*, 2008, **9**, 294–298.
57. K. Hashimoto, Y. Hayashi, T. Inuzuka and I. Hozumi, *Neuroscience*, 2009, **163**, 244–251.
58. M. M. Julia, Y. Bing, J. T. Ronald and P. Roger, *Am. J. Med. Genet. B Neuropsychiatr. Genet.*, 2007, **144B**, 885–890.
59. J. M. McCord and I. Fridovich, *J. Biol. Chem.*, 1969, **244**, 6049–6055.
60. B. F. Shaw and J. S. Valentine, *Trends Biochem. Sci.*, 2007, **32**, 78–85.
61. D. R. Borchelt, M. K. Lee, H. S. Slunt, M. Guarnieri, Z. S. Xu, P. C. Wong, R. H. Brown, Jr, D. L. Price, S. S. Sisodia and D. W. Cleveland, *Proc. Natl. Acad. Sci. USA*, 1994, **91**, 8292–8296.
62. A. G. Estevez, J. P. Crow, J. B. Sampson, C. Reiter, Y. Zhuang, G. J. Richardson, M. M. Tarpey, L. Barbeito and J. S. Beckman, *Science*, 1999, **286**, 2498–2500.
63. M. Chattopadhyay and J. S. Valentine, *Antioxid. Redox Signaling*, 2009, **11**, 1603–1614.
64. C. A. Ross and M. A. Poirier, *Nat. Rev. Neurosci.*, 2004, **S10–S17**.
65. R. Rakhit, P. Cunningham, A. Furtos-Matei, S. Dahan, X.-F. Qi, J. Crow, N. R. Cashman, L. H. Kondejewski and A. Chakrabartty, *J. Biol. Chem.*, 2002, **277**, 47551–47556.
66. L. M. Sayre, M. A. Smith and G. Perry, *Curr. Med. Chem.*, 2001, **8**, 721.
67. S. C. Barber, R. J. Mead and P. J. Shaw, *Biochim. et Biophys. Acta (BBA) – Molec. Basis Dis.*, 1762, 1051–1067.
68. B. F. Shaw, H. L. Lelie, A. Durazo, A. M. Nersissian, G. Xu, P. K. Chan, E. B. Gralla, A. Tiwari, L. J. Hayward, D. R. Borchelt, J. S. Valentine and J. P. Whitelegge, *J. Biol. Chem.*, 2008, **283**, 8340–8350.
69. J. Wang, G. W. Farr, C. J. Zeiss, D. J. Rodriguez-Gil, J. H. Wilson, K. Furtak, D. T. Rutkowski, R. J. Kaufman, C. I. Ruse, J. R. Yates, 3rd, S. Perrin, M. B. Feany and A. L. Horwich, *Proc. Natl. Acad. Sci. USA*, 2009, **106**, 1392–1397.
70. A. Kerman, H.-N. Liu, S. Croul, J. Bilbao, E. Rogaeva, L. Zinman, J. Robertson and A. Chakrabartty, *Acta Neuropathol.*, 2010, **119**, 335–344.
71. M. Chattopadhyay, A. Durazo, S. H. Sohn, C. D. Strong, E. B. Gralla, J. P. Whitelegge and J. S. Valentine, *Proc. Natl. Acad. Sci. USA*, 2008, **105**, 18663–18668.
72. L. Banci, I. Bertini, A. Durazo, S. Girotto, E. B. Gralla, M. Martinelli, J. S. Valentine, M. Vieru and J. P. Whitelegge, *Proc. Natl. Acad. Sci. USA*, 2007, **104**, 11263–11267.

73. J. S. Elam, A. B. Taylor, R. Strange, S. Antonyuk, P. A. Doucette, J. A. Rodriguez, S. S. Hasnain, L. J. Hayward, J. S. Valentine, T. O. Yeates and P. J. Hart, *Nat. Struct. Biol.*, 2003, **10**, 461–467.

74. S. Antonyuk, J. S. Elam, M. A. Hough, R. W. Strange, P. A. Doucette, J. A. Rodriguez, L. J. Hayward, J. S. Valentine, P. J. Hart and S. S. Hasnain, *Protein Sci.*, 2005, **14**, 1201–1213.

75. J. Wang, G. Xu, V. Gonzales, M. Coonfield, D. Fromholt, N. G. Copeland, N. A. Jenkins and D. R. Borchelt, *Neurobiol. Dis.*, 2002, **10**, 128–138.

76. C. M. Karch, M. Prudencio, D. D. Winkler, P. J. Hart and D. R. Borchelt, *Proc. Natl. Acad. Sci. USA*, 2009, **106**, 7774–7779.

77. P. Pasinelli, M. E. Belford, N. Lennon, B. J. Bacskai, B. T. Hyman, D. Trotti and R. H. Brown, *Neuron*, 2004, **43**, 19–30.

78. M. Cozzolino, M. G. Pesaresi, I. Amori, C. Crosio, A. Ferri, M. Nencini and M. T. Carri, *Antioxid. Redox Signaling*, 2009, **11**, 1547–1558.

79. J. D. Atkin, M. A. Farg, B. J. Turner, D. Tomas, J. A. Lysaght, J. Nunan, A. Rembach, P. Nagley, P. M. Beart, S. S. Cheema and M. K. Horne, *J. Biol. Chem.*, 2006, **281**, 30152–30165.

80. A.-L. Ström, P. Shi, F. Zhang, J. Gal, R. Kilty, L. J. Hayward and H. Zhu, *J. Biol. Chem.*, 2008, **283**, 22795–22805.

81. P. N. Lacor, M. C. Buniel, P. W. Furlow, A. Sanz Clemente, P. T. Velasco, M. Wood, K. L. Viola and W. L. Klein, *J. Neurosci.*, 2007, **27**, 796–807.

82. M. Synofzik, R. Fernández-Santiago, W. Maetzler, L. Schöls and P. M. Andersen, *J. Neurol., Neurosurg. Psychiatry*, 2010, 764–767.

83. K. Forsberg, P. A. Jonsson, P. M. Andersen, D. Bergemalm, K. S. Graffmo, M. Hultdin, J. Jacobsson, R. Rosquist, S. L. Marklund and T. Brannstrom, *PLoS One*, 2010, **5**, e11552.

84. K. Edor, N. V. Paul, D. Patrick and A. R. Guy, *Ann. Neurol.*, 2007, **62**, 553–559.

85. E. B. Dale, M. E. Lisa, P. J. Hart, W.-P. Martina and V. Joan Selverstone, *Ann. Neurol.*, 1997, **42**, 135–137.

86. A. Gruzman, W. L. Wood, E. Alpert, M. D. Prasad, R. G. Miller, J. D. Rothstein, R. Bowser, R. Hamilton, T. D. Wood, D. W. Cleveland, V. R. Lingappa and J. Liu, *Proc. Natl. Acad. Sci. USA*, 2007, **104**, 12524–12529.

87. S. Gagliardi, E. Cova, A. Davin, S. Guareschi, K. Abel, E. Alvisi, U. Laforenza, R. Ghidoni, J. R. Cashman, M. Ceroni and C. Cereda, *Neurobiol. Dis.*, 2010, **39**, 198–203.

88. H. Yi, W. Gang, Y. J. C. Grace, F. Xin and Q. Y. Shao, *Electrophoresis*, 2003, **24**, 1458–1470.

89. H. Y. Yoo, M. S. Chang and H. M. Rho, *Mol. Gen. Genet.*, 1999, **262**, 310–313.

90. H. L. Lelie, A. Liba, M. W. Bourassa, M. Chattopadhyay, P. K. Chan, E. B. Gralla, L. M. Miller, D. R. Borchelt, J. S. Valentine and J. P. Whitelegge, *J. Biol. Chem.*, 2011, **286**, 2795–2806.

91. Y. Furukawa, A. S. Torres and T. V. O'Halloran, *EMBO J.*, 2004, **23**, 2872–2881.
92. S. Z. Potter and J. S. Valentine, *J. Biol. Inorg. Chem.*, 2003, **8**, 373–380.
93. Y. Furukawa and T. V. O'Halloran, *Antioxid. Redox Signal*, 2006, **8**, 847–867.
94. C. K. Bruns and R. R. Kopito, *EMBO J.*, 2007, **26**, 855–866.
95. S. Z. Potter, H. Zhu, B. F. Shaw, J. A. Rodriguez, P. A. Doucette, S. H. Sohn, A. Durazo, K. F. Faull, E. B. Gralla, A. M. Nersissian and J. S. Valentine, *J. Am. Chem. Soc.*, 2007, **129**, 4575–4583.
96. J. M. Leitch, P. J. Yick and V. C. Culotta, *J. Biol. Chem.*, 2009, **284**, 24679–24683.
97. J. D. Rothstein, M. Dykes-Hoberg, L. B. Corson, M. Becker, D. W. Cleveland, D. L. Price, V. C. Culotta and P. C. Wong, *J. Neurochem.*, 1999, **72**, 422–429.
98. F. Ding and N. V. Dokholyan, *Proc. Natl. Acad. Sci. USA*, 2008, **105**, 19696–19701.
99. J. M. Leitch, L. T. Jensen, S. D. Bouldin, C. E. Outten, P. J. Hart and V. C. Culotta, *J. Biol. Chem.*, 2009, **284**, 21863–21871.
100. F. Arnesano, L. Banci, I. Bertini, M. Martinelli, Y. Furukawa and T. V. O'Halloran, *J. Biol. Chem.*, 2004, **279**, 47998–48003.
101. P. A. Doucette, L. J. Whitson, X. Cao, V. Schirf, B. Demeler, J. S. Valentine, J. C. Hansen and P. J. Hart, *J. Biol. Chem.*, 2004, **279**, 54558–54566.
102. M. J. Lindberg, J. Normark, A. Holmgren and M. Oliveberg, *Proc. Natl. Acad. Sci. USA*, 2004, **101**, 15893–15898.
103. Y. Furukawa, K. Kaneko, K. Yamanaka, T. V. O'Halloran and N. Nukina, *J. Biol. Chem.*, 2008, **283**, 24167–24176.
104. Z. A. Oztug Durer, J. A. Cohlberg, P. Dinh, S. Padua, K. Ehrenclou, S. Downes, J. K. Tan, Y. Nakano, C. J. Bowman, J. L. Hoskins, C. Kwon, A. Z. Mason, J. A. Rodriguez, P. A. Doucette, B. F. Shaw and J. S. Valentine, *PLoS One*, 2009, **4**, e5004.
105. S. V. Seetharaman, M. Prudencio, C. Karch, S. P. Holloway, D. R. Borchelt and P. J. Hart, *Exp. Biol. Med. (Maywood)*, 2009, **234**, 1140–1154.
106. J. A. Rodriguez, B. F. Shaw, A. Durazo, S. H. Sohn, P. A. Doucette, A. M. Nersissian, K. F. Faull, D. K. Eggers, A. Tiwari, L. J. Hayward and J. S. Valentine, *Proc. Natl. Acad. Sci. USA*, 2005, **102**, 10516–10521.
107. J. Wang, H. Slunt, V. Gonzales, D. Fromholt, M. Coonfield, N. G. Copeland, N. A. Jenkins and D. R. Borchelt, *Hum. Mol. Genet.*, 2003, **12**, 2753–2764.
108. P. C. Wong, H. Cai, D. R. Borchelt and D. L. Price, *Nat. Neurosci*, 2002, **5**, 633–639.
109. M. Prudencio, P. J. Hart, D. R. Borchelt and P. M. Andersen, *Hum. Mol. Genet.*, 2009, **18**, 3217–3226.
110. S. M. Lynch and W. Colon, *Biochem. Biophys. Res. Commun.*, 2006, **340**, 457–461.

111. V. K. Mulligan, A. Kerman, S. Ho and A. Chakrabartty, *J. Mol. Biol.*, 2008, **383**, 424–436.
112. T. D. Rae, P. J. Schmidt, R. A. Pufahl, V. C. Culotta and T. V. O'Halloran, *Science*, 1999, **284**, 805–808.
113. L. Banci, I. Bertini, S. Ciofi-Baffoni, T. Kozyreva, K. Zovo and P. Palumaa, *Nature*, 2010, **465**, 645–648.
114. M. T. Carrì, A. Ferri, M. Cozzolino, L. Calabrese and G. Rotilio, *Brain Res. Bull.*, 2003, **61**, 365–374.
115. J. P. Crow, J. B. Sampson, Y. Zhuang, J. A. Thompson and J. S. Beckman, *J. Neurochem.*, 1997, **69**, 1936–1944.
116. M. WiedauPazos, J. J. Goto, S. Rabizadeh, E. B. Gralla, J. A. Roe, M. K. Lee, J. S. Valentine and D. E. Bredesen, *Science*, 1996, **271**, 515–518.
117. J. R. Subramaniam, W. E. Lyons, J. Liu, T. B. Bartnikas, J. Rothstein, D. L. Price, D. W. Cleveland, J. D. Gitlin and P. C. Wong, *Nat. Neurosci.*, 2002, **5**, 301–307.
118. J. Wang, A. Caruano-Yzermans, A. Rodriguez, J. P. Scheurmann, H. H. Slunt, X. Cao, J. Gitlin, P. J. Hart and D. R. Borchelt, *J. Biol. Chem.*, 2007, **282**, 345–352.
119. J. Wang, G. Xu, H. Li, V. Gonzales, D. Fromholt, C. Karch, N. G. Copeland, N. A. Jenkins and D. R. Borchelt, *Hum. Mol. Genet.*, 2005, **14**, 2335–2347.
120. W. I. M. Vonk and L. W. J. Klomp, *Biochem. Soc. Trans.*, 2008, **036**, 1322–1328.
121. D. R. Brown, F. Hafiz, L. L. Glasssmith, B.-S. Wong, I. M. Jones, C. Clive and S. J. Haswell, *EMBO J.*, 2000, **19**, 1180–1186.
122. J. M. Christa, I. B. Ashley, L. M. Colin, C. Roberto and L. Qiao-Xin, *Int. J. Exp. Pathol.*, 2005, **86**, 147–159.
123. Y. Ogawa, H. Kosaka, T. Nakanishi, A. Shimizu, N. Ohoi, H. Shouji, T. Yanagihara and S. Sakoda, *Biochem. Biophys. Res. Commun.*, 1997, **241**, 251–257.
124. M. Kiaei, A. I. Bush, B. M. Morrison, J. H. Morrison, R. A. Cherny, I. Volitakis, M. F. Beal and J. W. Gordon, *J. Neurosci.*, 2004, **24**, 7945–7950.
125. C. Bertoni-Freddari, P. Fattoretti, T. Casoli, G. Di Stefano, B. Giorgetti and M. Balietti, *Exp. Gerontol.*, 2008, **43**, 389–393.
126. I. P. Ermilova, V. B. Ermilov, M. Levy, E. Ho, C. Pereira and J. S. Beckman, *Neurosci. Lett.*, 2005, **379**, 42–46.
127. Y. H. Gong and J. L. Elliott, *Exp. Neurol.*, 2000, **162**, 27–36.
128. C. W. Levenson, *Physiol. Behav.*, 2005, **86**, 399–406.
129. X. Yao, *NeuroToxicology*, 2009, **30**, 121–126.
130. J. D. Rothstein, L. J. Martin and R. W. Kuncl, *N. Engl. J. Med.*, 1992, **326**, 1464–1468.
131. J. D. Rothstein, M. Dykes-Hoberg, C. A. Pardo, L. A. Bristol, L. Jin, R. W. Kuncl, Y. Kanai, M. A. Hediger, Y. Wang, J. P. Schielke and D. F. Welty, *Neuron*, 1996, **16**, 675–686.

132. S. G. Carriedo, H. Z. Yin and J. H. Weiss, *J. Neurosci.*, 1996, **16**, 4069–4079.
133. M. L. Cucchiaroni, M. T. Viscomi, G. Bernardi, M. Molinari, E. Guatteo and N. B. Mercuri, *J. Neurosci.*, **30**, 5176–5188.
134. J.-H. Song, C.-S. Huang, K. Nagata, J. Z. Yeh and T. Narahashi, *J. Pharmacol. Exp. Ther.*, 1997, **282**, 707–714.
135. M. R. Celio, *Neuroscience*, 1990, **35**, 375–475.
136. C. Voisine, J. S. Pedersen and R. I. Morimoto, *Neurobiol. Dis.*, 2010, **40**, 12–20.
137. T. Gidalevitz, E. A. Kikis and R. I. Morimoto, *Curr. Opin. Struct. Biol.*, 2010, **20**, 23–32.

CHAPTER 12

Copper and Prion Protein Function: A Brief Review of Emerging Theories of Neuroprotection

G. L. MILLHAUSER

Department of Chemistry & Biochemistry, University of California Santa Cruz, 1156 High Street, Santa Cruz, CA 95064, USA

12.1 Introduction

Prion diseases result from accumulation of a misfolded form of the endogenous prion protein (PrP) and present an ongoing threat to human health, agricultural and wildlife.[1-6] Referred to as the transmissible spongiform encephalopathies (TSEs), these infectious, fatal diseases share pathologies with Alzheimer's and Parkinson's diseases.[6,7] As such, single instances of mad cow disease (bovine spongiform encephalopthay, BSE) create immediate concern with regard to the safety of our food supply and greatly affect the cattle industry.[8,9] Chronic wasting disease currently infects upwards of 15% of deer and elk in the Rocky Mountain regions of the United States and Canada.[10,11] Since the early 1990s, approximately 140 young individuals in the UK have died from variant Creutzfeldt–Jakob disease (vCJD) contracted from BSE-contaminated food.[12-14] Surgical procedures involving corneal transplants and dura mater grafts have transmitted disease to

RSC Drug Discovery Series No. 7
Neurodegeneration: Metallostasis and Proteostasis
Edited by Danilo Milardi and Enrico Rizzarelli
© Royal Society of Chemistry 2011
Published by the Royal Society of Chemistry, www.rsc.org

healthy individuals, as have contaminated surgical instruments, cadaver derived hormones[5] and blood transfusions.[15]

Prion diseases are triggered by conversion of the normal cellular form of the prion protein (PrP^C) to the scrapie form (PrP^{Sc}).[16,17] In parallel with Alzheimer's and Parkinson's diseases, which also arise form protein misfolding, most research on the TSEs focuses on mechanisms of PrP^{Sc} accumulation, often in the form of amyloid fibrils, and resulting neurotoxicity.[18] Despite great advances in our understanding of neurodegenerative processes, the specific mechanism by which fibrillar structures damage cells remains unclear. Moreover, many instances of CJD, the most common form of prion disease in humans, lack presentation of amyloid.[19,20] There are wide ranging efforts aimed at pharmaceuticals that interfere with amyloid,[21] but none so far have proven to be widely effective, and in some cases these new therapies may be deleterious.[22]

Much less is known about the normal function of PrP^C in healthy tissue. PrP^C is an abundant, glycophosphatidylinositol (GPI) anchored, glycoprotein (Figure 12.1) found in a wide range of tissues and concentrated at presynaptic membranes in the CNS.[23] Original work with mouse and hamster PrP knockouts revealed only

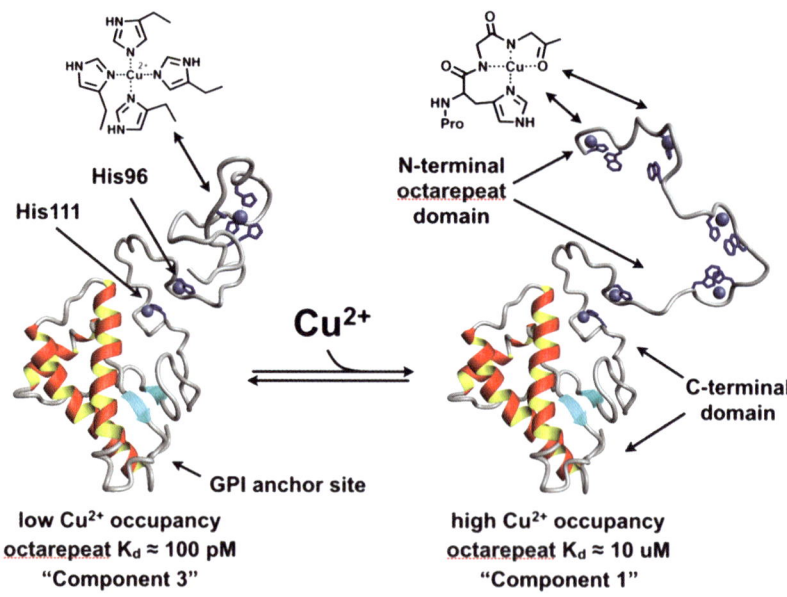

Figure 12.1 Coordination modes of PrP^C showing the folded C-terminal domain, the octarepeat (OR) domain (residues 60–91), the non-OR sites at H96 and H111 (human sequence) and location of the C-terminal GPI anchor that attaches to cellular membranes. Low Cu^{2+} concentration favors multi-His coordination, referred to as component 3. At higher concentrations, Cu^{2+} binds within each single repeat, coordinated to the HGGGW segment. Molecular features are shown above; K$_d$ values below. Copper uptake exhibits negative cooperativity.

mild phenotypic changes and initially suggested little about intrinsic PrP^C function.[24] Recent investigations, however, find that PrP plays an essential role in numerous stress responses. PrP^C influences sleep–wake cycles,[25] maintains proper levels of mitochondria,[26] supports myelin development,[27] helps regulate phagocytosis,[28] modulates excitability and seizures,[29] supports neuron differentiation,[30] suppresses apoptosis,[31,32] helps maintain memory in mature animals[33,34] and defends against oxidative stress.[35,36] PrP^C is now considered to be a neuroprotective protein, essential for long-term neuron stability.[24] Interestingly, in human brain trauma resulting in ischemia and hypoxia, PrP is dramatically up-regulated at the injury site.[37]

The prion protein is also implicated as protective against Alzheimer's disease. PrP^C undergoes endocytotic cycling and it is thought to play a role in clearance and regulation of a variety of extracellular species. The protein contains a highly conserved hydrophobic segment that binds specifically to toxic oligomers of the Abeta peptide, which constitutes the major protein component in extracellular senile plaques of Alzheimer's disease.[38] Moreover, PrP^C inhibits β-secretase, which is responsible for one of the two essential cleavage events that release Abeta from the amyloid precursor protein (APP).[39] Finally, PrP^C has been shown to protect against memory loss caused by Abeta plaques.[40]

The discovery of the role of PrP in neuron maintenance and protection is motivating new research avenues into its specific mechanism of action. It is very possible that prion neurodegeneration is exacerbated, or even caused, by loss of intrinsic PrP^C function through aggregation or alternative proteolytic processing resulting from the aggregated state. Investigations into the functional roles of proteins that misfold in Alzheimer's disease (Abeta and tau), Parkinson's disease (α-synuclein) and Huntington's disease (huntingtin) are accelerating and providing critical new insights across the spectrum of neurodegenerative disorders.[41]

PrP^C is a copper binding protein.[42–44] Studies aimed at identifying neuroprotective segments in PrP^C show that N-terminal residues responsible for copper uptake are essential. In most studies, simple mutations of copper coordinating His residues with isosteric species completely abrogate PrP^C neuroprotection.[32] The brain is rich in copper, which transports across neuronal membranes and is maintained at extracellular concentrations between 1.0 μM and 200 μM.[45] Copper binds to the PrP promoter region and up-regulates PrP expression.[46,47] Our laboratory endeavored to accurately identify the molecular features of the copper sites (see Figure 12.1),[48–50] characterize cooperativity in Cu^{2+} uptake,[51] assay affinity as a function of copper occupancy,[51] and evaluate the interplay with Zn^{2+},[52] another abundant species of the CNS.

This chapter considers two emerging theories of PrP^C function. The first considers PrP^C as protective against uncomplexed, extracellular Cu^{2+}. The second examines new findings suggesting that copper may participate in PrP^C processing, thus producing signaling fragments that regulate apoptosis. We begin with a brief review of the copper binding properties of PrP^C, with a focus on coordination features and affinity.

12.2 Molecular Features of the PrP^C Copper Sites

The $\sim$210 residue human prion protein possesses a folded, and largely helical, C-terminal domain and a flexible N-terminal domain. PrP^C is capable of taking up approximately six equivalents of Cu^{2+}, with all sites localized to the flexible segment of PrP, as shown in Figure 12.1. There are high affinity Cu^{2+} sites at His96 and His111, with dissociation constants of approximately 0.10 nM. The Cu^{2+} coordination sphere arises from the His imidazole and exocyclic nitrogen, as well as the amide nitrogens from the two amino acids immediately preceding the histidine. The polypeptide segment containing these sites is within the protease resistant core of PrP^{Sc}. Both experiment and computational studies suggest that copper may therefore inhibit conversion to the amyloid PrP.

The octarepeat (OR) domain, residues 60–91, is responsible for PrP's ability to respond to varying copper levels. This segment is composed of four tandem PHGGGWGQ units, and is almost perfectly conserved among mammals.[44] At low copper concentration, the OR domain wraps around a single Cu^{2+} with coordination through three to four His residues, so-called component 3 coordination. At high copper, each HGGGW unit within a single repeat coordinates Cu^{2+} though a single His and two Gly amide nitrogens (component 1), as shown in Figure 12.1.[49] The affinity varies dramatically between these two fundamental coordination states. Component 3 binds with a dissociation constant of $\approx$100 pM, whereas component 1 exhibits a substantially lower affinity (K$_d$ $\approx$ 10 μM).[51] The increase in K$_d$ of five orders of magnitude demonstrates that copper uptake proceeds with negative cooperativity, as demonstrated by electron paramagnetic resonance (EPR) competition experiments[51] and isothermal titration calorimetry (ITC).[53] The OR domain also binds Zn^{2+}, another highly abundant extracellular species of the CNS. And while zinc is not able to displace copper, it can shift the equilibrium towards component 1 coordination (Figure 12.1).[52]

The direct connection between PrP expression and copper regulation in the CNS has been tenuous. Initial reports suggested that PrP knockout mice exhibited significantly lower copper levels relative to wild type.[54] However, subsequent experiments performed on fractionated brain tissue found little variation in either copper levels or copper-dependent enzyme activity among animals expressing PrP at 0, 1× and 10× the normal level.[55] The disagreement between these assessments motivated a closer look at this critical issue. Using X-ray fluorescence, Pushie and colleagues performed imaging experiments on mouse brain slices (to be published). The results, shown in Figure 12.2, reveal the Zn, Fe and Cu levels in a coronal section. Interestingly, whereas Zn seems relatively constant across the tissue sections, and is independent of PrP expression, both Cu and Fe vary significantly. PrP increases Cu levels around the ventricles and Fe levels in the caudate putamen and several other regions. These findings not only support the connection between PrP and copper, but also motivate further investigation into iron regulation.

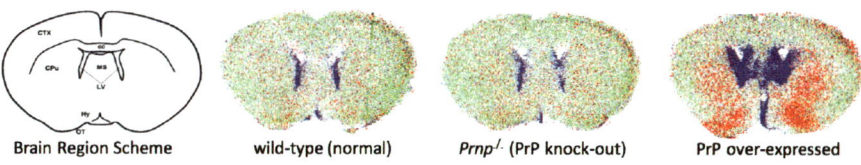

Figure 12.2 X-ray fluorescence imaging of iron (red), copper (blue) and zinc (green) in a select coronal section from mice expressing differential levels of PrP^C (Pushie *et al.*, submitted). Zinc is abundant throughout the brain. The highest levels of copper are located around ventricles (LV) and iron levels are significantly altered in the caudate putamen (CPu), olfactory bulb (not shown in this section), thalamus (not shown in this section) and hypothalamus (HY). Figure provided by Dr Jake Pushie, University of Saskatchewan.

12.3 Theories of Neuroprotection

Extracellular copper is found primarily in the Cu^{2+} oxidation state.[45] Without complexation, however, copper cycles between the Cu$^+$ and Cu^{2+} oxidation states, contributing to the production of reactive oxygen species (ROS) that may damage cellular components. Copper concentrations vary significantly in the synaptic space, a region of high PrP expression. At rest, synaptic copper concentrations are approximately 3.0 μM. However, upon neuronal depolarization, copper is released from the presynaptic surface and the concentration elevates significantly, perhaps as high as 250 μM. The role of released copper is not clear at the current time but it may participate in the modulation of synaptic receptors.[56]

The resting synaptic concentration is below the component 1 dissociation constant of 10 μM, indicating that component 3 coordination dominates. This coordination mode allows for redox cycling, suggesting perhaps that PrP^C may be a copper reductase. However, as the copper concentration rises, PrP^C reorganizes to take up Cu^{2+} in the component 1 mode. Component 1 coordination involves negatively charged backbone nitrogens, which confer a strong ligand field that suppresses redox activity. Thus, redox protection against copper mediated ROS production emerges at high copper levels. Cerebrospinal fluid has only low concentrations of proteins such as albumin and ceruloplasmin capable of complexing copper. Given the relative absence of these proteins, it is likely that PrP^C functions as an antioxidant, controlling the rate of ROS production.

The protective role of component 1 coordination is supported by a recent analysis of prion disease causing insert mutations in the OR domain, where individuals carry from one to nine additional octarepeats. Statistical analysis suggests that the number of inserts determines disease progression, with a remarkable threshold just beyond four inserts.[57,58] With one to four extra octarepeats, the average age of onset is above 60 years, whereas five to nine extra octarepeats results in early onset prion disease with symptoms starting between 30 and 40 years of age, a difference of almost three decades.

Given the profound influence of OR domain length in expansion disease, we explored whether the OR response to copper is altered by insertion number.[59] Specifically, we examined the balance component 1 *vs* component 3 coordination, as a function of OR length. Using a series of PrP derived constructs, we showed that domains with four to seven repeats (*i.e.*, zero to three inserts) behave much like wild type, reaching a maximum of component 3 at 1.0–1.5 equivalents. However, the eight OR construct (four inserts) exhibits persistent component 3 coordination that reaches a maximum at approximately 2.0–2.5 equivalents. The maximum observed for the nine OR construct is shifted to even higher copper concentration. Equivalent trends were observed with full-length recombinant protein, where we compared wild type with mutant PrP^C containing five OR inserts. These data reveal a remarkable relationship, where decreased age of onset and persistent component 3 coordination take place at a threshold of eight or more repeats. In turn, our findings suggest an important protective role for component 1 coordination that is lost in cases of OR expansion disease with four or more inserts. As noted above, component 1 may protect against redox cycling, and subsequent oxidative damage. However, expanded OR domains may also modulate PrP^{Sc} formation or OR-driven interactions with other cellular components or proteins.

PrP^C may also be a precursor to neuroprotective signaling peptides. In the normal course of PrP^C processing, extracellular PrP^C is cleaved into two fragments, referred to as N1 and C1, following residue 110 (human sequence), as identified in Figure 12.3.[60] Current findings suggest that endoprotolysis is carried out by the α-secretases ADAM10 and ADAM17, which are zinc metallproteases.[61] These α-secretases are also responsible for the protective alpha cleavage of the amyloid precursor protein (APP), which prevents release and build up of the Alzheimer's Abeta peptide. Because of the similarity between APP and PrP^C processing, the N1/C1 cut is often referred to as "alpha" cleavage. It is noteworthy that alpha cleavage takes place in a hydrophobic segment implicated in PrP toxicity.[62] Interestingly, the N1 fragment suppresses caspases and modulates p53 transcription, and is highly protective against apoptosis.[63] The receptor through which extracellular N1 mediates these intracellular activities has not been identified. The C-terminal fragment C1 likely

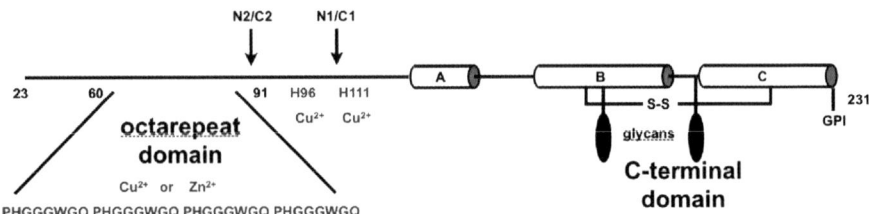

Figure 12.3 PrP^C schematic showing C-terminal helices, the disulfide bond, GPI anchor and N-linked glycans. Also shown are the cleavage sites that produce the N1/C1 and N2/C2 fragments.

remains anchored to the membrane and is weakly pro-apoptotic, but not enough to offset the protective effects of N1.

In the human prion disease CJD, PrP^C is found to be processed at another cut point, giving the N2/C2 fragments.[64] Unlike the N1 polypeptide, N2 is not protective.[63] Moreover, C2 constitutes a major component of CJD plaques. By analogy to the β-secretase cleavage of APP, giving the aggregation prone Abeta peptide, release of N2/C2 is referred to as beta cleavage.[65] N2/C2 cleavage takes place just following the OR segment, but the specific cut point is not well defined and searches for the responsible enzymes remain inconclusive. Interestingly, in cellular preparations, the longer N1 fragment does not undergo further processing to produce N2. Given the proximity to the OR domain, several laboratories suggest that copper may be responsible for N2/C2 cleavage, but the direct link to copper coordination has not been demonstrated *in vivo*.[66,67]

Recent experiments suggest that N1/C1 *vs* N2/C2 cleavage may be related to copper. Using RK13 cells, Haigh *et al.* found that N1/C1 cleavage is enhanced at higher copper levels.[65] The specific mechanism is not clear but may arise from copper initiated disruption of PrP-membrane raft interactions. From a different perspective, computational studies by Pushie and Vogel[66] suggest that low copper occupancy in the component 3 mode enhances beta cleavage through an electron transfer mechanism, which explains why the N2/C2 cut takes place adjacent to the OR domain. Interestingly, whereas the conserved OR Trp residues are not essential for either component 1 or component 3 coordination (see Figure 12.1), they are required as an electron source for the copper center, ultimately leading to the production of OH radicals. Regardless of the specific mechanism, the concepts above motivate a different look at copper mediated cellular damage.

There is no doubt that high levels of uncomplexed and weakly complexed copper are damaging to biological tissues. However, these studies of prion protein processing suggest that low copper levels are also problematic. The protective fragment N1 is likely only to be produced at copper levels that populate the component 1 mode. Perhaps PrP^C is a copper sensor designed to trigger apoptosis at low Cu^{2+} levels. If so, this points to a new role for copper, and places PrP^C as a central player in copper homeostasis and neuroprotection.

Acknowledgements

This work was supported by a grant from the National Institutes of Health (GM065790). The author thanks Dr Jake Pushie, University of Saskatchewan, who kindly provided Figure 12.2.

References

1. S. B. Prusiner, *Science*, 1991, **252**, 1515–1522.
2. S. B. Prusiner, *Science*, 1997, **278**, 245–251.
3. S. B. Prusiner, *Proc. Natl. Acad. Sci. USA*, 1998, **95**, 13363–13383.

4. C. Weissmann, *Nat. Rev. Microbiol.*, 2004, **2**, 861–871.
5. S. B. Prusiner, *Prion Biology and Diseases*, 2nd edn., Cold Spring Harbor Laboratory Press, Cold Spring Harbor, 2003.
6. A. Aguzzi and T. O'Connor, *Nat. Rev. Drug Discov.*, 2010, **9**, 237–248.
7. P. T. Lansbury Jr., *Neuron*, 1997, **19**, 1151–1154.
8. G. Pauli, *Cell Tiss. Bank*, 2005, **6**, 191–200.
9. J. L. Harman and C. J. Silva, *J. Am. Vet. Med. Assoc.*, 2009, **234**, 59–72.
10. M. W. Miller, E. S. Williams, C. W. McCarty, T. R. Spraker, T. J. Kreeger, C. T. Larsen and E. T. Thorne, *J. Wildl. Dis.*, 2000, **36**, 676–690.
11. R. C. Angers, H. E. Kang, D. Napier, S. Browning, T. Seward, C. Mathiason, A. Balachandran, D. McKenzie, J. Castilla, C. Soto, J. Jewell, C. Graham, E. A. Hoover and G. C. Telling, *Science*, **328**, 1154–1158.
12. P. Horby, *J. Paediatr. Child Health*, 2002, **38**, 539–542.
13. J. W. Ironside, *Semin. Hematol.*, 2003, **40**, 16–22.
14. R. C. Holman, E. D. Belay, K. Y. Christensen, R. A. Maddox, A. M. Minino, A. M. Folkema, D. L. Haberling, T. A. Hammett, K. D. Kochanek, J. J. Sejvar and L. B. Schonberger, *PLoS One*, 2010, **5**, e8521.
15. F. Houston, S. McCutcheon, W. Goldmann, A. Chong, J. Foster, S. Siso, L. Gonzalez, M. Jeffrey and N. Hunter, *Blood*, 2008, **112**, 4739–4745.
16. I. V. Baskakov, *FEBS J.*, 2007, **274**, 576–587.
17. A. Aguzzi, M. Heikenwalder and M. Polymenidou, *Nat. Rev. Mol. Cell Biol.*, 2007, **8**, 552–561.
18. P. Brundin, R. Melki and R. Kopito, *Nat. Rev. Mol. Cell Biol.*, 2010, **11**, 301–307.
19. J. G. Safar, M. D. Geschwind, C. Deering, S. Didorenko, M. Sattavat, H. Sanchez, A. Serban, M. Vey, H. Baron, K. Giles, B. L. Miller, S. J. Dearmond and S. B. Prusiner, *Proc. Natl. Acad. Sci. USA*, 2005, **102**, 3501–3506.
20. D. W. Colby, R. Wain, I. V. Baskakov, G. Legname, C. G. Palmer, H. O. Nguyen, A. Lemus, F. E. Cohen, S. J. DeArmond and S. B. Prusiner, *PLoS Pathog.*, 2010, **6**, e1000736.
21. L. D. Estrada and C. Soto, *Curr. Top. Med. Chem.*, 2007, **7**, 115–126.
22. B. Y. Feng, B. H. Toyama, H. Wille, D. W. Colby, S. R. Collins, B. C. May, S. B. Prusiner, J. Weissman and B. K. Shoichet, *Nat. Chem. Biol.*, 2008, **4**, 197–199.
23. A. Aguzzi, F. Baumann and J. Bremer, *Annu. Rev. Neurosci.*, 2008, **31**, 439–477.
24. A. D. Steele, S. Lindquist and A. Aguzzi, *Prion*, 2007, **1**, 83–93.
25. I. Tobler, S. E. Gaus, T. Deboer, P. Achermann, M. Fischer, T. Rulicke, M. Moser, B. Oesch, P. A. McBride and J. C. Manson, *Nature*, 1996, **380**, 639–642.
26. G. Miele, M. Jeffrey, D. Turnbull, J. Manson and M. Clinton, *Biochem. Biophys. Res. Commun.*, 2002, **291**, 372–377.
27. J. Bremer, F. Baumann, C. Tiberi, C. Wessig, H. Fischer, P. Schwarz, A. D. Steele, K. V. Toyka, K. A. Nave, J. Weis and A. Aguzzi, *Nat. Neurosci.*, 2010, **13**, 310–318.

28. C. J. de Almeida, L. B. Chiarini, J. P. da Silva, P. M. R. e Silva, M. A. Martins and R. Linden, *J. Leukoc. Biol.*, 2005, **77**, 238–246.
29. R. Walz, O. B. Amaral, I. C. Rockenbach, R. Roesler, I. Izquierdo, E. A. Cavalheiro, V. R. Martins and R. R. Brentani, *Epilepsia*, 1999, **40**, 1679–1682.
30. J. Kanaani, S. B. Prusiner, J. Diacovo, S. Baekkeskov and G. Legname, *J. Neurochem.*, 2005, **95**, 1373–1386.
31. M. J. Gains, K. A. Roth and A. C. LeBlanc, *Neuroreport*, 2006, **17**, 903–906.
32. B. Drisaldi, J. Coomaraswamy, P. Mastrangelo, B. Strome, J. Yang, J. C. Watts, M. A. Chishti, M. Marvi, O. Windl, R. Ahrens, F. Major, M. S. Sy, H. Kretzschmar, P. E. Fraser, H. T. Mount and D. Westaway, *J. Biol. Chem.*, 2004, **279**, 55443–55454.
33. A. S. Coitinho, R. Roesler, V. R. Martins, R. R. Brentani and I. Izquierdo, *Neuroreport*, 2003, **14**, 1375–1379.
34. J. R. Criado, M. Sanchez-Alavez, B. Conti, J. L. Giacchino, D. N. Wills, S. J. Henriksen, R. Race, J. C. Manson, B. Chesebro and M. B. Oldstone, *Neurobiol. Dis.*, 2005, **19**, 255–265.
35. W. Rachidi, D. Vilette, P. Guiraud, M. Arlotto, J. Riondel, H. Laude, S. Lehmann and A. Favier, *J. Biol. Chem.*, 2003, **278**, 9064–9072.
36. F. Klamt, F. Dal-Pizzol, M. L. Conte, DA Frota Jr., R. Walz, M. E. Andrades, E. G. da Silva, R. R. Brentani, I. Izquierdo and J. C. F. Moreira, *Free Radic. Biol. Med.*, 2001, **30**, 1137–1144.
37. N. F. McLennan, P. M. Brennan, A. McNeill, I. Davies, A. Fotheringham, K. A. Rennison, D. Ritchie, F. Brannan, M. W. Head, J. W. Ironside, A. Williams and J. E. Bell, *Am. J. Pathol.*, 2004, **165**, 227–235.
38. J. Lauren, D. A. Gimbel, H. B. Nygaard, J. W. Gilbert and S. M. Strittmatter, *Nature*, 2009, **457**, 1128–1132.
39. K. A. Kellett and N. M. Hooper, *Prion*, 2009, **3**, 190–194.
40. D. A. Gimbel, H. B. Nygaard, E. E. Coffey, E. C. Gunther, J. Lauren, Z. A. Gimbel and S. M. Strittmatter, *J. Neurosci.*, 2010, **30**, 6367–6374.
41. A. Aguzzi and C. Haass, *Science*, 2003, **302**, 814–818.
42. D. R. Brown, *Dalton Trans.*, 2009, 4069–4076.
43. G. L. Millhauser, *Acc. Chem. Res.*, 2004, **37**, 79–85.
44. G. L. Millhauser, *Annu. Rev. Phys. Chem.*, 2007, **58**, 299–320.
45. E. L. Que, D. W. Domaille and C. J. Chang, *Chem. Rev.*, 2008, **108**, 1517–1549.
46. S. A. Bellingham, L. A. Coleman, C. L. Masters, J. Camakaris and A. F. Hill, *J. Biol. Chem.*, 2009, **284**, 1291–1301.
47. K. Qin, L. Zhao, R. D. Ash, W. F. McDonough and R. Y. Zhao, *J. Biol. Chem.*, 2009, **284**, 4582–4593.
48. E. Aronoff-Spencer, C. S. Burns, N. I. Avdievich, G. J. Gerfen, J. Peisach, W. E. Antholine, H. L. Ball, F. E. Cohen, S. B. Prusiner and G. L. Millhauser, *Biochemistry*, 2000, **39**, 13760–13771.
49. C. S. Burns, E. Aronoff-Spencer, C. M. Dunham, P. Lario, N. I. Avdievich, W. E. Antholine, M. M. Olmstead, A. Vrielink, G. J. Gerfen, J. Peisach, W. G. Scott and G. L. Millhauser, *Biochemistry*, 2002, **41**, 3991–4001.

50. C. S. Burns, E. Aronoff-Spencer, G. Legname, S. B. Prusiner, W. E. Antholine, G. J. Gerfen, J. Peisach and G. L. Millhauser, *Biochemistry*, 2003, **42**, 6794–6803.

51. E. D. Walter, M. Chattopadhyay and G. L. Millhauser, *Biochemistry*, 2006, **45**, 13083–13092.

52. E. D. Walter, D. J. Stevens, M. P. Visconte and G. L. Millhauser, *J. Am. Chem. Soc.*, 2007, **129**, 15440–15441.

53. P. Davies, F. Marken, S. Salter and D. R. Brown, *Biochemistry*, 2009, **48**, 2610–2619.

54. D. R. Brown, K. Qin, J. W. Herms, A. Madlung, J. Manson, R. Strome, P. E. Fraser, T. Kruck, A. von Bohlen, W. Schulz-Schaeffer, A. Giese, D. Westaway and H. Kretzschmar, *Nature*, 1997, **390**, 684–687.

55. D. J. Waggoner, B. Drisaldi, T. B. Bartnikas, R. L. B. Casareno, J. R. Prohaska, J. D. Gitlin and D. A. Harris, *J. Biol. Chem.*, 2000, **275**, 7455–7458.

56. Y. H. Hung, A. I. Bush and R. A. Cherny, *J. Biol. Inorg. Chem.*, 2010, **15**, 61–76.

57. E. A. Croes, J. Theuns, J. J. Houwing-Duistermaat, B. Dermaut, K. Sleegers, G. Roks, M. Van den Broeck, B. van Harten, J. C. van Swieten, M. Cruts, C. Van Broeckhoven and C. M. van Duijn, *J. Neurol. Neurosurg. Psychiatry*, 2004, **75**, 1166–1170.

58. Q. Kong, W. K. Surewicz, R. B. Petersen, W. Zou, S. G. Chen, P. Gambetti, P. Parchi, S. Capellari, L. Goldfarb, P. Montagna, E. Lugaresi, P. Piccardo and B. Ghetti, in *Prion Biology and Diseases*, ed. S. B. Prusiner, Cold Spring Harbor Library Press, Cold Spring Harbor, NY, 2004, pp. 673–775.

59. D. J. Stevens, E. D. Walter, A. Rodriguez, D. Draper, P. Davies, D. R. Brown and G. L. Millhauser, *PLoS Pathog*, 2009, **5**, e1000390.

60. F. Checler and B. Vincent, *Trends Neurosci.*, 2002, **25**, 616–620.

61. B. Vincent, M. A. Cisse, C. Sunyach, M. V. Guillot-Sestier and F. Checler, *Curr. Alzheimer Res.*, 2008, **5**, 202–211.

62. C. N. O'Donovan, D. Tobin and T. G. Cotter, *J. Biol. Chem.*, 2001, **276**, 43516–43523.

63. M. V. Guillot-Sestier, C. Sunyach, C. Druon, S. Scarzello and F. Checler, *J. Biol. Chem.*, 2009, **284**, 35973–35986.

64. S. G. Chen, D. B. Teplow, P. Parchi, J. K. Teller, P. Gambetti and L. Autilio-Gambetti, *J. Biol. Chem.*, 1995, **270**, 19173–19180.

65. C. L. Haigh, S. Y. Marom and S. J. Collins, *Front. Biosci.*, 2010, **15**, 1086–1104.

66. M. J. Pushie and H. J. Vogel, *Biophys. J.*, 2008, **95**, 5084–5091.

67. N. T. Watt and N. M. Hooper, *Biochem. Soc. Trans.*, 2005, **33**, 1123–1125.

Subject Index

Note: References to figures are given in *italic type*; references to tables are given in **bold type**. Abbreviations: AD = Alzheimer's disease; ALS = amyotrophic lateral sclerosis; HD = Huntington's disease; PD = Parkinson's disease